Television and Political Communication in the Late Soviet Union

Television and Political Communication in the Late Soviet Union

Kirsten Bönker

LEXINGTON BOOKS
Lanham • Boulder • New York • London

Published by Lexington Books
An imprint of The Rowman & Littlefield Publishing Group, Inc.
4501 Forbes Boulevard, Suite 200, Lanham, Maryland 20706
www.rowman.com

6 Tinworth Street, London SE11 5AL, United Kingdom

British Library Cataloguing in Publication Information Available

Library of Congress Cataloging-in-Publication Data
Names: Bönker, Kirsten, author.
Title: Television and political communication in the late Soviet Union / Kirsten Bönker.
Description: Lanham : Lexington Books, [2020] | Includes bibliographical references and index. | Summary: "This study focuses on Soviet television audiences and examines their watching habits and the way they made use of television programs. Kirsten Bönker challenges the common misconception that viewers perceived Soviet television programming and entertainment culture as dull and formulaic. This study draws extensively on archival sources and oral history interviews to analyze how Soviet television involved audiences in political communication and how it addressed audiences' emotional commitments to Soviet values and the Soviet way of life. Bönker argues that the Brezhnev era influenced political stability and brought an unprecedented rise of the living standards, creating new meanings for consumerism, the idea of the "home," and private life among Soviet citizens. Exploring the concept of emotional bonding, this study engages broader discussions on the durability of the Soviet Union until perestroika"— Provided by publisher.
Identifiers: LCCN 2019052422 (print) | LCCN 2019052423 (ebook) | ISBN 9781498526883 (cloth) | ISBN 9781498526890 (epub)
Subjects: LCSH: Television viewers—Soviet Union. | Television programs—Soviet Union. | Communication in politics—Soviet Union. | Soviet Union—History—1953-1985.
Classification: LCC HE8700.66.S65 B66 2020 (print) | LCC HE8700.66.S65 (ebook) | DDC 302.23/45094709045—dc23
LC record available at https://lccn.loc.gov/2019052422
LC ebook record available at https://lccn.loc.gov/2019052423

To Mats and Maja

Contents

Abbreviations

ch.	chast' = part
CPSU	Communist Party of the Soviet Union
d.	delo = file
f.	fond = collection
GARF	Gosudarstvennyi arkhiv Rossiiskoi Federatsii (State Archive of the Russian Federation)
GARO	Gosudarstvennyi arkhiv Rostovskoi oblasti (State Archive of Rostov Province)
Gosplan	Gosudarstvennyi planovyi komitet (State Planning Committee, 1957–1991)
Gosteleradio	Gosudarstvennyi komitet Soveta Ministrov SSSR po televideniiu i radioveshchaniiu, from 1957 to 1970 it was named Gosudarstvennyi komitet po radioveshchaniiu i televideniiu pri Sovete Ministrov SSSR (State Committee for Television and Radio by the Council of Ministers of the USSR)
l.	list = page
ll.	listy = pages
NMO	Nauchnii-metodicheskii otdel = Soviet TV's Scientific-Methodological Department
op.	opis' = inventory
RGAKFD	Rossiiskii gosudarstvennyi arkhiv kinofotodokumentov (Russian State Film and Foto Archive)
RGANI	Rossiiskii Gosudarstvennyi Arkhiv Noveishei Istorii
TsAOPIM	Tsentral'nyi arkhiv obshchestvenno-politicheskoi istorii Moskvy (Central Archive of Social-Political History of Moscow)

TsGA SPb	Tsentral'nyi gosudarstvennyi arkhiv Sankt-Peterburga (Central State Archive, St. Petersburg)
TsGAIPD SPb	Tsentral'nyi gosudarstvennyi arkhiv istoriko-politicheskikh dokumentov Sankt-Peterburga (Central State Archive of Historical-Political Documents, St. Petersburg)
TsGALI SPb	Tsentral'nyi gosudarstvennyi arkhiv literatury i iskusstva Sankt-Peterburga (Central State Archive of Literature and Arts, St. Petersburg)
TsNSIO	Tsentr nezavisimikh sotsial'nykh issledovanii i obrazovaniia, Irkutsk (Center of Independent Social Research and Education, Irkutsk)

Acknowledgments

It is not easy to name all the colleagues and friends who supported and enhanced my research and encouraged me in the writing of this book. All their suggestions, corrections, and comments greatly helped to make a book out of my first vague idea to explore the way Soviet people watched television and made use of the medium. I am particularly obliged to Stephan Merl (Bielefeld University), who was confident enough to give me the chance to develop my research project within the Collaborative Research Center (*Sonderforschungsbereich*: SFB) "The Political as Communicative Space in History" at the Bielefeld History Department. Stephan generously shared with me his broad expertise in the history of late socialism and his knowledge about the relevant archival holdings in the Russian party and state archives. He was always ready to read pieces and provided countless insightful comments and advices. Apart from our academic discussions, I really appreciate our friendship that has developed over the last decade including joint research trips to the United States, tourist excursions, and many seafood diners. Finally, Stephan always supported our family attempt to harmoniously combine raising kids, doing research abroad, teaching, and administrative duties. His attitude is still no self-evident experience for a woman in German academia.

The colleagues of the Bielefeld History Department and especially those engaged in the SFB "The Political as Communicative Space in History" provided a highly inspiring intellectual environment for research and discussion crossing the boundaries of Russian and Soviet history. Moreover, the SFB funded by the German Research Foundation (DFG) allocated generous funds. The financial resources enabled me to set up a broad sample of oral history interviews with the help of Russian and Russian-German colleagues and to travel to Russia for many research stays. I always took this as a family-friendly privilege, because it gave me the opportunity to reconcile family life and academic research in a time when our kids, Mats and Maja, were still quite young.

The time of being a postdoctoral researcher is always precarious. When the funding of the SFB had run out, Angelika Epple (Bielefeld University) "adopted" me at her professorship for the "History of the 19th and 20th Centuries" and gave me the opportunity to work with her as an assistant professor.

I owe a great deal to many other colleagues and friends who helped to develop the research concept and to test arguments and who provided feedback or supported me by just being there: Bettina Brockmeyer; Sandra Dahlke; Larissa Dufaud-Zakharova (1977–2019), a wonderful person, an exceptional colleague and friend, who unfortunately passed away far too early in March 2019; Christine E. Evans; Stefan Gorissen; Sven Grampp; Tobias Kies; Sandra Maß; Jurij Murašov; Julia Obertreis; Kristin Roth-Ey; and Meike Vogel. In the long-term perspective starting with my study at Bielefeld University, Jurij Murašov is kind of "vinovat" that I finally wrote this book about Soviet TV viewers, as he was the first academic teacher in my career who broached the issue of television and explored with us students the way television impacted on the viewers' perceptions. At a later stage, the same is true for many colleagues who (co-)organized inspiring conferences, edited volumes and special issues, invited me to talk in their colloquia, or offered important feedback in one way or another. In this context, I thank among others Susanne Schattenberg, Anke Hilbrenner, Klaus Gestwa, Jan Kusber, Dietmar Neutatz, as well as the "Swiss connection" Nada Boškovska, Angelika Strobel, Daniel Ursprung, Ekaterina Emeliantseva, Carmen Scheide, Ulrich Schmid, and Peter Collmer. Also, Manfred Hildermeier, who was my doctoral supervisor at an earlier stage, was available to provide letters of recommendation at any time, when I still needed them. Alevtina I. Gavrilova not only hosted me during my stays in Moscow but became a dear friend of our family. Gudrun Lehmann helped much with all the administrative issues, while I was teaching at Bielefeld University. Robert Bache, Tim Schaefer, and Alexander Zalojnov were supportive and smart student assistants who were much more committed to assist than one could them expect to.

This book is the outcome of my postdoctoral thesis, the traditional German *Habilitation* that took place at Bielefeld University in 2017. I am especially grateful to my reviewers Stephan Merl, Thomas Welskopp, and Ulrich Schmid (University of St. Gallen), who accepted the efforts to be part of the academic committee. Thomas Welskopp did not hesitate to head the committee and to manage the habilitation procedure, although he had surely never expected that I may ask him to accept this responsibility. Besides that, I am much obliged to him, as he entrusted his professorship "History of Modern Societies" to me to act in place for him for two terms.

I am also grateful to all interviewees who agreed to contribute to this project by sharing their experiences, emotions, and thoughts about Soviet television, the Soviet regime, and often even about their Soviet lives. I am equally grateful to the Russian and Russian-German colleagues Elena Bogdanova (St. Petersburg), Alexander Ermakov (Irkutsk), Ol'ga Galanova (Bielefeld), E. Ivanova (Irkutsk), Sof'ia Kontorovich (Rostov-on-Don), Vadim Korolik (Göttingen), Elena Mingaleva (Göttingen), Galina Orlova (Rostov-on-Don), E. Peters (Göttingen), Michail Rozhanskii (Irkutsk), Julia Schmidt (Göttingen), Anna Tikhomirova (Bielefeld/Yaroslavl'), Oksana Zaporozhets (Samara), and Elena Zhidkova (Samara), who recruited most of the interviewees, and conducted and transcribed the interviews. It was a pleasure to work on the interview guide together with Elena Bogdanova and Galina Orlova. I thank them very much for their comments, efforts, and time.

As parts and revised chapters of this book have been published elsewhere first, I am grateful for the permission to reproduce material from the following publications: "'Dear television workers . . .': TV consumption and Political Communication in the Late Soviet Union", *Cahiers du monde russe* 56 (2015), 2, pp. 371–399' and "'Talking with the Consumer': Consumer Issues on Soviet TV," *Laboratorium: Russian Review of Social Research* 8 (2016), 1, pp. 30–57'.

I am profoundly indebted to the anonymous peer reviewers and the editors of the special issues of *Cahiers du monde russe* and *Laboratorium* who disclosed several weaknesses and faults in my argumentation. They helped me much to improve these parts.

However, I owe the greatest debt to my family. My parents, Birgit and Eckhard Bönker, have supported us by caring a lot for our kids. Mats and Maja always let me travel to do the archival research or to participate in conferences: I know that I have been often away from home. But all this would not have been possible without the loving and generous support of my husband, Lutz Häfner.

Introduction

Watching Television, Emotional Bonding, and Political Communication in the Late Soviet Union

A WINDOW TO THE WORLD

"Soviet television, well, I cannot say that it was a propaganda sphere for me because I just did not watch this propaganda. Well, let's say, it was to a greater extent, nevertheless, a source of knowledge, a small window to the world."[1] With these words forty-nine-year-old Elena retrospectively recounted her contemporary assessment of Soviet television.[2] Elena was born in Leningrad in 1961 and graduated in business economy in the late 1980s. She was asked about her TV-watching habits during the Soviet period of her life in an interview in November 2010. This is part of a sample of eighty interviews covering television uses and watching habits in the late Soviet Union. Interestingly enough, many of the respondents, who were mainly interviewed in 2010 and 2011, would have agreed with Elena.[3] Thus, one of the first impressions from the interviews was that Elena was far from the only former Soviet citizen at the time of the interview who retrospectively did not unilaterally condemn Soviet television programming but, on the contrary, really appreciated certain programs and specificities of the Soviet system.

Although Elena represented herself as a critical media consumer, she recalled having never basically rejected Soviet television. Her receptivity was not least due to the peculiarities of a televisual medium that easily changed the perceptions of the immediate surroundings, of the Soviet society, and somehow of the "world": Elena was not the only respondent contemporarily considering Soviet TV as a "window to the world." Her namesake Elena Sergeevna, who was born in 1957 on Sakhalin, remembered television as

> a vent, so to speak, as a window to the big world. (. . .) of course, we watched TV, well, how shall I say, that the TV set [was] like a dreamland, it's a bit like

> an escape from reality. (. . .) That means that all your immediate problems, any hardships, they move aside, you forget a bit about it. You leave for the dreamland, there, you watch and, well, everything directly gets better.[4]

These ambivalent perceptions of TV's propagandistic communication strategies on the one hand and its realistic, constructivist, and from time to time escapist representations of reality and the outside world on the other not only triggered new relations to time, space, and distance among TV viewers. The respondents' statements also demonstrate the different emotional conditions the viewers could engage in by including television into their everyday life and the sociopolitical environment they lived in. By way of preliminaries, I quote the reminiscences of one more respondent to illustrate this panorama television opened for many contemporaries: Georgii was born in Leningrad in 1953 and raised in the family of an established scientist. He frankly admitted that, in Soviet times, he had "sincerely believed to live in the best country, everything [was] fine." Concerning his perception of Soviet TV broadcasts, Georgii presented an equally open-minded view confirming that "I believed that everything was true."[5]

These remarks lie far away from Western clichés about tiresome party meetings and boring speeches by CPSU general secretaries, and do not reduce TV to a mere propagandistic mouthpiece of the Communist Party. The respondents' memories about Soviet television have very little in common with prejudices about unglamorous TV films and endless documentaries propagating Soviet achievements, worker and peasant heroes, and the recent record harvest.

Even more, the interviewees' statements raise several questions this book seeks to trace about watching TV in the late Soviet Union. During the 1960s, TV became the most important technical and cultural device that rendered home a place and an object of negotiation between "ordinary citizens" and the regime about new meanings of everyday life.[6] The medium merged home, consumption, and leisure into a continuum of the public and private spheres whose boundaries increasingly overlapped, yet were also permeable and shifting.[7] There are many reasons to suppose that Soviet television quickly became the new key factor of public communication processes by setting, constructing, and interpreting societal issues. The book must not least consider how television impacted on structures and mechanisms of the public sphere constrained by authoritarian censorship, taboos, and political control. These constituted obvious differences compared to liberal models of the public sphere. However, socialist technical mass media hardly engaged in conveying such societal realities and, moreover, grew in a similar way to their Western counterpart over the decades. Thus, mass media pervaded socialist societies and their public spheres as much as they did on the capitalist side of the Iron Curtain. However, the central issue of this book is not about the mechanisms of governing and controlling the media but people's practices of consuming them. The particular focus on the viewers' use of television should heighten awareness of changes in political communication, of discursive linkages between the private and the public spheres, and perhaps even between the different segments of the public sphere (*Teilöffentlichkeiten*), such as scientific, literary, artistic, political, and religious.[8]

The parallels within Europe and across the Iron Curtain regarding the rapid spread of television should not conceal the extent to which public communication processes via mass media in socialist countries differed. We need to inquire into the range of criticism—whether directed or not—and the way television enhanced collective communities, political legitimacy, and emotional bonding. This perspective involves the viewers and leans on the assumption of the classic study of Siegfried Kracauer on "totalitarian propaganda" that recent studies have further confirmed: Any official communication and messages under authoritarian, illiberal conditions needed to address the audience's emotions, to "captivate the souls of the masses" and "to mentally influence" the people, as Kracauer put it.[9] The "technologies of the soul" that Swiss slavist Ulrich Schmid recently highlighted with regard to Vladimir Putin's regime adapt successful communication strategies of the Soviet regime, as both are based on cognitive binaries such as true and false, friend and foreign.[10] As a stronger consideration of the viewer practices would help to deconstruct them, we should further assume that political propaganda often merged with pop-cultural features. Recent propaganda studies therefore draw on approaches from British cultural studies and take the audience's preferences into account. This should at least gradually shift common frames of interpretation of everyday life.[11]

Against this backdrop, I will explore Soviet citizens' material and immaterial TV consumption and the communicative changes the TV set brought about in family life and in attitudes toward television and the Soviet regime. This book is interested in the way the new medium transformed the negotiations on the new Soviet lifestyle and how it embedded them within the shifting continuum of private and public life. To what extent did TV allow for ambiguities and ambivalences in "authoritative discourse" that framed and interlaced the private-public continuum on the basis of performing ritualized acts?

Alexei Yurchak has established the way ritualized acts perpetuated the late Soviet system without simply mirroring official values and discourses.[12] Television was a far-reaching medium that was predestined to engage in the staging of the new Soviet way of life and thereby needed to represent authoritative discourse; this book explores how the TV viewers responded to it. How did television impact on the way TV viewers participated in the reproduction of the form of the ritualized acts constituting authoritative discourse? How did viewers use television and its constative meanings in the context of these ritualized acts? Did they possibly adapt the medium to engage in or did they rather transform the performative essence of authoritative discourse by their somehow unpredictable use of television in everyday life?

Against this background, the book analyzes the way viewers embedded television, and the language, topics, and presentations of the world and the Soviet society it produced, into their everyday life. To put it more concretely, it asks how viewers engaged in negotiations about demands for pleasure, entertainment, and rest in front of the screen. In what way did television and its viewers become objects, as well as subjects of intellectual constructions and political negotiations within authoritative discourse?

In television's early days, TV employees and viewers achieved a relatively broad scope of action, as party leaders had kept a distance from the new medium during the 1950s and early 1960s. Only from the 1960s onward did they realize the propaganda potential of television and started to invest a lot of money to develop the technical infrastructure.[13] Thus, television could enhance its cultural impact boosting the basic changes in post-Stalinist material culture. Kristin Roth-Ey has strongly stressed television's ambivalent influences on political communication, the cultural education of the "masses," and the reluctant attitude of established cultural elites toward TV from its early days in the 1950s. She explores the interaction and negotiations between TV producers and the party in the process of establishing television as the Soviet key medium. Christine E. Evans has focused on TV program politics, on the refinement of the schedule, and of TV shows and the launching of new genres. In considering negotiations between the audience and TV producers in the context of musical holiday programming and game shows, she convincingly demonstrates how changes in the format of these shows made them more contest-based entertainment genres. These changes mirrored the regime's search for new ways of engaging Soviet citizens. The aim was to increase citizens' and particularly young citizens' commitment to the Soviet sociopolitical order.[14]

Thus, there is much ground for this book to opt for a somewhat different research perspective and to be rather selective about TV producers' concepts of programming, schedule developments, and aesthetical considerations. This book, therefore, aims to present the viewers' side of the story. It traces how viewers purchased their first TV set, arranged their domestic life according to the new technical device, embedded watching television in their daily life, and, not least, established such emotional bonds to the medium and its use that they are still figuring in today's often-nostalgic narratives about the lost Soviet life.

Many respondents were still concerned about these changes in their private life, as Liudmila Vasil'evna, a sixty-four-year-old teacher of Russian language and literature, established:

> television is simply an open window to the world, that is, however, diverse, this world. But, nevertheless, in the beginning this, this world, that opened for all people, it was possible to drop in, it was possible to travel around, because it was very sincere, very trustful and you trusted in it, . . . we were not scared of it. It always attracted us somehow. Television gave the opportunity to know the world, the customs, and habits of people, about whom you read, but there you could see them. There, the visualization did its work, very big and very positive. That it became a medium of education. (. . .) Certainly, that it started to bear such an educational character, this called an even more respectful attitude to television. Then television also became a medium of pleasure for us, just pleasure. In this connection it was not simply pleasure, let's say, also the discussion about movies, yes? Movies, we could watch them free of charge, we didn't need to pay money, yes, but we had the possibility to watch television, that tells about living people in such programs like *A nu-ka, devushki!* (Let's Go, Girls!), *A nu-ka, parni!* (Let's Go, Guys!) and *KVN*.[15]

Such statements point to the fact that we obviously do not know much about viewers' perceptions of Soviet television as a pleasure-giving medium as well as a source of information at the same time. The fact that TV programs brought about tremendous changes in leisure habits raises questions about the TV audience, its consumption practices, and its habits. More generally, this means that the role of TV consumption in the social, political, and cultural communication processes under the conditions of authoritarian regimes requires explanation.

TELEVISION AND THE SOVIET WAY OF LIFE

It is clear that censorship was more invasive and communicative restrictions were undoubtedly stronger in state socialist than in democratic countries. Nonetheless, the etatist concept of government-related television channels in Charles de Gaulle's France or similar ideas of restrictive media politics in the Federal Republic of Germany in the 1950s and 1960s suggest that bipolar models contrasting the "dictatorial" East with the "democratic" West are too undifferentiated and should at least be carefully refined.[16] Apart from a period of slight relaxation during the 1970s, such bipolar models of the "good," that is, liberal Western media system versus the "bad," that is, illiberal media system of Soviet style dominated the research on mass media in socialist Europe at least until the early 1990s.[17] This approach has long been inspired by the now-obsolete concept of totalitarianism and thus based on a dichotomous model of party-state versus population. Dichotomous models tended to highlight the regime's monopoly on mass media and described them as seemingly persuasive propaganda channels. The research in times of the early Cold War especially showcased the press and the radio as instruments of thought control and brainwashing. Obviously still inspired by Aldous Huxley's dystopia *Brave New World*, Cold War scholars suspected socialist media consumers of being merely passive recipients—if they considered them at all.[18] Even though the media consumer remained rather hidden in these investigations, contemporary Western social and media scientists made considerable efforts to explore especially the Soviet system of political communication during the Cold War that had been exported to the Eastern bloc states. After the fall of the Iron Curtain, media and social scientists drew their attention to the post-socialist changes in communication, staging of power, and the democratization processes in which mass media played a crucial role.[19] Thereby, post-socialist television, the restructuring of ownership, political influence, and changes within the profession of journalists gained explicit research interest.[20] For some time now, state socialist television has captured growing attention in the fields of history and historically working media studies.[21] Recent research on socialist television in the GDR,[22] the East Middle European states,[23] and the Soviet Union[24] at least touches upon TV consumption and has given initial insights into how TV was involved in the vertical communication among the audience, the TV staff, and the party. Thus, it is now state of the art to suppose the production and consumption of TV programs as being negotiated among media producers, viewers, state, and party institutions.

The new approaches have already borne fruit in the broader context of recent research on the period of "developed socialism," a discussion with which this book also aims to engage. It is interested in the stability of the late socialist regimes and focuses on societal processes and practices. Consumerism, leisure-time activities, tourism, and sports in state socialist societies have become very productive fields of research.[25] They bring to light complex relations and negotiations between state policies, activities of mass organizations, and individual and group practices. Well-known contributions to socialist popular culture studies have posed the question how Western influences and the "imagined West" were perceived and adapted in socialist societies.[26] Obviously, a strict juxtaposition of Western influences and the validity of socialist values cannot be made. Rather, the adoption of elements of Western culture easily coexisted with established norms and power structures on the Eastern side of the Iron Curtain. This is an important insight for historical television studies as well: the consumption of Western films, of foreign radio stations like *Radio Free Europe* or *BBC World*, and of television programs did not automatically lead to a rejection of socialist culture or the socialist system of governance.[27] However, the question remains in which ways television contributed to the decomposition of socialist rule—as a medium perceived as state-controlled and coming under fire during the

1980s, or as a medium providing critical information that accelerated tendencies of erosion.[28]

Tracing the trajectories of television and radio as possible subversive forces, Kristin Roth-Ey takes a clear position and suggests that the Soviet Union "lost the cultural Cold War," as it had finally failed to establish an attractive media empire.[29] However, the continuing success of the present Russian TV channel *Rossiia Kul'tura* among an audience aged fifty years and older demonstrates that Soviet popular culture still resonates with these people's tastes and interests. Since its founding in 2004, it has a small but steady share in audience preferences. It obviously benefits from nostalgic trends in Russian society, as it mainly serves its audience by broadcasting popular Soviet films, serials, and features.[30]

This phenomenon should lead us to be more cautious about interpretations that tend to perceive media cultures as binary black boxes. Program exchanges among the socialist states and between Eastern and Western TV stations, broadcasts of common features within the transmission range of *Intervision,* or live linkups between *Eurovision* and *Intervision* let us suppose that cross-references and amalgamations prospered beyond and across the Iron Curtain.[31] Television not only created national "imagined communities" but also set in motion complex transnational circulation of cultural products that almost naturally involved transnational comparisons and negotiations about aesthetic, narrative, and cultural standards.[32]

There are some historical and current reasons not to share somewhat missionary assumptions about cultural Americanization as the downside of an ultimately failed cultural Sovietization represented by Soviet mass media. At least, we need to explore how people embedded television in social and cultural discourses and how they perceived its pop-cultural potential for identification and for emotional and personal ascriptions. How did people perceive themselves as viewers, how did they use the new medium, and how did they communicate about and with Soviet television? Which demands did the audience communicate toward the television as information and entertainment medium? In the end, television only gains sociopolitical and cultural significance through viewers' usages and their communication about or with the medium. A program that is ignored by the audience has no societal relevance. Thus, Soviet television opened a new site of communication, sociocultural negotiations, and societal self-observation not only from the perspective of the regime but also for the audience. It did so because the failure to implement economic reforms during the Khrushchev period urged the regime to search for other strategies of mobilizing the population to win people's acceptance and support. With enhancement of the concept of "peaceful coexistence" television became a central part of fighting out the "cultural wars."[33] Each superpower aimed to offer an engaging popular culture that should win people's hearts and minds in order to convince them of the advantages of the respective sociopolitical system providing a special way of life. The Brezhnev era was not least molded by an internal struggle for cultural integration. This nonmilitary side of the Cold War made material and cultural consumption, taste, and moral attitudes into important sites of competition. It exposed the ideological confrontation between the capitalist and the communist systems, as practices and norms of soft power and emotional bonding aimed to boost the home front.

Thus, the cultural education of the Soviet television viewer partly renewed the missionary project of creating the New Soviet Man that dated back to the early days of the young Soviet state and had been adapted throughout its history. The project represented a central means to overcome self-perceived backwardness and to surpass the capitalist West. The New Soviet Man was supposed to be the bearer of the future communist society. But before this idea might come true, she or he was expected to do more than just develop a cultured behavior in the sense of politeness and proper appearance in public. Khrushchev's promise to overtake the United States in terms of consumption issues and the standard of life gave the propagandized *novyi byt* (the new way of life) new drive. Henceforth, the New Soviet Man was not only supposed to behave in a certain way; in the context of the expanding media landscape and the increasing use of media, he was also supposed to improve his cultural consumption and leisure activities.[34]

Private life and leisure time gained even further legitimation, as the constitution of 1977 explicitly allowed Soviet citizens to perceive cultural consumption as an official right and a legitimate aspect of the new Soviet lifestyle. The constitution granted Soviet citizens not only the right to rest and leisure (Article 41) but also the right to enjoy cultural benefits (Article 46). Among other aspects, this should be ensured by "developing television and radio broadcasting." Moreover, the constitution related the freedom of speech, of the press and so on to "the opportunity to use the press, television, and radio" (Article 50).[35]

Indeed, the Soviet 1960s and 1970s witnessed a reinterpretation of the relation between the "public" and the "private." Contemporaries perceived that people were retreating from the politicized public sphere into "their" private spheres.[36] Recent research has already successfully questioned the conventional binary model that emphasized an antagonistic demarcation of these spheres of action. This shift in research perspective has also drawn attention to the mechanisms that contributed to the stability of the sociopolitical system which has otherwise been understood from the point of view of its collapse.[37] Referring to Alexei Yurchak's interpretations, I would argue that Soviet stability had much to do with new or revised communication practices during the Brezhnev era. Television played an important role because it became a pleasure-giving medium that changed ordinary people's leisure and communication practices as well as the material and medial consumer culture. The above-quoted memories of the respondents strongly suggest that television by this time gradually gained the capacity to evoke an affirmative emotional commitment among viewers toward their Soviet life.

WINNING THE HEARTS AND MINDS OF THE TELEVISION AUDIENCE OR "WHAT DO PEOPLE DO WITH MEDIA?"

Exploring this field of emotional bonding and the way television addressed and involved viewers in communication, the book also ties into the much broader discussion on how to explain the durability of the Soviet Union until perestroika. To make this approach clearer, some methodical assumptions must be highlighted.

Contemporarily, television was officially charged with many tasks, not least among them "educating Soviet youth to purposefulness, love for the country and socially useful work, trust and devotion to our party" Television was furthermore supposed "to develop the fraternal friendship of multinational Soviet youth and of the whole world, to develop . . . high moral quality, smartness for live, uprightness, honesty, diligence towards comrades and irreconcilability towards enemies of communism."[38]

The question of to what extent television involved the audience emotionally refers to Martha Nussbaum's concept of political emotions. Nussbaum argues that political principles always need an emotional basis to maintain the society. In her perspective, any society needs emotions like compassion and affection. A liberal, democratic society is faced with the challenge of fostering societal engagement and negotiating how to include the weak. This contributes to a kind of "civil religion." Referring to Rousseau, Locke, and Kant, it includes ceremonies and rituals that support a public culture of emotions that reinforces the bonding to societal norms. According to Nussbaum, these "public emotions" contain cognitive assessments. In her view, it makes a categorical difference regarding the stability of a society whether its political principles are secured via people's active engagement or rather through passive approval. Nussbaum supposes that a society's sustainability is much more lasting if people stand up for values and ideas and not only agree on a compromise for pragmatic reasons.[39] Although this leaves out the question of force and repression, one could transfer this idea to authoritarian societies, as these regimes also aim to trigger people's emotional commitment to values, symbols, or traditions, even if these explicitly relate to enemy images and thus do not adhere to morally "good" concepts. Nussbaum emphasizes that solely emotionally connoted linkages between individual perceptions and memories on the one hand, and societal principles based on "high" ideals on the other, may strongly bond people to the order.[40] In our case, the questions then are about how television was able to evoke such public emotions backing up the Soviet regime and to what extent they were based on commitment or rather routinized approval ensured by authoritative discourse.

There are many theories concerning the question of what emotional effects television has on the viewer.[41] However, historical research has still much empirical investigation to do in order to explain these effects. This is especially true regarding state socialist societies. Although emotional responses to concrete TV programs are undoubtedly important and need to be examined in a historical perspective, I am interested here in the more general emotional bonding of viewers to the TV and the impact of this process on viewers' relations to the Soviet system. The investigation of state socialist television needs to explore the scope between Elihu Katz's classical questions: "*what do people do with media?*" and "*what does media do to people?*"[42]

To conceptualize the question of emotional bonding by asking for the use of television, I draw on a basic assumption of cultural media theories. They aim to explain how mass media contribute to societal integration. With this, mass media appear as the central agent of societal reproduction and communitarization: they can generate social commitment and credibility, initiate communication, and diversify as well as standardize cultural conventions. Thus, the ambivalent character of television flows promotes the synchronization of individual and collective perceptions of contents.[43]

Manfred Zeller, focusing on Soviet soccer fans, profoundly examines how television not only impacted on perceptions but especially built emotional communities.[44] Some media scholars try to grasp this kind of societal medialization and community building in the two dimensions of cognition and emotionality. To analyze the affective-emotional involvement of viewers in relation to cultural codes of interpretation, they introduced the term "media culture of emotions" (*Medien-Gefühlskultur*).[45] Emotions constitute a highly complex phenomenon and unsolved interdisciplinary problem incurring competing and sometimes even opposite answers to the question of how to explore them. Historians find it difficult to detect them beyond their linguistic, discursive, or symbolic representations. Based on a predominantly social-constructivist approach, historical research mostly does not differentiate between the inner self and the outside representation of emotions. This means that historians quite disregard emotions as a psychological state of mind that may exist without language. The obvious reason is that we have no or at least only few sources to analyze the psychological dimension of emotions in past times in contrast to their external representations.[46]

For similar pragmatic reasons, I will explore the emotional setting while watching television especially based on oral history interviews. Therefore, I prefer the social-constructivist approach to trace, on the one hand, how the respondents retrospectively represent their emotional state watching Soviet television. On the other, I'm aiming to relate these emotional attitudes to the sociocultural factors possibly shaping and framing them. For this purpose, I draw on the concept of "media culture of emotions" to describe the process of emotional commitment that referred to the sociopolitical subtext of the TV program representing the Soviet way of life. Thus, the "media culture of emotions" was immediately entangled with an officially routinized and socially accepted popular culture.

Among other aspects, the negotiation about "shared emotions" contributes to this highly complex process of communitarization.[47] Thus, the sociopolitical context framed TV users' practices and emotional bonding within both an authoritative discourse and a societally accepted popular culture that might lead to everyday hybridization of official meanings.[48] Hybridization is in Mikhail Bakhtin's sense a dialog of different voices representing cultural differences. This dialog sets a process in motion that could undermine authoritative discourse. Hybridization thus refers to the interacting and merging of representations without erasing one of them. It takes place in an intermediate space.[49]

Taking up Bakhtin's idea, I assume that television not only constituted but also framed an intermediate space in which possible cultural differences, interpretations, meanings, and, in the end, sociocultural balances of power may be negotiated. This approach somewhat echoes Homi K. Bhabha's postcolonial approach to social hierarchies and balances of power. Borrowing Mikhail Bakhtin's term, Bhabha calls this intermediate space of negotiation "third space." Bhabha's idea of hybridization is, however, somewhat different from that of Bakhtin, who supposes that any hybridization contests authority. Bhabha attributes a hybrid potential to authoritative discourse itself.[50]

The speed and scope of television rather complicated the staging of authoritative discourse and allowed for a variety of voices and representations. Further, viewers watch and discuss meanings at home, and they must transgress any boundaries between the private and the public spheres to enter the "third space" and negotiate with the regime. As viewers could hardly be forced to respond to televised authoritative discourse, it makes sense to subscribe to Bhabha's more dynamic understanding of hybridization and to suppose that television potentially might hybridize authoritative messages. This should be seen as a dynamic, rather ambivalent process of achieving people's acceptance of and perhaps even commitment to central Soviet values and attitudes.

These negotiations were surely not without constraints. To consider possible constraints. we find a stimulating perspective in Umberto Eco's semiotic approach. Like British cultural studies but unlike Bakhtin and Bhabha, it explicitly focuses on television: Eco determines that one of the most important characteristics of television is that of an "open text."[51] In this sense, one could understand television's "third space" as nondetermined. Cultural studies subscribe to this basic assumption but have plausibly established the way hegemonic rules of framing make certain interpretations more likely than others.[52]

Hegemonic constraints and asymmetric powers of interpretation were not exclusive to authoritarian regimes, but it seems very likely that the authoritative discourse of Soviet style highly affected the official side of television. For our purposes, the way it affected viewers' uses should be considered an open question. Nevertheless, the access to television's sources of knowledge differed depending on the actors' sociopolitical background. Thus, Soviet television had—like any other mass media—to refer to broadly familiar interpretation frames that ordinary viewers might have accepted or even felt committed to. In Stuart Hall's interpretation, television could be a source of social consensus that again might be accepted by viewers without totally agreeing.[53]

The question of building societal consent within the "third space" of television is rather central to our interests here, as it is closely related to the question of the stability of the social order. A further complementary view is raised by Niklas Luhmann. In contrast to British cultural studies, Luhmann is much more skeptical about the relation between societal consent probably produced by mass media and the stability of a social order. He argues that it is not that any consensus or any social contracts are keeping a social order stable but rather what Luhmann calls "objects": these are issues generated by the mass media.[54] In his view, these are topics that gain familiarity among the public via the mass media, so that the public is able to negotiate with them without necessarily agreeing with them. Thus, Luhmann rather aims at the variety of representations displayed by television. Like Stuart Hall, he argues that nobody is obliged to agree to the reality constructed by the mass media.[55] From a theoretical point of view, these negotiations do not need to satisfy democratic or liberal requirements. Luhmann's perspective on societal systems does not depend on the actual political order, because his basic assumption refers to a functional differentiation of "modern society." As Luhmann not only limits the role of an acting subject in his analysis but rather rejects the notion, it theoretically does not matter which actors, with what kind of political standpoint, operate within the media system.[56] This is a challenging assumption for historical research; however, the missing actor in Luhmann's theoretical concepts does not decrease the tempting offer of the way he describes the mechanisms working within the media system. As Luhmann's own theoretical connection between the media system and the other social systems is not totally clear, it is in one way or another left to research to analyze the relations between the subsystems. Thus, the question for those who do not deny agency is who negotiated upon which kind of "objects," and in what way these "objects" might have contributed to a consensus-building communication between the audience, television, and the regime. Or, to expand the question in Luhmann's sense, to what extent the negotiations on objects created familiarity and perhaps even a possible variety of interpretations.

Against this background, different viewers' practices, such as viewers' letters to Soviet television, discussing about broadcasts with family and friends, switching off the TV set, and so on, were performative acts engaging in these negotiations. These practices turned out to be irreducible to the order of discourse. The crucial heuristic point seems to be the connections between discursive and other practices and how the latter might contradict, reject, or "vampirize" (Michel de Certeau) the first. This is a central argument of French cultural historian Roger Chartier to explain historical change, the relation of actors' perceptions, practices, and structures and of actors' subjectivity and structural framings.[57]

Roger Chartier is especially interested in the impact of books and texts on their readers in bringing about changing practices and perceptions. He attributes high relevance to the materiality of the text for its interpretation. Reading and hearing the text or, more generally speaking, using the medium becomes a concrete practice that may differ according to the sociocultural and political disposition of the "interpretative communities." Media practices always depend on both discourses and social conditions.[58] To analyze their correlations, Chartier takes up Michel Foucault's approach and considers discourses as practices that are submitted to formative rules. These regulating pressures often remain unconscious for the subjects and might contradict their ideas and intentions. Although the control of these rules is normally out of reach for the majority, Chartier rejects the idea that all practices could be coupled to discourses and that meaning is automatically produced or determined by structural conditions. The question of the actors' intentionality versus routinization and practices' potential for change or for maintenance remains, however, crucial. These aspects are even more essential with regard to the Soviet system strongly aiming to reduce people's agency on the one hand, while simultaneously triggering people's involvement on the other.

Chartier distinguishes his notions of the subject and agency from Foucault's and, to a certain extent, also from Pierre Bourdieu's concept of habitus. Without supposing a completely unrestrained subject, Chartier—in explicit contrast to Foucault—stresses the variety of uses, utterances, and understandings an individual may opt for. However, to prevent an arbitrary attribution, they shall be set in relation to dominant discourses, because their connection allows for sociopolitical and cultural dynamics.[59] This perspective helps with investigating the shared cultural constructions and perceptions about society which can be understood as the result of power relations between dominant, obedient, and resistant interpretations.[60]

For our purposes, we can, thus, go a step back and understand television as a medium of social reproduction, mutual observation, and self-reassurance. Television bundled the respective practices of individuals within the "third space"—a process probably resulting in groupings and classifications.

Based on this conceptualization, I suppose that television entailed new interpretative communities in front of the screen, with which members then entered into negotiations about sociocultural classifications. Chartier objects to the research practice that attributes certain characteristics of popular or highbrow culture to certain cultural products and then deduces social classifications to the people consuming these products. He also rejects the idea that the social background of media consumers predetermines their interpretations or the classification of the respective cultural product. According to Chartier, the creation of sociocultural power relations is not an unrestrained process but a much more fluid one based on consumption practices of texts and media products.[61]

To understand the sociopolitical and emotional context of Soviet TV practices, I draw on John Fiske's idea of TV as "bearer/provoker of meanings and pleasures."[62] The mentioned insights into the retrospective assessments of some former Soviet citizens about their TV uses let us suspect that watching TV, everyday life, and politics were not only closely entangled. I also assume that the Soviet TV indeed provoked meanings and pleasures. Many structural elements of the medium coincided on both sides of the Iron Curtain. In the capitalist West as well as in the socialist East, the television linked the private to the public sphere in a new way. All over the world, viewers became able to consume the "outside" and public world at home in their living rooms. The shifting of space boundaries boosted at least superficially domesticity and the private sphere of the people. And much more banal: Even an authoritarian regime could not force the viewers to either switch on or off the TV set, so that anyone could decide on her or his own when to watch television.

Thus, the summing up of this methodological and theoretical spectrum leads to the idea that watching TV at home and negotiating on meanings of the Soviet lifestyle might have been a highly political practice as it perpetuated the Soviet system by people's "doing and living Soviet." My use of the term "political" derives from the constructivist approach of the new political history. This concept focuses on the historically changing boundaries, mechanisms, media, and people constituting a "communicative space of the political" correlating to our "third space." It pays special attention to practices and discourses apart from decision-making processes concerning questions of the social order, rules of living together, power relations, or the limits of what can be said or done. The communication covering these aspects should transgress individual interests. It refers to imagined or real collectives and aims at having, or even has, a wider impact on others. In this sense, the "political" should be recognized as sustainable or obligatory.[63] The question of how audience letters may have contributed to the regime's stabilization concerns these political implications of communication. As a result, "political" never simply means "oppositional." The notion of "political" also exceeds the communication of political decisions, talks, or audience responses to broadcasts of party or trade union meetings. Communication should be considered as political as well, if it reveals that people did not refuse or—on the contrary—accept the regime within the limits of authoritative discourse. My assumption is that this kind of communication contributed to the regime's durability, as it benefited from television's capacity to trigger affirmative emotions and commitment among viewers. Driving this idea forward, I suggest that most of the TV viewers emotionally accepted the way the medium presented Soviet popular culture, societal values, and the Soviet way of life in its entertainment programming. Already several contemporary sociological studies of the late 1960s and early 1970s indicated that a large part of the audience regarded television as a medium of after-work entertainment. One of these Soviet studies underlined that most viewers reported that they especially appreciated those programs that satisfied their "emotional-aesthetic" demands.[64]

The renewal of popular culture and the visualization of new consumerist lifestyles across time zones and urban borderlines on Soviet screens showed how strongly TV consumption meshed with material consumption: the first step in becoming a TV consumer was to buy a TV set. This effect is interesting insofar as Nikita Khrushchev's proclamation to overtake America made consumption a central factor of legitimizing the Soviet regime. From this time on, consumer practices and private interests belonged to the core set of themes on which the Soviet regime invited people to communicate to gain their acceptance.[65] Recent research convincingly suggests that the consumer culture of the 1960s and 1970s—embracing goods, services, taste, and mass media—became a cohesive factor that held the regime and society together rather than divided it. It did so not least because late Soviet consumer culture offered a new field of communication.[66] Television became perhaps the most constitutive part of this ambiguous change. Officially, it remained the "mouthpiece of communist education"—a designation that seemed to contradict a consumerist lifestyle.[67] Challenging this contradiction, I further argue that the popularity of televised Soviet lives somehow promoted viewers' adherence to a societal consensus and lastly to the political order. The stabilization of an authoritarian regime is certainly much more complex than this argument alone, but it offers a component of this process. The crucial point is not the achievement of consent, as the cultural studies have proposed, but rather the effect of raising topics by television. It thus acted as a medium amplifying, simplifying, and familiarizing certain issues, ideas, values, or feelings that viewers could relate to in their daily lives. To address the audience by providing views defining the "normal" life was the main way for the regime and TV producers to convey messages, offer identification models, and evoke commitment.

"Normality" is historically closely related to private life, everyday practices, and seemingly reliable future life prospects. Therefore, it should be of interest for us how "ordinary" Soviet citizens—contemporarily and looking back—included these practices that were supposed to create normality and the changes they underwent into discursive constructions of the "normal" Soviet life. "Normal," "normality," and "normalization" are rather hazy notions and would deserve much deeper considerations than I'm able to provide here. As many scholars relate the generation of normality to everyday rituals and habits, it makes sense for my purpose of a historical exploration based on interviews to focus on discursive constructions of normal everyday life. On what kind of television did the respondents seem to agree? With which kind of values presented on the screen did they comply? Did viewers appreciate news and documentary programs as much as entertainment broadcasts to be included into the regular Soviet TV program? To what extent did possibly heterogeneous constructions of "normal" realities create solid interpretation frames that stabilized the social order?[68]

TRACING THE SOVIET TV VIEWER AND PRACTICES OF WATCHING TV

I trace the different aspects of watching TV on two different paths. On the one hand, I explore a range of materials of the central and regional television committees, as well as of Communist Party organizations originating from several national and regional archives in Moscow, St. Petersburg, Rostov-on-Don, Irkutsk, and Samara. These materials include correspondences between the party organizations and television stations negotiating, for example, the financial support and infrastructural developments, television's audience research, reports of the letter desks or internal documents on the programming, and the staff.

On the other hand, I approach the Soviet TV viewer based on a sample of eighty oral history interviews that explicitly cover TV-watching habits in the late Soviet Union. Most of the interviews were conducted by experienced Russian sociologists and historians between 2010 and 2012. That means they cover the years after the heavy economic crisis of 2008 and mostly before

Putin took up his third presidency in Mai 2012. Thus, the interviews are still not that much affected by propagating the renewed style of foreign policy fed by the reenactment of attitudes of a superpower and the enforcement of patriotism and Western enemy images.

The sample is transcribed and mostly available as audio files so that I could get an impression of the way the respondents spoke and reacted to certain questions. The majority of interviewees was explicitly informed that their narrations would be analyzed within a West European research project. Some respondents displayed some irritations about this kind of Western curiosity. They were much more reluctant in answering questions on everyday life and Soviet television when I myself conducted the interviews than those respondents who were interviewed by my Russian colleagues. This gap seemed to be due to the fact that the course of an interview is much influenced by the way the interviewer reacts to the interviewee's responses and which issues she or he further tracks. Certainly, my Russian colleagues sometimes listened and responded differently to the interviewees than I was able to because of my differing sociocultural socialization.[69]

The interviewers applied a guide for semistructured qualitative interviews that I had first refined in my own interviews.[70] They referred to the guide in a flexible way to possibly take up the topics the respondents touched upon. They should remember any watching habits, the invention of certain watching traditions, or any family quarrels about the TV program. The questions should stimulate them to talk rather freely about their leisure practices, their media consumption, and the TV set as consumer good. At best, they should remember the first time they saw a TV set, the purchase of TV sets, and a TV's domestication by rearranging the living room or placing it on the refrigerator in the kitchen. Thus, the questions should stir the interviewees to think of their earliest memories of television, to look back on concrete practices and changes of media consumption during their life. Did they read newspapers and listen to the radio? How did their media consumption change with the emergence of television? The respondents should narrate about the way TV impacted on their leisure practices, which programs they preferred, how they assessed them, and how they perceived entertainment, documentary, and news programs and made up their idea of "Soviet television." I also intended to let them highlight the changes perestroika brought about and to compare their Soviet and post-Soviet media uses.

These questions should enable us to trace central narratives people shaped to make sense of their Soviet life with the specific view to their leisure practices and media use. The perspective of their current life and the way they tackled the transition period and coped with the changes surely highly impacted on the development of the narratives. Although interviews are to a certain extent a snapshot of the historical trajectory at that specific moment, they also give insights into patterns of remembering as they relate different layers of time. As historians analyze interviews primarily the same way they explore other textual sources, there is little interest in linguistic and text-structuring aspects of narratives but in the creation, arrangement, and framing of past practices. The question as to which practices the respondent remembered in which way is closely connected to the past and present social, cultural, and economic contexts. Paul Ricœur established the interesting argument that the memory work of the individual is complementary to collective memories. The sociocultural contextualization of memory makes it possible to attribute one's own mnemonic aspects to others and the whole society as soon as people recognize the value of the memories of the others.[71]

People's relations always facilitate and constrain their making of meanings and the life stories they tell. Narratives based on remembrances thus constitute an active process of putting the past into a personally meaningful trajectory linking past and present. They themselves practice giving diverse, ambivalent, and not seldom contradictory experiences and developments a temporal and topical order. It seems that people often aim to imbue their life stories with certain coherence, perhaps even continuity. This strategy not seldom means to be selective toward the past, to leave some points out or to forget them.[72] This tendency becomes rather obvious, as many interviewees tended to perceive the reforms of perestroika as a catastrophe that came over them from the outside.[73] Many respondents struggled to integrate their perceptions of perestroika and the demise of the Soviet Union into their life stories. Interestingly enough, however, these decisive moments often did not cause a clear temporal structuring of the narration in periods before and after. On the contrary, many respondents merged time layers and needed to be reminded by the interviewers to define the years they were talking about. This way of remembering and narrating the Soviet life story proved at least partly the respondents' striving for coherence. Like a trauma, it also showed that even a catastrophe does not prevent the dropping of memories and the tendency to oblivion.

The sample of respondents does not thoroughly represent the social structure of European Russia. It comprises forty-eight women and thirty-two men born between 1929 and 1965. Thus, the focus is on an age group of people who grew up before perestroika and experienced a profound change of the political system as adults. This upheaval often specifically impacted on the way they related their Soviet and post-Soviet life stories and made the interviews even more interesting regarding the political use of history and nostalgia.

Women were overrepresented for pragmatic reasons, as they were much more willing to be interviewed. Instead of striving for representativity, the sample collected a variety of ideal types that should help to assess different

Soviet media users based on a qualitative approach. The ideal types were drawn from those whom media scholar Michael Meyen identifies as audience for the GDR TV, stressing that he always found a mixture of different TV users: the compliant, the affirmative, the indifferent, the reserved, and the subversive TV viewer.[74] Although the ideal-typical categories should not limit the interpretation of the respondents' narrations, the review of the first interviews indicated that many narrations roughly revealed similar practices and assessments. It seemed to be justified to accept an overrepresentation of higher-educated interviewees—64 percent among women, 70 percent among men—in relation to the Soviet populace, because they tended to reveal a greater variety of viewing habits and TV content assessment than less-educated people. In addition, higher-educated people were often more talkative and willing to engage themselves in the questions and issues raised by the interviewers.

Most of the respondents were urban dwellers, although rural areas were well represented. During the Soviet era, they lived in the cities and counties of Moscow, Leningrad, Kuibyshev (Samara), Rostov-on-Don, Yaroslavl, Irkutsk, and Kemerovo. Some moved from villages to the city at a later stage of life and were able to witness the changes of urban and rural lifestyles.

Most interviewees represented themselves as having belonged to the "middle" of Soviet society and thus met my claim to interview ordinary people, at least in their own perception. Indeed, they figured as former ordinary Soviet citizens, as only a few of the interviewees were rather distinguished in their purchasing power, privileges, and social networks in Soviet times. Nevertheless, they conveyed a certain variety of social groups, as some of them were blue-collar workers, whereas others taught at universities.

Further, the interview sample included a small number of experts. These persons—three women and three men—worked at regional TV stations since the 1960s. Ėvelina E. and Roman R. at Rostov TV, Liudmila at Irkutsk TV station, and V. V. and Valentina V. at Kuibyshev television. Boris M. Firsov is an outstanding personality who was director of Leningrad TV station from 1962 to 1966 and then turned to an academic career. These experts gave valuable insights into the daily work and let us trace professional perceptions of Soviet television.

It is, however, generally a heuristic challenge that interviews certainly do not immediately disclose contemporary perceptions of emotional bonding, confidence, and the normalization of Soviet lifestyles. The respondents' narratives are highly fractured by the personal course of life during the sociopolitical, cultural, and economic changes, as well as by the current situation at the time of the interview. Nevertheless, the narrative amalgamation of several temporal layers is the especially interesting potential of interviews, as it helps to get a better understanding of how the Russian society deals with its Soviet legacy. Hence, we often meet affirmative emotional ascriptions to Soviet values embedded in nostalgic narratives about the lost Soviet order. Nostalgia is a very widespread phenomenon in contemporary Russia involving different groups of societal actors. It must be considered as a very influential social and political interpretation frame in post-Soviet Russia. Understood as part of narrative strategies making sense of the past, it reveals how people compare their current and past lives, on which basis they judge the changes in social practices and political values they perceive as important.

Looking back, former ordinary Soviet citizens currently include their Soviet everyday practices and the changes they underwent into discursive constructions of the "normal" Soviet life. From today's point of view, the "normality" of leisure practices, such as watching TV, reading journals, going to the cinema, even queuing for consumer goods, was closely related to their private life and seemingly reliable future life prospects. These subjective dimensions of normality provide ground for current nostalgic representations of socioeconomic stability and political security during the 1970s and pre-perestroika years. In the post-Soviet years, stability and security, having allegedly characterized the "Golden" 1970s retrospectively, became categories legitimizing and appreciating the lost Soviet model of society.[75] This observation corresponds to the findings of Russian sociologists who have recently established that certain Soviet social values like social justice and equality are still widespread in contemporary Russia.[76] The actual strong increase of social diversity in the post-Soviet society and the retrospective knowledge about the unjust distribution of privileges in Soviet times further underline the importance of nostalgia as narrative strategy in the current Russian political culture. This impact on the political culture is, not least, crucial as nostalgia in the sense of a "yearning for yesterday" refers to a state of transformation, of uncertainty, and of adjustment. The remembered lost order appears to have satisfied certain emotional needs that have become contested in the present. Russian-American comparative literature scholar Svetlana Boym conceptualizes nostalgia as consisting of two ideal-typical main narratives, the reflective and the restorative one. The restorative strategy aims to reconstruct the lost order. It is pinned to tradition, tries to construct terms of belonging, and does not reflect upon cultural ambivalences. It seems to preponderate in many contemporary societies of former state socialist countries.[77]

In contrast, the reflective narrative strategy allows for ambivalence, for shattered dreams of the past to become visions of the future, even for ironic interpretations of the lost order or cultural forms.[78] People thus may take up several places and stances in time, picking out certain cultural elements, social forms, or political values of the late Soviet era without necessarily appreciating the Soviet order as a whole. The reflective and restorative strategies are often so deeply entangled that the question to what extent the restorative strategy implies concrete demands for the restoration of the lost order is hard to answer.

To take account of inconsistencies, ambivalences, and antagonistic representations of the past in the respondents' narratives, I suggest a third narrative ideal-typical strategy that I propose to call "synthesizing." It emphasizes the process of synthesizing memories and reflects the observation that people often merge conflicting interpretations about the past. In comparison to Boym's concept of the reflective strategy, the synthesizing narrative is less intellectual and more down-to-earth. It leaves little space for irony or utopian ideas. However, it may express open sympathies for the lost certainties, values, and practices of the former Soviet life. But unlike the restorative, the synthesizing strategy is characterized by rather pragmatic considerations, comparisons, and differentiations between the Soviet past and the post-Soviet present.[79]

Basically, all nostalgic narratives offer ground for current political strategies of reviving Soviet practices, values, and concepts. In addition, the different cultural and political contexts of the various post-socialist states determine the actual meanings, appropriations, the particular impacts on the historical consciousness, and the use of the Soviet past.[80]

Apart from official memory politics, we should consider nostalgic practices and emotions as subjective ways of bringing the lost state and certain Soviet experiences to mind again. Nostalgia represents at least a symbolic, perhaps even mental return to the past and has a strong transfiguring or obscuring potential.[81] Certainly, an important aspect to explore is the extent to which nostalgic narratives detach the lost material, social, and cultural practices, things, and symbols from their original political and ideological context. In any case, the longing for them gives us a notion of how the interviewees attribute specific sense to their past Soviet lives by contextualizing their memories.[82] It is as important to indicate the possible historical shallowness as to explore the ingredients and motives of the nostalgic narratives: Do they refer to biographical cuts, to a lost—maybe retrospectively constructed—feeling of a collective belonging, to concrete symbols or practices that were common and familiar? And if so, to what extent are they imbued with political meanings and let TV programs and media use figure as an influential "soft power" of stabilizing the Soviet order?

However, interviews stimulate the respondents to update memories by entangling temporal and spatial layers of the Soviet past with current perceptions. Although nostalgia differs from contemporary contentedness, the respondents' narratives may reveal the communicative foundation of the society. They give us a notion as to how the interviewees attribute sense to their past Soviet lives by contextualizing their memories.

Further, Soviet television has itself interlaced different time layers and dimensions of space. From its beginnings in the mid-1950s, viewers perceived television as having opened a "window to the world," as some respondents put it. Current television complicates these layers even more by rerunning Soviet features today. Against this background, the interviews perfectly reveal self-perceptions, individual practices, motives, and interpretations on TV watching and media use.

This book is more thematically than chronologically organized, although the five chapters internally follow a chronological narrative. It approaches television consumption from different angles. The arguments referring to the interviewees' remembrances sometimes leave a circumscribed time frame, as the respondents tended to merge temporal structures.

Chapter 1 focuses on quantitative, technical, and aesthetic dimensions of Soviet television consumption in order to highlight the way the television sets impacted on domestic interiors and consumer desires. It explores how the TV set gained sociocultural importance as an object of consumer desire, how it yielded specific consumer practices, and how it changed aesthetics and hierarchies in the domestic home space. The first part of this chapter asks for the quantitative aspects of the changing consumer world and looks at the rising distribution of TV sets. To relate this statistical site of the material culture to subjective meanings and cultural practices, I take up narratives of former Soviet TV viewers about technical developments, finesses, and flaws of Soviet TV sets. The second part of the chapter analyzes how the TV set impacted on consumer desires. When and why did the desire emerge to buy a TV set? Which family member initiated the purchase and decided on the model? How did ordinary consumers handle them and found ways to buy a TV set? Which were the challenges that might have obstructed the purchase? Which were the distribution channels, and to what extent did personal relations and money matter to get hold of a TV set?

The second chapter is, on the one hand, interested in the way television emotionalized the viewers. On the other hand, it asks to what extent television created a credible world of information to the audience and how it was challenged by foreign radio programs. Therefore, I examine how television impacted on privacy, interpersonal relations, and family ties. This means that I trace quantitative aspects of television's impact on the changing leisure practices. It should be highlighted in what way gender, education, social, cultural, and economic aspects determined watching television. I highlight these developments based on contemporary sociological studies and mirror them with my findings derived from the oral history interviews: How did the respondents make sense of the socio-significance of the new media use? How did they embed watching TV into their everyday family life? How did the respondents use the foreign radio programs? Thus, it should be established to what extent viewers perceived television as a "window to the world."

Chapter 3 explores how television modified the hierarchy of performing arts in the perception of contemporaries. How did it change the assessment of cultural professionals, creative artists, and journalists of the established hierarchy? How did the new medium change the debates on Soviet mass culture and the claim to educate the New Soviet man? To what extent did television shift the established boundaries between pop culture and entertainment on the one hand and the classical genres like ballet, opera, and drama on the other. To what extent did contemporaries classify television as cultural facilitator which let "ordinary" citizens gain access to the so-far dominating cultural canon? To what extent did the television broadcasts of classical performing arts impact on the new medium's standing?

This chapter focuses on the contemporary elite debates about television and brings them together with retrospective perspectives of former Soviet television viewers based on interviews. The contemporary observers and representatives of the established arts dominated the important feuilleton and culture sections of print media and radio that kept an eye on the new medium. They were keenly and somehow unrelentingly engaged in drawing the line between legitimate and illegitimate pleasure, between lowbrow entertainment and highbrow arts. The debates aimed to define the social, political, and cultural significance of the new medium and its programs. Thus, this chapter tells about the contemporary perceptions of television as medium and is not interested in the artistic and aesthetic arrangements of the television programs. Also, it does not concentrate on how television affected the audience but, following Elihu Katz, on how the viewers used the medium. This opens a perspective on viewers' motivation to watch TV, on their ideas of entertainment and pleasure. Entertainment and pleasure are cognitive effects that are not easy to grasp, as they depend on time, social, political, and cultural contexts, and so on. They were closely related to those debates that aimed to define sociocultural boundaries.[83] I apply a broad notion of entertainment and pleasure that include laughter, relaxation, and distraction felt by viewers watching a film, a philharmonic concert, as well as historical documentaries, animal features, or Central Television's nightly news program.[84] We will take a closer look at how contemporaries reinterpreted concepts of popular culture and of the established hierarchy of performing arts considering the new technical capacities of television.

Chapter 4 examines the communicative relationship between audience and television. This refers to the way television workers imaged the audience, the way they got into contact, as well as the specificity of letters sent to television's editorial desks and the demands of viewers toward the medium. What do these viewer letters tell us about the role television played in Soviet people's leisure habits and about their viewing practices? What do they reveal about the question of how "ordinary" citizens made use of television programs and how did they assess the television content? What does the mail reveal about viewers' emotional bonding toward the medium within an officially and societally accepted popular culture and representation of late Soviet lifestyle? To what extent did television engage its audience in political communication? How might the involvement of the audience have facilitated the communicative stabilization of the Soviet regime up to 1985?

Chapter 5 is an inquiry into the communication between Soviet television and its audience about consumer issues. It refers to the local TV stations of Rostov and Leningrad, as well as to Moscow Central Television. I highlight the communicative strategies the television stations employed to address viewer-consumers. How did Soviet TV aim to involve the audience? How did TV offer viewer-consumers information on consumption issues? How did viewer-consumers take up this discourse, and to what extent did they claim their rights on consumption via letters to the television stations?

The chapter explores how TV programs like Rostov Television's *Pogovorim o vashikh pis'makh* and Central Television's *Bol'she khoroshikh tovarov* (More good products) visualized consumer issues, how viewers reacted to them, and how television interlaced concrete cases with broader debates on consumption. I chose Rostov-on-Don because it is located about 1,000 kilometers from Moscow and does not belong to its coverage area. Furthermore, Rostov's regional TV station was on air as early as 1958. Rostov-on-Don has traditionally been a prosperous city and an economic, cultural, and educational center of European Russia's southwest. Leningrad's TV station was the most important local one and had potentially more liberties than smaller TV stations.

The book concludes with some thoughts on the way television normalized the Soviet way of life, on television's lasting impacts on Soviet popular culture, and the idea of Soviet cultural superiority. Searching for the communicative changes perestroika brought about, I finally touch upon the ambiguous role television played in the last years of the Soviet Union as it started to televise people's voices on the street and to stage the unexpected economic and political transformation.

NOTES

1. The interview with Elena was conducted by Elena Bogdanova in St. Petersburg in October 2010.

2. When I speak of "Soviet television," I mean a complex setting of different interest groups that influenced television programming and contents: the different professional groups such as journalists, editorial staff, artists, producers and directors, anchor people, presenter, and technical staff on the one hand and party members

trying to impact on television on the other. Where necessary to clarify the argument, I try to name the concrete actors.

3. I cite the names of the respondents according to their agreement. Most consented to give their first name; some also agreed to give their patronym.

4. The interview with Elena Sergeevna was conducted by Oksana Zaporozhets in Samara in October 2010. She grew up in a small village in the Astrakhan region, moved to Kuibyshev (Samara) and graduated from the local history department. She worked as a teacher and today she is a saleswoman.

5. The interview with Georgii was conducted by Elena Bogdanova in September 2010. Georgii graduated from university but works now as a public employee in St. Petersburg. Like his father, he was a member of the Communist Party.

6. In my understanding, an "ordinary" citizen was someone who did not enjoy access to the elites' system of distribution. He or she was defined by his or her lack of formal political power. By contrast, "regime" refers to members of organizations such as the party, government, and trade unions.

7. Susan E. Reid, "The Meaning of Home: 'The Only Bit of the World You Can Have to Yourself,'" in *Borders of Socialism: Private Spheres of Soviet Russia*, ed. Lewis Siegelbaum (New York, Houndmills, Basingstoke: Palgrave Macmillan, 2006), 145–170.

8. For a still-stimulating conceptualization of researching communication mechanisms, the public sphere, and the role of mass media, see: Jörg Requate, "Öffentlichkeit und Medien als Gegenstände historischer Analyse," *Geschichte und Gesellschaft* 25 (1999): 5–32.

9. Siegfried Kracauer, *Totalitäre Propaganda*, ed. Bernd Stiegler (Berlin: Suhrkamp Verlag, 2013), 29–30.

10. Ulrich M. Schmid, *Technologien der Seele. Vom Verfertigen der Wahrheit in der russischen Gegenwartskultur* (Berlin: Suhrkamp Verlag, 2015).

11. Cf. Jörg Requate, "Kommunikation und Propaganda in den Medien der DDR. Möglichkeiten und Grenzen des kommunikativen Austausches zwischen den audiovisuellen Medien der DDR und ihrem Publikum," *Siegener Periodicum zur Internationalen Empirischen Literaturwissenschaft (SPIEL)* 14, 2 (1995): 230–243; Thymian Bussemer, *Propaganda und Populärkultur. Konstruierte Erlebniswelten im Nationalsozialismus* (Wiesbaden: Deutscher Universitätsverlag, 2000).

12. Alexei Yurchak has borrowed the notion of "authoritative discourse" from Mikhail Bakhtin. Cf. Alexei Yurchak, *Everything Was Forever, Until It Was No More: The Last Soviet Generation* (Princeton, NJ: Princeton University Press, 2005), 24–29, 36–76.

13. Kristin Roth-Ey, *Moscow Prime Time. How the Soviet Union Built the Media Empire that Lost the Cultural Cold War* (Ithaca, NY and London: Cornell University Press, 2011), 208–222.

14. Roth-Ey, *Moscow*; Christine E. Evans, *Between Truth and Time. A History of Soviet Central Television* (New Haven, CT and London: Yale University Press, 2016); idem, "Song of the Year and Soviet Culture in the 1970s," *Kritika: Explorations in Russian and Eurasian History* 12, 3 (2011): 617–645; idem, "The Soviet Way of Life as a Way of Feeling: Emotion and Influence on Soviet Central Television in the Brezhnev Era," *Cahiers du monde russe* 56, 2–3 (2015): 543–69.

15. The interview with Liudmila Vasil'evna was conducted by Galina Orlova in Rostov-on-Don in October 2010. Liudmila was born in the small town of Matveev-Kurgan, Rostov-on-Don region, in 1946. She is higher educated.

16. Jean-Pierre Esquenazi, "Télévision et espace public sous De Gaulle," *Cahiers d'histoire. Revue d'histoire critique* 86 (2002): 49–61; Aude Vassallo, *La télévision sous de Gaulle. Le contrôle gouvernemental de l'information (1958/1969)* (Bruxelles: De Boeck Supérieur, 2005); Meike Vogel, *Unruhe im Fernsehen. Protestbewegung und öffentlich-rechtliche Berichterstattung in den 1960er Jahren* (Göttingen: Wallstein Verlag, 2010), 39–47.

17. Mark Hopkins was one of the more leftist Anglophone social scientists who drew on the idea of convergence between the ideological systems with regard to communication and media mechanisms. Cf. Mark W. Hopkins, *Mass Media in the Soviet Union* (New York: Pegasus, 1970).

18. Gayle Durham Hollander, *Soviet Political Indoctrination. Developments in Mass Media and Propaganda since Stalin* (New York, Washington, London: Praeger, 1972); Alex Inkeles, *Public Opinion in Soviet Russia. A Study in Mass Persuasion* (Cambridge, MA: Harvard University Press, 1951); Paul Roth, *Die kommandierte öffentliche Meinung. Sowjetische Medienpolitik* (Stuttgart: Seewald, 1982); with regard to the GDR rather exemplary for a strict sender-receiver model: Gunter Holzweißig, *Zensur ohne Zensor. Die SED-Informationsdiktatur* (Bonn: Bouvier Verlag, 1997); idem, *Die schärfste Waffe der Partei. Eine Mediengeschichte der DDR* (Köln, Weimar, Wien: Böhlau Verlag, 2002).

19. Barbara Thomaß and Michaela Tzankoff eds., *Medien und Transformation in Osteuropa* (Wiesbaden: Westdeutscher Verlag, 2001); John Downey and Sabina Mihelj eds., *Central and Eastern European Media in Comparative Perspective. Politics, Economy and Culture* (Farnham/Surrey, Burlington: Ashgate, 2012); Julie A. Cassiday and Emily D. Johnson, "Putin, Putiniana and the Question of a Post-Soviet Cult of Personality," *The Slavonic and East European Review* 88 (2010): 681–707; Kaarle Nordenstreng, Elena Vartanova and Yassen Zassoursky eds., *Russian Media Challenge* (Helsinki: Kikimora Publications, 2002).

20. Cf. for example: Stephen Hutchings and Natalia Rulyova eds., *Television and Culture in Putin's Russia. Remote Control* (London, New York: Routledge, 2009); Ellen Mickiewicz, "Excavating Concealed Tradeoffs: How Russians Watch the News," *Political Communication* 22, 3 (2005): 355–380; Sarah Oates, "The Neo-Soviet Model of the Media," *Europe-Asia Studies*, 59, 8 (2007): 1279–1297.

21. This proliferation has already led to the publication of some edited volumes containing empirical and conceptual contributions. Cf. Alexander Badenoch, Andreas Fickers and Christian Henrich-Franke eds., *Airy Curtains: European Broadcasting during the Cold War* (Baden-Baden: Nomos-Verlag, 2013); Aniko Imre, Timothy Havens and Katalin Lustyk eds., *Popular Television in Eastern Europe During and Since Socialism* (Abingdon, New York: Routledge, 2012); idem, "Audience History as a History of Ideas: Towards a Transnational History," *European Journal of Communication* 30, 1 (2015): 22–35.

22. To mention just a few studies: Michael Meyen, *Einschalten, Umschalten, Ausschalten? Das Fernsehen im DDR-Alltag* (Leipzig: Leipziger Universitätsverlag,

2003); Uwe Breitenborn, *Wie lachte der Bär? Systematik, Funktionalität und thematische Segmentierung von unterhaltenden nonfiktionalen Programmformen im Deutschen Fernsehfunk bis 1969* (Berlin: Weißensee Verlag, 2003); Claudia Dittmar, *Feindliches Fernsehen: Das DDR-Fernsehen und seine Strategien im Umgang mit dem westdeutschen Fernsehen* (Bielefeld: Transcript, 2010); Claudia Dittmar and Susanne Vollberg eds., *Zwischen Experiment und Etablierung. Die Programmentwicklung des DDR-Fernsehens 1958 bis 1963* (Leipzig: Leipziger Universitätsverlag, 2007); idem eds., *Alternativen im DDR-Fernsehen? Die Programmentwicklung 1981 bis 1985* (Leipzig: Leipziger Universitätsverlag, 2005); Heather Gumbert, *Envisioning Socialism: Television and the Cold War in the German Democratic Republic* (Ann Arbor, MI: University of Michigan Press, 2014).

23. Paulina Bren, *The Greengrocer and His TV. The Culture of Communism after the 1968 Prague Spring* (Ithaca, NY and London: Cornell University Press, 2010); Irena Reifová, "A Study in the History of Meaning-Making: Watching Socialist Television Serials in the Former Czechoslovakia," *European Journal of Communication* 30, 1 (2015): 79–94; Patryk Wasiak, "The Great Époque of the Consumption of Imported Broadcasts: West European Television Channels and Polish Audiences during the System Transition," *VIEW. Journal of European Television History and Culture* 3, 5 (2014): 68–68; Michael Zok, "Das polnische Fernsehen in den 1980 Jahren. Polska Telewizja als Gegenstand und Austragungsort politischer Konflikte," *Rundfunk und Geschichte* 39, 3–4 (2013): 25–34; idem, *Die Darstellung der Judenvernichtung in Film, Fernsehen und politischer Publizistik der Volksrepublik Polen 1968–1989* (Marburg: Verlag Herder-Institut, 2015); Dana Mustata, "'The Revolution Has Been Televised . . .': Television as Historical Agent in the Romanian Revolution," *Journal of Modern European History*, 10, 1 (March 2012): 76–97; Sabina Mihelj, "Negotiating Cold War Culture at the Crossroads of East and West: Uplifting the Working People, Entertaining the Masses, Cultivating the Nation," in *Comparative Studies in Society and History* 53, 3 (2011): 509–539; idem, "The Politics of Privatization: Television Entertainment and the Yugoslav Sixties," in *The Socialist Sixties: Crossing Borders in the Second World*, ed. Anne E. Gorsuch and Diane P. Koenker (Bloomington, IN: Indiana University Press, 2013), 251–267.

24. Evans, "The 'Soviet Way of Life'"; idem, *Between Truth*; Roth-Ey, *Moscow*.

25. Cf. recently Nada Boškovska, Angelika Strobel and Daniel Ursprung eds., *'Entwickelter Sozialismus' in Osteuropa. Arbeit, Konsum und Öffentlichkeit* (Berlin: Duncker & Humblot, 2016); Luminita Gatejel, *Warten, hoffen und endlich fahren. Auto und Sozialismus in der Sowjetunion, in Rumänien und der DDR (1956–1989/91)* (Frankfurt, New York: Campus, 2014); Anne E. Gorsuch and Diane P. Koenker eds., *The Socialist Sixties: Crossing Borders in the Second World* (Bloomington, IN: Indiana University Press, 2013); Natalya Chernyshova, *Soviet Consumer Culture in the Brezhnev Era* (London, New York: Routledge, 2013); Milena Veenis, *Material Fantasies. Expectations of the Western Consumer World among East Germans* (Amsterdam: Amsterdam University Press, 2012); David Crowley and Susan E. Reid eds., *Pleasures in Socialism: Leisure and Luxury in the Eastern Bloc* (Evanston, IL: Northwestern University Press, 2010); Patrick Hyder Patterson, *Bought and Sold: Living and Losing the Good Life in Socialist Yugoslavia* (Ithaca, NY: Cornell

University Press, 2011); Breda Luthar and Maruša Pušnik eds., *Remembering Utopia: The Culture of Everyday Life in Socialist Yugoslavia* (Washington, DC: New Academia Publishing, 2010).

26. Yurchak, *Everything was forever*; Sergei I. Zhuk, *Rock and Roll in the Rocket City: The West, Identity, and Ideology in Soviet Dnipropetrovsk, 1960–1985* (Baltimore, MD: Johns Hopkins University Press, 2010).

27. Cf. the GDR: Michael Meyen, "Haben die Westmedien die DDR stabilisiert? Zur Unterhaltungsfunktion bundesdeutscher Rundfunkangebote," *SPIEL* 20, 1 (2001): 117–133.

28. James Dingley, "Soviet Television and Glasnost,'" in *Culture and the Media in the USSR Today,* eds. Julian Graffy and Geoffrey Hosking (London/New York: St. Martin's Press, 1989), 6–25. Cf. also: Ellen Mickiewicz, *Split Signals: Television and Politics in the Soviet Union* (New York, Oxford: Oxford University Press, 1988).

29. Roth-Ey, *Moscow*, 7, 23.

30. Alaina Lemon, "Sympathy for the Weary State?: Cold War Chronotopes and Moscow Others," *Comparative Studies in Society and History* 51, 4 (2009): 832–864; Sergei A. Oushakine, "'We're Nostalgic But We're Not Crazy': Retrofitting the Past in Russia," *The Russian Review* 66 (2007): 451–482; Kevin M. F. Platt, "Russian Empire of Pop: Post-Socialist Nostalgia and Soviet Retro at the 'New Wave' Competition," *The Russian Review* 72 (2013): 447–469; Ekaterina Kalinina, "Multiple Faces of the Nostalgia Channel in Russia," *VIEW. Journal of European Television History and Culture* 3, 5 (2014): 108–118; Anna Arutunyan, *The Media in Russia* (Maidenhead/Berkshire: McGraw Hill Open University Press, 2009), 19.

31. Cf. the recent contributions in Kirsten Bönker, Julia Obertreis and Sven Grampp eds., *Television Beyond and Across the Iron Curtain* (Cambridge: Cambridge Scholars Publishing, 2016). Katalin Lustyik, "From a Socialist Endeavour to a Commercial Enterprise: Children's Television in East-Central Europe," *Popular Television in Eastern Europe During and Since Socialism*, eds. Anikó Imre, Timothy Havens and Katalyn Lustyik (New York, London: Routledge, 2013), 105–122; Thomas Beutelschmidt and Richard Oehmig, "Connected Enemies? Programming Transfer between East and West during the Cold War and the Example of East German Television," *VIEW. Journal of European Television History and Culture* 3, 5 (2014): 60–67; Yulia Yurtaeva, "Intervision. Searching for Traces," *VIEW. Journal of European Television History and Culture* 3, 5 (2014): 23–34; Heather L. Gumbert, "Exploring the Transnational Media Exchange in the 1960s," *VIEW. Journal of European Television History and Culture* 3, 5 (2014): 50–59; Michael Meyen, *Denver Clan und Neues Deutschland. Mediennutzung in der DDR* (Berlin: Christoph Links Verlag, 2003); Wasiak, "The Great Époque."

32. Andreas Fickers and Catherine Johnson, "Transnational Television History: A Comparative Approach. Introduction," *Media History* 16, 1 (2010): 1–11.

33. Richard Stites, *Russian Popular Culture: Entertainment and Society Since 1900* (Cambridge: Cambridge University Press, 1992), 148–149.

34. Susan E. Reid, "Cold War in the Kitchen: Gender and the De-Stalinization of Consumer Taste in the Soviet Union under Khrushchev," *Slavic Review* 61 (2002): 211–252; Kirsten Bönker, "Der sowjetische Fernsehzuschauer als ‚neuer Mensch':

Die sozialwissenschaftliche Freizeitforschung und das Problem der Rückständigkeit," in *Die Zukunft der Rückständigkeit. Festschrift für Manfred Hildermeier*, ed. David Feest and Lutz Häfner (Köln, Weimar, Wien: Böhlau, 2016), 373–397.

35. Cf. http://www.constitution.org/cons/ussr77.txt (accessed September 1, 2014). The 1936 constitution had also provided a right to rest and leisure (Article 119). Radio was not included. The press was mentioned, comparable with the 1977 constitution press, in the context of the freedom of speech (Article 125).

36. See my own attempt to describe the new dynamics that characterized private life after the Khrushchev era: Kirsten Bönker, "Depoliticalisation of the Private Life? Reflections on Private Practices and the Political in the Late Soviet Union," in *Writing Political History Today*, ed. Willibald Steinmetz, Ingrid Gilcher-Holtey and Heinz-Gerhard Haupt (Frankfurt a. M.: Campus, 2013), 207–234.

37. Alexei Yurchak was one of the first researchers who rejected binary models and strived to explain the Soviet Union's longevity: Cf. *Everything Was Forever, Until It Was No More. The Last Soviet Generation* (Princeton, NJ: Princeton University Press, 2006). See also Lewis H. Siegelbaum, "Introduction," in *Borders of Socialism. Private Spheres of Soviet Russia*, ed. ibid (New York: Palgrave Macmillan, 2006), 1–21, esp. 3; Ekaterina Emeliantseva, "The Privilege of Seclusion: Consumption Strategies in the Closed City of Severodvinsk," *Ab Imperio: Studies of New Imperial History and Nationalism in the Post-Soviet Space* 2 (2011): 238–259, esp. 238–244; see on the CSSR, Bren, *The Greengrocer*; on the GDR, Mary Fulbrook, *The people's state. East German Society from Hitler to Honecker* (New Haven, CT: Yale University Press, 2005).

38. GARF, f. 6903, op. 1, d. 667, 1960: Dokumenty o rabote obshchestvennogo soveta telezritelei pri tsentral'noi studii televideniia (spravki, predlozheniia), ll. 30–31.

39. Martha C. Nussbaum, *Politische Emotionen* (Berlin: Suhrkamp, 2014), 13–24.

40. Nussbaum, *Emotionen*, 25.

41. A fundamental overview is provided by Katrin Döveling, *Emotionen—Medien—Gemeinschaft. Eine kommunikationssoziologische Analyse* (Wiesbaden: VS Verlag, 2005); Katrin Döveling, Christian von Scheve and Elly A. Konijn eds., *The Routledge Handbook of Emotions and Mass Media* (London, New York: Routledge, 2010).

42. Elihu Katz, "Mass Communication Research and the Study of Popular Culture: An Editorial Note on a Possible Future for This Journal," *Studies in Public Communication*, 2, 1–6 (1959): 2.

43. Cf. the inspiring ideas of Siegfried J. Schmidt, "Medien, Kultur: Medienkultur. Ein konstruktivistisches Gesprächsangebot," in *Kognition und Gesellschaft. Der Diskurs des Radikalen Konstruktivismus*, ed. idem (Frankfurt a. M.: Suhrkamp, 1992), 425–450. In relation to liberal systems, it is admittedly not uncontested. Some media scholars note that current media cultures also create contradictions and social disintegration. These phenomena are surely not new. See the critical remarks of Andreas Hepp, *Medienkultur. Die Kultur mediatisierter Welten* (2nd ed. Wiesbaden: VS Verlag, 2013), 16–21.

44. Manfred Zeller, *Das sowjetische Fieber. Fußballfans im poststalinistischen Vielvölkerreich* (Stuttgart: *ibidem*-Verlag, 2015), 145–180.

45. Cf. Ulrich Saxer and Martina Märki-Koepp, *Medien-Gefühlskultur. Zielgruppenspezifische Gefühlsdramaturgie als journalistische Produktionsroutine* (München: Oelschläger, 1992), 51. This term has also not remained uncontested. Cf. a critical review: Andreas Hepp, *Netzwerke der Medien: Medienkulturen und Globalisierung* (Wiesbaden: VS Verlag, 2004), 74–75.

46. Rüdiger Schnell, *Haben Gefühle eine Geschichte? Aporien einer History of emotions*, part 1 (Göttingen: V&R, 2015), 15–59.

47. Döveling, *Emotionen—Medien—Gemeinschaft*, 183.

48. Knut Hickethier argues that East German television offered viewers a cozy image of the GDR characterized by a specific sentimental culture of familiarity. He states that sports programs, in particular, might have ensured "identification" with the state. See Knut Hickethier, "Das Fernsehen der DDR," in *Wie im Westen, nur anders: Medien in der DDR*, ed. Stefan Zahlmann (Berlin: Panama-Verlag, 2010), 123, 128.

49. Michail M. Bachtin, "Das Wort im Roman," in idem, *Die Ästhetik des Wortes*, ed. Rainer Grübel (Frankfurt a. M.: Suhrkamp, 1979), 244–247.

50. Homi K. Bhabha, *The Location of Culture* (London: Routledge, 1994), 33, 56, 312; Silvia Sasse, *Michail Bachtin zur Einführung* (Hamburg: Junius, 2010), 136–140.

51. Umberto Eco, *Das offene Kunstwerk* (Frankfurt a. M.: Suhrkamp, 1996), 201–203.

52. Stuart Hall, "Encoding/Decoding," in *Media and Cultural Studies: Key Works,* eds. Meenakshi Gigi Durham and Douglas M. Kellner (Malden, MA: Blackwell, 2006), 163–173.

53. Stuart Hall, "Die strukturierte Vermittlung von Ereignissen," in *Grundlagentexte zur Fernsehwissenschaft. Theorie—Geschichte—Analyse*, ed. Ralf Adelmann et al. (Konstanz: UVK, 2001), 357.

54. Niklas Luhmann, *Die Realität der Massenmedien* (4th ed. Wiesbaden: VS Verlag für Sozialwissenschaften, 2009), 121–122.

55. Luhmann, *Realität*, 77, 112, 115.

56. Niklas Luhmann, *Die Wissenschaft der Gesellschaft* (Frankfurt a. M.: Suhrkamp, 1990), 566; idem, *Soziale Systeme. Grundriß einer allgemeinen Theorie* (Frankfurt a. M.: Suhrkamp 1987), 111.

57. Roger Chartier, "The Chimera of the Origin: Archaeology, Cultural History, and the French Revolution," in *Foucault and the Writing of History*, ed. Jan Goldstein (Oxford, Cambridge, MA, 1994), 167, 174, 184–185.

58. Roger Chartier, *Lesewelten. Buch und Lektüre in der frühen Neuzeit*, Frankfurt a. M., New York, Paris, 1990, 7–24; idem, "Le Monde comme representation," *Annales ESC* 6 (1989): 1505–1520.

59. Chartier, "Chimera," 186; idem, "Die Welt als Repräsentation," in *Alles Gewordene hat Geschichte. Die Schule der ANNALES in ihre Texten* 1929–1992, ed. Matthias Middell, Steffen Sammler (Leipzig: Reclam, 1994), 330–331.

60. Chartier employs the notion of "representation" to connect classifications, practices, and structures. Cf. Chartier, Welt, 329–343.

61. See Roger Chartier, *Forms and Meanings: Texts, Performances, and Audiences from Codex to Computer* (Philadelphia, PA: University of Pennsylvania Press,

1995), 83–97; idem., *Lesewelten. Buch und Lektüre in der frühen Neuzeit* (Frankfurt a. M., New York, Paris: Campus, 1990), 7–24.

62. John Fiske, *Television Culture* (London, New York: Routledge, 1987), 1.

63. Cf. Collaborative Research Centre (Sonderforschungsbereich) 584: *The Political as Communicative Space in History. Research Programme* (Manuscript: University of Bielefeld, 2007) http://www.uni-bielefeld.de/%28en%29/geschichte/forschung/sfb584/research_program/index.html (accessed June 12, 2012).

64. Valerii G. Sesiunin, "Funktionen der Massenkommunikationsmittel," in *Massenkommunikation in der UdSSR. Sowjetische Beiträge zur empirischen Soziologie der Journalistik*, ed. Hansjürgen Koschwitz (Freiburg, München: Verlag Karl Alber, 1979), 46–47.

65. Stephan Merl, "Konsum in der Sowjetunion: Element der Systemdestabilisierung?" *Geschichte in Wissenschaft und Unterricht*, 58 (2007): 519–536; Reid, "Cold War in the Kitchen."

66. Concerning the question of how consumption contributed to stability, see the convincing arguments of Emeliantseva, "The Privilege of Seclusion: Consumption Strategies in the Closed City of Severodvinsk."

67. GARF, f. 6903, op. 1, d. 612, 1959: Stenogramma zasedaniia obshchestvennogo soveta telezritelei pri tsentral'noi studii televideniia, l. 2.

68. Cf. for a discussion on methodological approaches to normality: Herbert Willems, "Normalität, Normalisierung, Normalismus," in *›Normalität‹ im Diskursnetz soziologischer Begriffe*, ed. Jürgen Link, Thomas Loer and Hartmut Neuendorff (Heidelberg: Synchron, 2003), 51–83; Hannelore Bublitz, "Diskurs und Habitus. Zentrale Kategorien der Herstellung gesellschaftlicher Normalität," in *›Normalität‹ im Diskursnetz soziologischer Begriffe*, ed. Jürgen Link, Thomas Loer and Hartmut Neuendorff (Heidelberg: Synchron, 2003), 151–162.

69. Cf. Anke Stephan, "Erinnertes Leben: Autobiographien, Memoiren und Oral-History-Interviews als historische Quellen," in *Digitales Handbuch zur Geschichte und Kultur Russlands und Osteuropas*. https://epub.ub.uni-muenchen.de/627/1/Stephan-Selbstzeugnisse.pdf, (accessed February 14, 2015), 17–18; with special regard to women's voices Kathryn Anderson and Dana C. Jack, "Learning to Listen: Interview Techniques and Analyses," in *The Oral History Reader*, ed. Robert Perks and Alistair Thomson (3rd ed. Abingdon/Oxon and New York: Routledge 2016), 129–142; for the reflection on the conversations with soccer fans: Zeller, *Das sowjetische Fieber*, 37–40.

70. I lean on the concept of the "interview of remembrance." Cf. Gabriele Rosenthal, *Erlebte und erzählte Lebensgeschichte. Gestalt und Struktur biographischer Selbstbeschreibungen* (Frankfurt a. M., New York: Campus, 1995), 70–98; Roswitha Breckner, "Von den *Zeitzeugen* zu den *Biographen*. Methoden der Erhebung und Auswertung lebensgeschichtlicher Interviews [1994]," in *Oral History*, ed. Julia Obertreis (Stuttgart: Franz Steiner, 2012), 131–151.

71. Roger Chartier, "Mémoire et oubli. Lire avec Ricoeur," in Christian Delacroix, François Dosse and Patrick Garcia eds., *Paul Ricoeur et les sciences humaines* (Paris: La Découverte, 2007), 241–242.

72. See for the diverse ways and structural patterns with which people order the past in order to create a sense of collective belonging in the present time: Eviatar

Zerubavel, *Time Maps: Collective Memory and the Social Shape of the Past* (Chicago, IL: University of Chicago Press, 2003).

73. Cf. Kirsten Bönker, "*Perestroika* and the Loss of Certainties: The Post-Soviet Revaluation of Soviet Money Practices and Social Equality," *Journal of Modern European History* 15, 3 (2017): 367–394.

74. Cf. Michael Meyen, *Einschalten, Umschalten, Ausschalten? Das Fernsehen im DDR-Alltag* (Leipzig: Leipziger Universitätsverlag, 2003), 98–109.

75. Cf. for example the revealing judgments of some respondents taking part in Donald Raleigh's study on Soviet baby boomers: Donald J. Raleigh, *Soviet Baby Boomers. An Oral History of Russia's Cold War Generation* (Oxford: Oxford University Press, 2012), 237.

76. Liudmila Khakhulina, "The Persistence of Mass Conceptions of Justice," *Sociological Research* 54, 2 (2015): 71–90, 74; Anna A. Mironova, "Trust, Social Capital, and Subjective Individual Well-Being," *Sociological Research* 54, 2 (2015): 121–133, 121, 127, 130–132.

77. For nostalgia in contemporary Russia, see recently Kalinina, "Multiple faces." Cf. also the discussion of nostalgia with regard the GDR in Katja Neller, *DDR-Nostalgie. Dimensionen der Orientierungen der Ostdeutschen gegenüber der ehemaligen DDR, ihre Ursachen und politischen Konnotationen* (Wiesbaden: Springer, 2006), 41–42. With regard to nostalgic TV consumption in the Czech Republic: Veronika Pehe, "Responses to *The Thirty Cases of Major Zeman* in the Czech Republic," *VIEW. Journal of European Television History and Culture* 3, 5 (2014): 100–107.

78. Svetlana Boym, *The Future of Nostalgia* (New York: Basic Books, 2001), xviii; idem, http://monumenttotransformation.org/atlas-of-transformation/html/n/nostalgia/nostalgia-svetlana-boym.html (accessed April 15, 2015).

79. I refer here to my deliberations on nostalgia and the memories of late Soviet money practices in Bönker, "*Perestroika*."

80. See for nostalgic practices of music competitions playing in the Soviet retro style in the Baltic region and Georgia: Platt, "Russian Empire of Pop."

81. Fred Davis, *Yearning for Yesterday. A Sociology of Nostalgia* (New York, London: Free Press, 1979), 1–29.

82. See for the effect Cold War chronotopes still have today: A. Lemon, "Sympathy for the Weary State?: Cold War Chronotopes and Moscow Others," *Comparative Studies in Society and History* 51, 4 (2009), 832–864; S. A. Oushakine, "'We're Nostalgic But We're Not Crazy': Retrofitting the Past in Russia," *The Russian Review* 66 (2007), 451–482.

83. I take up here the considerations of David Crowley and Susan E. Reid, "Introduction: Pleasures in Socialism?" in *Pleasures in Socialism. Leisure and Luxury in the Eastern Bloc*, ed. David Crowley and Susan E. Reid (Evanston, IL: Northwestern University Press, 2010), 4–8; Susan E. Reid, "Communist Comfort: Socialist Modernism and the Making of Cosy Homes in the Khrushchev Era," *Gender & History* 21, 3 (2009): 469.

84. Communication scientist Elisabeth Klaus proposes not to contrast the spheres of "entertainment" and "information," but to view them as entangled elements of journalism and media products. See Elisabeth Klaus, "Der Gegensatz von Information ist Desinformation, der Gegensatz von Unterhaltung ist Langeweile," in *Medien—Politik—Geschlecht. Feministische Befunde zur politischen Kommunikationsforschung*, ed. Johanna Dorer, Brigitte Geiger and Regina Köpl (Wiesbaden: Springer, 2008), 51–64.

Chapter 1

TV Sets Capturing Soviet Homes

As early as in the late 1950s, the TV set became a symbol of the new Soviet lifestyle. The Soviet TV set reflected the fact that the Cold War competition pushed the regime to pay greater attention to material culture and the standard of living on the domestic front. During the so-called kitchen debate on the opening of the American trade exhibition in Moscow in July 1959, Soviet Premier Nikita S. Khrushchev and U.S. vice president Richard Nixon propagated the peaceful competition between the two power blocs. Focusing on consumer goods and lifestyle, both Khrushchev and Nixon deemed television to be a vital element to gain the future victory in the Cold War. Both verbally competed over which political system would offer the more attractive lifestyle based on labor-saving and recreational devices for ordinary people.[1] The "kitchen debate" gained momentum with Khrushchev claiming that "in rockets we've left you [the United States] behind, and also in this technology [he referred to color television] we've outstripped you!"[2]

It was not without logic that material culture gained more and more importance after the mass housing campaign enabled millions of people to move into their own four walls. The considerably improving housing supply during the 1960s paved the way for a specific Soviet domesticity and previously unknown possibilities of shaping one's private lifestyle.[3]

This chapter explores how the TV set gained its considerable sociocultural importance as an object of consumer desire and further as a source of information and entertainment. It focuses on the question of how television changed people's domestic life, its material aesthetics, and its hierarchies of space. It is inspired by Lynn Spigel's analysis of the way people placed the TV set and, in doing so, not only rearranged the postwar American living room but also adapted their leisure practices to the new medium.[4]

The first part of this chapter highlights the quantitative spread of local TV stations and the rising distribution of TV sets all over the country. The second section links this statistical site of the changing material culture to subjective meanings and practices of consumption. It analyzes the narratives of former Soviet TV viewers to explain how the TV set impacted on consumer desires. The interviewees talked about when and why the desire to buy a TV set emerged, or which family member initiated the purchase and decided on the model.

Having purchased a TV set, the proud owners had to find an inviting place in their usually rather small flats to rest in front of the screen. Therefore, the third section of this chapter is dedicated to the question of how the TV set changed the domestic space and the Soviet way of life. The way people spent their leisure time, consumed culture and mass media, clothed, prepared meals, or furnished their living rooms had been the object of thorough ideological observance and official instructions. The Soviet regime promoted a whole bunch of advice literature, housekeeping books, and advice programs on television. These programs specifically addressed the Soviet housewife and, sometimes more generally, the Soviet consumer as such.[5] Mass media, however, observed not only themselves but also sociocultural practices of everyday life and media consumption. These media representations of the Soviet way of life helped to frame people's attributions to their own lifestyle and domestic space. I trace these interpretation frames within contemporary advice literature and explore representations of the TV set in selected Soviet movies. Consulting the interview sample, I further analyze how the respondents remember the way they placed the TV set in their apartments.

THE EXPANSION OF TELEVISION

The quantitative and geographical spread of TV sets, as well as the institutional consolidation of central and local TV stations, told a story about the success of Soviet television from the early 1950s. The early years of Soviet television were characterized by internal administrative conflicts, as different interest groups struggled about technical, aesthetical, and political supervision. Until 1957, Central Television was subordinated to the Ministry of Culture and to the Ministry of Communications, which provided the technical infrastructure for radio and television. Both were not very eager to promote the new medium at the expense of theater and cinema, which meant that cinema, for example, was much better funded than television. The regional TV stations especially had to cope with major challenges in the early days, not least because their technical equipment left much to be desired. Riga TV director Ialynskii's complaint about his TV studio in 1956 was not unusual:

It was so small that only one person could be filmed at once. Riga TV staffers had literally no place to perform and were, in addition, highly affected by a temperature of 42 degrees Celsius.[6] The provincial TV studios were, however, no real exceptions to the rule, as even the employees of Moscow Central studio had to content themselves with a building of only 300 square meters in the beginning.[7]

Due to competing authorities, television could not easily rely on better equipment, even if cinema studios offered free capacities. Therefore, the Radio and Television Board successfully urged the responsible ministers to settle competences and bring the management of television together. In 1957 the administrative responsibilities were assigned to the newly founded State Committee for Television and Radio by the Council of Ministers of the USSR (*Gosteleradio*).[8]

This administrative restructuring fostered the quantitative and qualitative development of television that picked up speed afterward. There are no thorough statistics about the distribution of TV sets in the early years but only dispersed data in archive documents of party origin or related to *Gosteleradio*. In 1950, the Moscow and Leningrad programs were only aired a few hours per week and could merely be received by approximately 15,000 TV sets in the areas of the two capitals.[9] The number of TV sets, however, quickly increased toward more than 225,000 until 1953 and almost 3 million in 1958. This rising distribution of TV sets corresponded with the increasing numbers of TV stations: Starting with three stations in 1953, nine stations in 1955 and reaching a number of already fifty-three stations as early as 1958, regional and local TV stations quickly spread all over the country.[10] Regional TV stations aired daily in Leningrad since October 1955 and, for example, in the Latvian capital of Riga at least twice a week. Thus, viewers living in several big cities received two programs since the mid-1950s.[11] Television's expansion was further reflected by the increasing daily transmission period: From January 1955, Moscow television had started broadcasting a few hours daily. Within only four years Central Television extended its airtime from 15.3 hours per week in 1955 up to almost 54 hours per week in 1959.[12] The audience, however, was still small and lived primarily in Moscow, Leningrad, and the republican capitals. In 1956, Moscow television workers estimated that more than three million inhabitants of Moscow and Tver region watched TV each day.[13]

The Main Board of Radio broadcasting projected a number of six to eight million TV sets served by at least seventy-five stations for 1961, whereas Brezhnev announced in his speech at the XXIII party congress that the number of TV sets would account for forty million in 1970.[14] At least, the effective development of the TV stations even surpassed any optimistic forecasts, as by 1960 the number of TV stations officially funded by state organs rose to 84, by 1965 to 121, reaching 131 in 1975. After the late 1960s, the medium provided at least technically a nationwide reception.[15] In the mid-1960s, the number of TV sets was estimated at 12.5 to 15 million and the audience was considered to comprise 60 million people. In several districts, the broadcasts were further retransmitted by nearly 50 state-run and more than 300 smaller amateur relay stations.[16]

Many of the local studios and most of the local relay stations were established by enthusiastic tinkerers interested in the new medium. They had the necessary technical skills to set up operable TV stations with simple means. The infrastructural and technical conditions under which these studios and stations operated were challenging. In the beginning, the range of many local stations was rather limited. Tomsk television station, for example, had a scope of only 120 to 140 km in 1956.[17] In 1959, Sverdlovsk studio was even more difficult to receive, as its steady broadcasting zone only covered a radius of 70 km, although it sometimes might have reached up to 120 km.[18] The range of local stations gradually improved. Thus, allegedly almost 90 percent of Kyrgyz people could receive the broadcasts from the regional TV station in Frunze (Bishkek) in 1969. Half of the people could watch Moscow Central Television, the regional programs of the neighboring republics of Uzbekistan and Kazakhstan aired from Tashkent and Alma-Ata (Almaty). In this sense, a representative of the Kyrgyz party claimed as early as 1966 that all districts were covered by TV signals and that 11 TV sets were spread among 100 families.[19]

The rather uncontrolled spread of stations all over the country constituted a problem as Moscow Central Television initially had no influence on their schedules, the selection of employees, and so on. Depending on the city and region, local enthusiasts sometimes maintained the studios and relay stations during their leisure time, sometimes backed up by local party committees. Although they usually needed the support of local party committees, Central Television in Moscow only achieved a constant institutional influence over them in the late 1960s. From the late 1950s on, the new central institutional structure was gradually expanded to the regions. As in Kazan, these committees took over the amateur studios in December 1959.[20] Nevertheless, Moscow State Committee for Television and Radio often complained about local impediments and the repeated lack of support from local party organizations in its attempt to centralize Soviet broadcasting.[21]

The 1960s, however, were a decade of homogenization, centralization, and professionalization of the television staff. Central Television aimed to set up a nationwide program addressing all nationalities and ethnic groups. An important step on this way was to connect the far-off cities and regions to Moscow central channels. In 1965, viewers in such distant cities as Baku, Yerevan, Perm, or Kurgan could receive programs of Central Television.

The next step was to expand the cable and new satellite network to studios in Middle Asia and Siberia. In the mid-1960s, the big regional TV stations such as in Leningrad, Kiev, Riga, or Tiflis not only received broadcasts from Moscow, but themselves fed the nationwide channels with regional programs. Due to the technical progress, the exchange of regional live broadcasts was already complemented by exchanged video recordings. The satellite *Molniia–I* secured the exchange of programs between Moscow and Vladivostok. Moscow programs were aired even to foreign countries via the transnational broadcasting networks of *Intervidenie* and *Eurovision*, which conversely enabled Soviet viewers to watch foreign broadcasts. The commissioning of the new television center *Ostankino* in 1967 made it possible to launch a fourth nationwide channel. Like West European TV station *Gosteleradio* also started to broadcast regular color TV programs in 1967. Viewers in many regions could receive up to six channels including those from Moscow in the late 1960s. Non-Russian regional stations aired national language programs few hours each day.[22] This ethnic component of regional television became an important element of Central Television's program policies.

The number of employees considerably grew during these years: Central TV started with 402 employees in 1955, and employed 8,295 people in 1960 and more than 17,800 in 1965.[23] Correspondingly, the number of channels rose. At first, viewers in Moscow and Leningrad could watch three channels from 1965 on.[24]

Many foreign correspondents considered the new Soviet lifestyle as an important issue for media coverage that they supposed to catch the interest of their Western audience from the mid-1950s onward. The correspondents thoroughly depicted the changes in Soviet leisure habits and the new significance of watching television. Thereby, they conveyed a picture of Soviet viewers who enthusiastically welcomed and requested the expansion of television. As early as in 1954, the *New York Times* foreign correspondent, Harrison E. Salisbury, reported about people's excitement: "Muscovites are frankly wild about TV," he contended.[25] According to Salisbury, Moscow viewers seemed to be quite mad about television, although—in contrast to American TV—it was still of a rather experimental nature. It was a sensation for most Soviet people to watch live images on the TV screen at home, as not more than one million households owned a TV set at this time. The price of TV sets amounted to several times an average monthly salary. Salisbury was astounded that Muscovites "queue up for weeks, if necessary, to buy a set and don't seem to mind paying prices up to about 2,400 rubles, which is $600, for receivers whose viewing screen is quite small by American standards."[26]

Typically, foreign correspondents reporting from Moscow in the 1950s tell about the way television intruded into Soviet life. Like Salisbury, American journalist Marguerite Higgins was torn about the impression that TV sets were distributing so quickly and that any family needed to set up their own antenna on the one hand. On the other hand, she was surprised that the Soviet Union was not able to produce even more TV sets at smaller prices. TV antennas were the most visible signs of television's dissemination. It seemed to contemporaries that each day hundreds of new antennas were protruding from the roofs of the houses. Foreign correspondents were especially eager to let their audience know about the profound changes of Soviet daily life and illustrated their stories with photographs depicting roofs on which a lot of antennas were lined up.[27]

Some of the interviewees remember how they were impressed by the spread of antennas. They were, however, of social significance, as fifty-nine-year-old historian Yuri Vasil'evich depicted with a view to his childhood in the small town Otradnyi in Samara region: "Yes, one more important memory from early childhood. Well, Otradnyi was a town of oil worker. Well, people were pretty rich, if this notion is possible. And there in Otradnyi one considered as the first sign of wealth the numbers of antennas on the house. The houses were quickly built. And there the antennas sprang up like mushrooms. Because there were no collective antennas but for any single TV set, yes."[28]

The images of spreading TV antennas all over the country constituted its own normative power and pushed even the Communist Party to consider television as an ambitious propaganda medium.[29] Indeed, the distribution of TV sets jumped up so quickly that Soviet journalist Sergei A. Muratov claimed that the depiction that "we live in a 'forest of television antennas'" was a typical narrative of newspapers in 1965.[30]

As early as in the mid-1950s, people's desire for TV sets was so great that the Ministry of Trade started to complain that it could not satisfy the consumer demand.[31] Until the mid-1960s, the Soviet industry was not able even roughly to fulfill the target figures. In particular, consumers living outside the capitals met considerable difficulties to buy a TV set. For example, West Siberian Tomsk Television workers noted in 1956 that TV sets were promptly sold out as soon as they were available in the local shops. A total of 1,200 TV sets assigned to Tomsk region for the period from October to January were sold within one day.[32] This does not come as a surprise considering that the city of Tomsk had 224,000 inhabitants, whereas the region accounted for more than 700,000. The same was true for other cities, where the supply never met the consumer demand for TV sets. According to local accounts, Sverdlovsk (Ekaterinburg) trade network sold 12,800 TV sets to the population of the regional capital and 8,000 to people living in the region in 1958. This meant it was ranked fourth within the Soviet Union with regard to the amount of sold TV sets, but these numbers do not speak for themselves, as almost 780,000 people lived in the city of Sverdlovsk, and in the whole region even more than four million people.[33] Nevertheless, the number of TV sets was increasing and local TV workers were eager to document this spread of TV sets all over the country. Employees of Tomsk TV station counted almost 17,000 TV sets in the whole region and—figuring as a telling sample—550 TV sets based in collective farms in one district located 120 kilometer away from the regional capital in 1956. Viewers in only five of sixteen districts of the region were able to receive TV signals from Tomsk. Tomsk employees were optimistic that the number of TV sets in the regional capital would soon reach 15–15,500.[34] These numbers make very clear that TV sets were rare pieces in the regions beyond the capitals. Still, the distribution of TV sets jumped up and basically spread—as Table 1.1[35] illustrates—as quickly as in the GDR and West Germany. The mere numbers were impressive. They did not, however, say much about the actual functioning of the TV sets, as many respondents remembered that the new devices were rather tricky and needed to be handled cautiously. Complaints about technical problems and missing spare parts were omnipresent.[36] The first mass-produced Soviet TV set, the KVN–49, was named after the initials of its constructors Kenigson, Varshavskii, and Nikolaevskii. This abbreviation, however, also turned out to unintentionally represent the real functioning of this TV set, as it was often ironically equated with "*Kupil—Vkliuchil—Ne rabotaet*" (Bought—Switched on—Doesn't work).[37]

Further, Soviet statistics reflect only the output figures and not the numbers of sold TV sets. Considering, however, the contemporary and retrospective reports about the high consumer demand, they, nevertheless, document the rising amounts of TV sets all over the country and accounted for 15,693

Table 1.1 Distribution of TV Sets Per Family (USSR) or Household in Percent[1]

USSR		*GDR*		*FRG*	
1960	8.0	*1960*	18.5	*1960*	17.2
1965	24.0				
1970	51.0	*1970*	73.6	*1970*	67.9
1975	74.0				
1980	97.0	*1980*	105.0	*1980*	83.7
1988	103.0	*1989*	129.3	*1987*	87.8

[1]The data are compiled from the following: *Narodnoe khoziaistvo SSSR za 60 let. Iubileinyi statisticheskii ezhegodnik* (Moscow: Statistika, 1977), 513; *Narodnoe khoziaistvo SSSR*, 1922–1972 (Moscow: Statistika, 1972), 588; *Narodnoe khoziaistvo SSSR v 1988 g. Statisticheskii ezhegodnik* (Moscow: Statistika, 1989), 119; Leonid A. Gordon, Eduard V. Klopov, Leon A. Onikov, *Cherty sotsialisticheskogo obraza zhizni: Byt gorodskikh rabochikh vchera, segodnia, zavtra* (Moscow: Izdatel'stvo Znanie 1977), 156; Thomas Lindenberger, "Geteilte Welt, geteilter Himmel? Der Kalte Krieg und die Massenmedien in gesellschaftsgeschichtlicher Perspektive," in *Zwischen Pop und Propaganda. Radio in der DDR*, ed. Klaus Arnold and Christoph Classen (Berlin: Christoph Links, 2004), 37; Stephan Merl, "Staat und Konsum in der Zentralverwaltungswirtschaft. Rußland und die ostmitteleuropäischen Länder," in *Europäische Konsumgeschichte, Zur Gesellschafts- und Kulturgeschichte des Konsums (18.-20. Jahrhundert)*, ed. Hannes Siegrist, Hartmut Kaelble, and Jürgen Kocka (Frankfurt a.M., New York: Campus, 1997), 227.

million TV sets in 1965 and 57.2 million, including 1.7 million color TV sets, in 1976.[38]

This quantitative development depicted that people's opportunity to watch TV statistically considerably improved after the late 1960s. Statistics counted 1 TV set per 7 persons in 1970, 1 per 4.5 persons in 1976, 1 per 4 persons in 1980, and 1 per 3.2 persons in 1988.[39] Thus, in the mid-1970s the audience comprised according to official estimates more than 75 percent of the whole Soviet population.[40] This internal figure corresponded with the observations of American correspondents. *Washington Post* journalist Robert G. Kaiser depicted the rising desires for consumer goods and the gradual transformation of people's attitude toward a new kind of consumerist lifestyle. Kaiser observed in the early 1970s that at this time television reached "most of the population." Like his Western colleagues, he presented the rapid spread of television as key part of the story of "the embourgeoisement of Soviet society." One does not need to share this interpretation with its slightly complacent satisfaction to recognize the interest and sometimes even fascination Soviet television aroused among Soviet citizens, as well as among foreign observers.[41]

ABOUT DESIRES AND PURCHASES OF A TV SET

After the late 1950s, the standard of living considerably improved for many people who could move into apartments of their own. It constituted new consumer experiences and required new consumer skills to furnish the apartment and equip it with technical devices such as refrigerator, radio, and TV set.[42] Thus, purchasing and maintaining TV sets gained in importance in a rather similar way to improvements of the living standard West European families experienced.[43] Indeed, hardly any other consumer good and medium changed everyday life and impacted on leisure and media practices of so many people in the industrialized countries of the East and the West as the television set. West German bestsellers like *Blaupunkt Tirol*, *Grundig Zauberspiegel*, and *Telefunken PAL Color* 708 corresponded with Soviet *KVN–49*, *Rekord–68*, and *Rubin C–201*. Apart from the *KVN–49*, which appeared rather coarse and simple like many Soviet devices, the external design of Soviet TV sets did not dramatically differ from West European televisions. *Rubin–102*, for example, resembled *Grundig Zauberspiegel* in its form and design: both had wooden housings and screens with rounded edges, a 43-centimeter tube and two controllers at the bottom—one on the left and one on the right side. In the beginning, the design and basic technology developed in a similar way, as the number of buttons and the screen diagonal gradually increased on both sides of the Iron Curtain.[44]

The year of 1967 further pushed the development of TV sets in the East and the West, as the launch of color TV programs in November 1967 required to provide the viewers with color TV sets. The first Soviet mass-produced color TV on the market was *Rubin–401*, which was launched in October 1967. Like sixty-four-year-old Natal'ia Petrovna from Kuibyshev, some respondents were at the time of the interview still fascinated by Soviet television's technical development. Like many others, Natal'ia Petrovna recapped the improvements: "It was interesting and surprising how our technology strode ahead. And the TV sets themselves changed; they became more cultured. The screens grew. The rear wall became smaller."[45]

Although the external resemblance of Soviet and Western TV sets said nothing about their technical qualities, they both were equally expensive and nevertheless much-desired consumer goods. Despite their high prices, TV sets quickly became a socially rather equally distributed popular consumer good in the Soviet Union. According to an account of 1970 from Ural city of Nizhnii Tagil, nearly 83 percent of workers and blue-collar specialists and 91.5 percent of white collars owned a TV set.[46] This seems to be surprisingly high rates, if we consider the relation between the average prices for a TV set and the average salaries. Table 1.2 shows how the average prices of some durable consumer goods and the average salaries developed from 1970 to 1985. TV sets were the most expensive household devices. In 1970, a color TV set cost more than six times the average monthly salary, a black-and-white device was considerably less expensive, but still cost more than two and half times the average salary. Tables 1.2 and 1.3 make clear that TV sets slightly cheapened in relation to salaries over the decade from 1970 to 1980. In 1980, a color TV set cost only four times an average monthly salary, a black-and-white one 1.3 times.[47]

Many interviewees remembered that TV sets were usually sold out in the short term, as soon as the news a shop would offer TV sets had spread. This was especially true for the regions beyond the capitals and huge cities, although TV sets cost more than two average monthly salaries. Only few

Table 1.2 Monthly Salaries and Prices of Durable Consumer Goods in Rubles

	1970	*1975*	*1980*	*1985*
Average salary	121.2	148.7	174	199.2
TV set (color)	744.34	650.22	706.87	643.99
TV set (b/w)	329.84	313.8	229.36	229.1
Radio	71.25	88.71	105.55	110.63
Tape recorder	167.01	181.64	232.69	275.52
Photo camera	33.83	37.58	63.36	71.76
Refrigerator	214.51	244.58	249.66	288.11
Washing machine	92.16	87.04	88.61	94.61

Source: Tsentr ėkonomicheskogo analiza i ėkspertizy: http://www.ceae.ru/SSSR.htm, accessed January 5, 2017.

Table 1.3 The Amount of a Durable Consumer Good in Relation to the Average Monthly Salary

	1970	*1975*	*1980*	*1985*
Average salary	121.2	148.7	174	199.2
TV set (color)	0.16	0.23	0.25	0.31
TV set (b/w)	0.37	0.47	0.76	0.87
Radio	1.70	1.68	1.65	1.80
Tape recorder	0.73	0.82	0.75	0.72
Photo camera	3.58	3.96	2.75	2.78
Refrigerator	0.57	0.61	0.70	0.69
Washing machine	92.16	87.04	88.61	94.61

Source: Tsentr ėkonomicheskogo analiza i ėkspertizy: http://www.ceae.ru/SSSR.htm, accessed January 5, 2017.

respondents were old enough to remember how people purchased the first TV sets in the mid-1950s. Seventy-three-year-old teacher Zinaida Mikhailovna from Tol'iatti depicted how the purchase of a TV set represented a strong social distinction in the early years. She moved to Moscow in 1954 to enter a pharmacological institute and saw the first TV set in her life, a *KVN*-49, when she visited relatives after arriving in the capital. She affirmed that on the one hand only the well-to-do who lived a prosperous life could afford to buy a TV set in the mid-1950s. On the other hand, people had needed some relations or privileges to obtain a television, because they would have not simply been available at shops. Thus, the purchase of a TV set was a prestigious matter, as "the TV set demonstrated a certain standard of living," as Zinaida recalled.[48]

Even a well-off respondent like former Leningrad biologist Georgii (born in 1953), whose parents had extraordinarily high salaries of 700 to 800 rubles in the 1960s and 1970s, considered the first TV sets "not to be luxury goods, but almost on the border to be" such luxury items. The income strongly influenced the purchase of the first TV set for many of the respondents. Engineer Nadezhda Ivanovna, who was born in 1950 in a small city near Irkutsk, like Zinaida Mikhailovna contended that in the 1950s only the better off bought TV sets.

Apart from the financial side, in many provincial places consumers interested in purchasing a TV set had to refine their practices. Zinaida Mikhailovna reported about her strenuous efforts to buy the first TV set for her family after she had moved from Moscow to Arkhangelsk in 1959. It was not only her initially difficult material situation, as her first wage only amounted for 94, somewhat later then for 160 rubles. After marrying, her situation considerably improved as her husband was a sailor and earned decent money. Nevertheless, they had no TV set in the early 1960s, as

> there were none available on free sale, actually nobody had one. I remember we got the first sometime in the mid-1960s, I brought it from Riazan, bought it

> there, brought it by train, it was a black and white device. Some tricolor membrane stuck on it. It should imitate a color TV. It got permanently broken. You bang the TV set with the fist in order to get the picture.

Zinaida Mikhailovna was motivated to accept this effort to purchase a TV set that did not even work properly because

> we had a dream for a long time: to buy a TV set. Because already sort of our friends started to buy one, at work they talked about how they bought a TV set, and we very much wanted to have a TV set [she emphasized the word "very"]. You couldn't buy a TV set in Archangelsk, because there were no shops offering TV sets, i.e. they didn't have enough. So, the idea appeared to bring a television, when I go on vacation. That means, I went on vacation to a village 200 km from Riazan. And on a wonderful day I got on a bus, went to Riazan, purchased a TV set . . . yes, in Riazan they were on free sale. Well, free, yet again the brand *Rekord*, and in my view, it was a *Rekord*. It was not very expensive, even as measured by that money. There . . . I purchased it, brought it with a taxi to the main station, put it in the luggage storage . . . it was heavy, somebody helped me. . . . that means, when we went back to Arkhangelsk after vacation, we grabbed it with us, brought it by train. I don't remember, if there was an antenna or if we simply switched it on and all . . . we lived on the fifth floor, somehow it received a picture. We brought it home and, certainly, it was a very, very, very great pleasure. Well, only the poor picture.

Apart from such strenuous consumer performances, other practices than purchasing in a state shop gained importance during the 1960s. Alternative ways were to obtain a TV set via the employer, to take a place on the waiting list of a state shop, or, in particular, to buy one on the black market.[49] State shops, however, also sold TV sets on credit—a practice that became rather popular and enabled consumers to own more than one TV set. Natal'ia Petrovna recounted that the employer of her parents offered the opportunity to buy a new color TV set at preferential rate, so that the black-and-white set was moved to the kitchen.[50] Thus, only in the 1970s had the supply and product range quantitatively and qualitatively improved to such an extent that almost any family had a TV set.

Many respondents reported that a TV set was among the most important purchases of a young married couple as soon as they could move into a flat of their own from the mid-1960s onward. Zinaida Mikhailovna described a "feeling of satisfaction" that the purchase of the first TV set evoked: "In my view, it was after the washing machine the second thing we bought for our home. First. Second, somehow a feeling of value."[51]

The fact that TV sets became one of the most popular consumer items was no retrospective construction. Among other contemporary observers, cartoonists and illustrators reflected the multiple meanings linked to the TV set and thus made it a core element of the iconography of Soviet consumerism. If one considers caricatures in authoritarian systems as "persuasive devices" that are used to bring up issues and to reiterate topics, the messages should guide the viewer to rightly judge the different aspects of the new consumerist lifestyle.[52] The caricature of Figure 1.1 showcases the official promise of the regime to provide a rising material living standard and the expectations of ordinary citizens about a better future symbolized by the TV set and a refrigerator as essential technical devices. The propagation of the new Soviet way of life also implied an ideological balancing act brought about by legitimizing socialist consumerist wishes on the one hand and criticizing the capitalist consumerist orientation on the other hand. Further, the caricature made fun of the gender conflicts consumption desires brought about, as newly married women expected their husbands to guarantee a decent way of living what including a TV set and other household devices.

Many respondents—female as well as male—would have agreed on this consumerist outlook on life. According to some narratives, the purchase of a TV set was even more important than to buy a refrigerator or a washing machine. Higher-educated Oleg Igor'evich, who was born in a village in 1958, emphasized the impact of television on communitarization and emotional

Figure 1.1 "But What, Couldn't You Get Hold of Such a Car?" *Source: Krokodil*, 12 (April 1975), 5.

bonding within families: "You were no family . . . without a TV set."[53] Oleg talked about the late 1970s, a time when almost 90 percent of Soviet households owned a TV set. He himself married at this time and moved from his native village to Kuibyshev, today's Samara. Like Oleg, other respondents, who had lived in villages during the 1960s, clearly remembered that only few people had owned a TV set in their social environment.

In fact, some of the respondents, who originated from villages or small towns, saw a TV set for the first time when they visited friends or relatives living in larger cities. Liudmila, a nurse born in Chapaevsk in Samara region in 1951, recounted that she regularly accompanied her parents, who went to Moscow in the early 1960s in order to buy those goods which were hardly available in Samara. They always stayed with Moscow friends who had already owned a TV set.[54]

Although many of those villagers who went to cities brought the wish to own a TV set back home, it took a certain time until watching TV became a private family habit. In contrast to urban spaces, watching TV in small towns and villages remained a convivial practice of family, friends, relatives, and neighbors partly even until the mid-1970s. Many of these respondents, who represented village experiences, emphasized that it was very popular to leave home to watch TV and spend time with others. Not before the mid-1970s, when the supply with TV sets had considerably improved, urban and village practices of watching TV gradually converged. From television's early days on, the fundamental enthusiasm about watching television, however, was a common feature of village and urban people, as the tractor driver and crane operator Valerii Nikolaevich recollected. Valerii Nikolaevich was born in 1958 and raised in a village near Irkutsk. He recollected this change of watching practices because of the rapid spread of TV sets very well: "When people had initially experienced watching TV and channel-surfing . . . , well, then virtually one hundred percent started to buy and to buy and to buy. Within one year practically anyone owned the first TV set, you set up an antenna, that was all."[55]

The purchase of the first TV set constituted for many respondents regardless of their place of residence a great family event. This was not least due to the high prices of TV sets. Talking about the purchase of the first TV set, many interviewees mentioned the financial expenditure and set it in relation to their parents' or their own incomes. A sample of Leningrad viewers cannot be considered as representative of the whole Soviet audience. Nevertheless, these contemporary Soviet audience studies, which gathered data from interviews, presented similar results already in 1967. Although only about 15 percent did not own a TV set, almost 60 percent of them stated that this was because they could not afford to buy one. However, more than 70 percent of them wanted to buy a TV set, as soon as they could afford to do so.[56] This desire was further encouraged by the regime promoting the new technical devices as core parts of the Soviet lifestyle. Thus, as early as the late 1950s, the regime was eager to demonstrate in the so-called commodity dictionaries that the Soviet consumers could experience a certain choice and advice. Therefore, mass media distributed images visualizing such new consumer habits as early as in the late 1950s.[57] In the beginning, these images often represented clear gender roles of customers and salesmen: Both were supposed to be male and interested in technical aspects.[58] Indeed, many respondents report that the father or the husband initiated the purchase of the first TV set. If there was any choice, the male members of the families were remembered to have opted for a certain model. The impact of these gendered practices seemed to have changed only slightly over the years, although women increasingly at least participated in the act of purchasing the TV set.

Some respondents clearly remembered how much they desired a TV set. A TV set of a young person's own would have symbolized a certain independence of the parents: "And generally, I dreamt of buying a TV set on my own," remembered the then sixty-four-year-old, higher-educated Masha in 2010. Masha read a lot of books as her mother worked in the library of the Leningrad Academy and regularly brought books home. Nevertheless, television quickly captured her curiosity in the mid-1950s, when she was a little school child. At this time, her family lived in one room of a communal apartment and could not afford to buy a TV set. Masha recalled that she visited her girlfriend on Sundays in order to watch cartoons. Some day in the late 1950s, her parents bought a *KVN–49* after winning 1,000 rubles from a lottery. *KVN–49* was one of the first TV models. As can be seen on the Figures 1.5 and 1.6 below, it needed a magnifying lens filled with water in front of its quite small screen to improve the TV picture. Masha remembered that the TV set was not easy to handle because of technical deficiencies. *KVN–49* could receive three channels, which was enough in the early years of Soviet television. Although *KVN–49* became technically rather quickly outdated during the 1960s, the family could not afford to buy a new television and replace it. Insufficient financial means forced them to watch still on the old black-and-white TV set even after color TV sets had become widespread and Moscow Central Television regularly aired color programs in the late 1960s: "Color TV sets were incredibly expensive. All people used black-and-white televisions."

Not seldom, the desire to have a TV set of one's own was intensified by quarrels about the program to be watched. Masha's father decided on the program and prohibited his daughter from watching during the day even if he was not at home. In such cases, television might have strained the family communication. Masha stayed with her parents until she married and could not watch the TV programs she wished to. Her father was a keen football fan and constantly preferred to watch sports broadcasts that her mother disliked as much as Masha did. However, both did not dare to rebel against the father's decisions about the program. The restrictions distressed Masha so much that her idea to own a separate TV set was due to the "endless quarrels with my father over the program."[59] Masha wanted so much a TV set of her own to become a self-determined viewer that she even accepted a relatively high financial burden to realize her dream. She purchased the TV set on credit with a term of two years and

> some devious methods. In order not to pay the down payment, this was too expensive, there appeared such a variant that one could buy a TV set without the first down payment within two years, well, I don't remember exactly, that's the way I did it. I worked as lab technician; I had a small salary. But I bought my own TV set. I fit it in a small space that I found between something and something. And I could watch what I wanted to!

Many respondents referred to their strained financial situation that made it difficult to purchase a TV set. Sixty-four-year-old Antonina described a rather unusual way to acquire a TV set, as her husband constructed one with his own hands:

> We build it ourselves! We fetched the basis from *Iunogo technika* [Young technician, a monthly journal]. He [her husband] was a gifted radio-technician, a gifted engineer! I don't know what he was not able to do, he could do anything! . . . And we knocked together this TV set, because we didn't have the money to buy one. It worked very well. I don't know where it ended up, when we gave it away.[60]

Another common consumer practice, particularly of students, was to look for used TV sets. Seventy-year-old Al'bina recounted that she had only very little money when she studied at the technical college. Her monthly grant amounted to 33 rubles, but her mother still supported her so that she even took her first stipend and managed somehow to buy an old *KVN–49* with a lens just as seen on Figures 1.5 and 1.6. Interestingly enough, her mother also did not have a TV set, as according to Al'bina she did not have much money and enjoyed other leisure practices like theater and fashion more than watching television. The second-hand *KVN–49*, however, pleased her. She tuned the device so that it gave a "wonderful picture" and still worked for a long time.

The narratives about purchasing a TV set had much in common, as most respondents did not belong to the better off. Therefore, they had to save money, to look for used TV sets, to buy one on credit, or to be clever and construct it themselves. The state, however, deliberately stimulated this consumer practice in the 1970s by offering generous credits for durable and expensive goods that were supposed to characterize the new Soviet lifestyle: especially technical devices like TV sets and furnishings. Many respondents remembered to have had highly welcomed this offer. They found that it became easy to buy a TV set on credit in Soviet times. Sixty-four-year-old Natal'ia Petrovna, who managed a trade company, reported how even buying a color TV set became rather unspectacular in the 1970s:

> It seems to me that probably all had already their own TV set, because they were already affordable, there were credits, again, I tell you, there were credits and people made very much use of it. You only needed to get a certificate from your work, that you work at that organization and your salary, that's all. Well, thus, there were no difficulties. Already. When I told that my mother bought a woman's jacket for 560 rubles, then [she referred to the early 1960s] you had to save money, there were no credits then. You borrow money from you neighbors, purchase, but the state didn't give any credits, they started to give them later. And people made use of them, bought furnishings and carpets, that's true, carpets were not resolved in gold, but purchased. And they bought TV sets, that was not difficult. Well, . . . yes, all had the opportunity to buy, even pensioners, although they had retired, it was all the same, they purchased, because they had the opportunity.[61]

The rise of television and consequently the spread of TV set, thus, entailed new consumer practices as state and consumers closely interacted. People's desire for TV sets not only brought about a new appreciation of consumer goods and lifestyle but also of the private sphere, as it became much more attractive to stay at home and watch TV. Moreover, the diversification and normalization of purchasing such an expensive and valuable good as a TV set contributed much to shape demanding and ambitious consumers. The narratives of the respondents reveal less dissatisfaction and complaints about obstacles to buy a TV set than one might assume. While this may be attributable to the distance in time and probably to some nostalgic sentiments about the lost Soviet reliabilities, the narratives document how strongly the consumption of lifestyle goods impacted on people's private life and how much the TV set became a symbol of the new Soviet life in the 1970s. Its material side was characterized both by the technical progress integrating more and more "ordinary" people into the Soviet consumer society and by the systemic failures and deficiencies.

THE TV SET AND THE CHANGES OF THE SOVIET DOMESTIC SPACE

Already in 1959, Irving R. Levine, the first permanent American radio and television correspondent after the war, assured his readers that Soviet television "sometimes provides instructive insight into the Soviet way of life."[62]

He had movies in mind, in which the young heroes envisioned their future lives in an apartment of their own. The story of people dreaming of a decent furnishing, including a TV set and a refrigerator, turned into a story of the increasing privatization of Soviet everyday life. If one believes the contemporary observers, this kind of retreat into the private sphere was not Soviet, particularity within the Eastern bloc countries. Many correspondents working beyond the Iron Curtain equated this alleged escape into the private sphere along with a depoliticalization of private life.[63] Not surprisingly, television figured as the excellent symbol of this new consumerist lifestyle.

Unlike the foreign observers, Soviet sociologists remained virtually vague about this interpretation but established in the late 1960s that many people were "still not able to spend leisure reasonably."[64] According to the renowned Soviet sociologist Boris Grushin, it depended on the individual, if the leisure activity contributed to personality development or rather brought about human beings who might "only passively 'consume' culture."[65] Other observers like the literary critic and publicist Lev A. Annenskii also engaged into the social and cultural critical discourse on television perfectly showing how the same arguments contemporarily impacted on the debates across and beyond the Iron Curtain. Annenskii, who published in renowned periodicals such as *Literaturnaia gazeta* and *Ogonek*, referred to a theorem of influential Austrian media philosopher Günther Anders. In 1956, Anders argued that television generated a certain type of man, namely an "isolated mass hermit." Moreover, television became the center of any family replacing the dining table; in his words, television became the "negative family table."[66] Annenskii deplored that "people don't sit vis-à-vis but look towards the TV set" and that television changed internal family communication practices for the worse.[67]

However, the Soviet intellectual discourse was diverse. Observers raised different perspectives on television's impact on people's personality and on the way television changed the private communication. Contrary to Annenskii, the journalist Sergei A. Muratov supported the opposite position arguing that television did not separate people as a matter of principle but also united them as soon as they had a common issue to talk about.[68]

These intellectual arguments also impacted on the broader public discourse on television use and mixed with the regime's fear of losing access to the private communication. Even if the regime had never been able to totally control how people perceived the mass media, the spread of TV sets raised significant questions and problems. Television not only challenged the traditional way the regime addressed the people, but also entailed new pop-cultural ambivalences and provided a seemingly depoliticized private communication. The fact that TV consumption quickly individualized proved that the truth lay somewhere in the middle between Annenskii's and Muratov's arguments. Considering the interviews, it is apparent that nearly all who directly remembered the purchase of the first TV set found it especially exciting to gain the opportunity to watch TV at any time. Respondents who represented themselves as reserved and critical media consumers, like seventy-two-year-old Yuri from Leningrad, were just as excited about the emergence of TV sets as interviewees who regularly switched on the television. Yuri depicted how the apartments of TV set owners became social meeting points, sometimes even for complete strangers:

> You went to visit someone to watch TV. . . . Until then (the time when his father bought a TV set] I went to visit someone myself. One of my friends lived right at the corner of Sovetskaia 8. There was a hostel and someone had a TV set. Up to ten people went there, they stood, watched, it was miracle. When my father purchased a TV set, the guys came to me to watch. . . . You had dinner and all watched TV.

Looking back, not all but many of the interviewees appreciated the socializing effect television had on family, friends, and neighbors. In many interviews, the aspect of coming together figured as a general characteristic of the lost Soviet life in contrast to the current life. Fifty-seven-year-old Georgii from Leningrad rather typically depicted the practice of watching jointly as Soviet specificity, even if not all respondents recalled such amicable relations to the neighbors like he did. His memories were perhaps not quite accurate regarding the chronology of launching color TV sets and further TV channels, but they reveal several intriguing aspects. Among other things, he made clear that the actual TV program hardly mattered as much as the practice of socializing. Although some friends came visiting in order to watch TV, because they did not have a TV set of their own, he observed more interaction in Soviet times:

> They came visiting, they came visiting. Then it was usual, moreover, it was for a long time like this. Afterwards already other televisions appeared. Later on, we had a more or less modern TV set. A modern one—such, probably, exist until today—you could already switch to another program, but even then, they came visiting. This means it was a usual phenomenon. And even not only during my childhood, but in my view even still in the 1970s, if not in the 1980s, they came to visit and watch TV. I think even color TV sets did not exist. There were simply more channels. Nevertheless, they were few, probably three. And the program directory of what you could watch was poor, but they came to watch, I remember, how we were sitting in front of the screen. . . . Perhaps it [i.e., television] was some means of interaction. Then people generally socialized more than before. It means that there was not such an unsociable world of one family or one person. People knew their neighbors round the yard and the street, and the staircase and they socialized more. This meant to drop in the

> apartment, ask there for something, watch TV or, in general, it was usual to ask literally for salt. Now I don't go to the neighbor because of this. It would appear strange. But then it was usual, we dropped in, the TV set was switched on—take a seat, let's watch. Hey, I wanted to watch it myself. It was all right. Schoolmates, with whom I learnt at school, came to me to watch TV. My parents reacted very well.[69]

Some respondents did not similarly experience the socializing effect or perhaps simply did not remember as strongly as Georgii that watching TV was a typical joint leisure habit. Seventy-three-year-old teacher Zinaida Mikhailovna from Tol'iatti argued in a more differentiated way. First, she regarded the time she was studying in Moscow after the mid-1950s; and second, she recollected her life in Arkhangelsk in the 1970s. In any case, she documented the variety of Soviet watching practices and probably the way they changed with the generally increasing living standard. She first talked about the 1950s:

> There, where, as we say, people were good and lenient towards the neighbors, neighbors met to watch TV. I watched to a greater extent there in the student dormitory. Then, when we had one in the dormitory . . . in Moscow, yes, in Moscow. Probably, in 1955, 1956. And there in the relaxation area, there one met, well, as many people as possibly found room in there—and watched, watched. And they also watched very many sports programs, some interesting youth programs, we watched *Goluboi ogonek* (Little Blue Flame)—it was very intriguing . . . well . . . there were some interesting youth broadcastings, there was *Allo, my ishchem talanty* (Hello, we look for talents), afterwards the couple *A nu-ka, parni!* (Let's go, guys!), in my opinion. Well, it means these were programs for which young people met . . . met and watched. . . . and in case of talking about individual TV sets—that was prestigious. And what I know, as I said, . . . that one simply met in great groups in order to watch some programs.

After graduating from university in 1959, Zinaida moved to Arkhangelsk, where she taught at a pharmacological school. In the mid-1960s, her family purchased the first TV set. At this time, the number of TV sets in Arkhangelsk were still comparatively small, which was why their TV set had, nevertheless, a certain socializing effect on their environment:

> Friends came and watched. Neighbors—you did not socialize much with neighbors, because the apartment was new, we did not know, perhaps the neighbors had TV sets, well . . . [Friends] came to visit and watch. But not especially because of that, no, no. We purposely went to the theater, to concerts. . . . You did not purposely go to watch TV, you mostly switched on the television at 8 p.m., one watched something, afterwards one had dinner and talked; yes, you switched off . . . we would rather communicate.

It was rather seldom that people did not feel attracted by television at all, like the nurse Liudmila: "In principle, TV is and was not my passion, not at all, to stare all day at the screen, no. I don't like television, I prefer to read."

In contrast to her, not a few respondents attached a considerable part of their self-perception and representation to the TV set. The talks at work, at school, at university, and with friends about the program of the previous evening often even motivated the purchase of the first television. Thus, the higher-educated Katia, who was born in Leningrad in 1955, regretted the late purchase of a TV set only in 1967. This was surprising insofar as her family especially valued reading books and took a rather reserved stance toward the new medium. Katia, however, wanted to take part in the discussions at school about movies.[70] Others remembered that their fathers and brothers desperately wanted to watch international sport events like Olympic Games or Soccer championships on TV and hence purchased a TV set.

It is important to point out that villagers remember the change of leisure activities television brought about in a rather similar way to urbanites. The professor of sociology Vladimir Iakovlevich, who was born in a village in Cheliabinsk region in 1947, observed that until the mid-1960s people spent more time together doing sports and gaming. Afterward, adults started to watch television in the evening, what in Vladimir Iakovlevich's words meant that "with the spread of television [. . .] one sat more at home."[71]

Sitting at home and watching television also created some uncertainties about the effect of the new leisure habit among the viewers. Especially in television's early days, viewers wondered if excessively watching TV might cause any health problems, for example affecting the eyes. Television workers certainly reassured concerned viewers that television would not harm the eyesight. Nevertheless, they used such questions to advise the audience how to place the TV set and organize the environment to reach a convenient watching experience: The room should not be obscured but sufficiently illuminated; any darkening would tire the eyes.[72] Moreover, parents worried about how to educate children to be responsible viewers who watched TV reasonably and chose adequate programs. Television producers were also eager to promote their medium as enlightening and very appropriate for children's leisure time.[73]

Thus, the regime was eager to teach people how to use the new medium and how to adapt it to Soviet domestic spaces. Guiding people to rearrange their private life was, however, not a one-way communication, but let them engage in negotiations on privacy and lifestyle.

The contemporary advice literature on housekeeping and furnishing published in the late 1950s and early 1960s contrasted the experiences from villages and small towns the respondents recounted retrospectively. The official advice literature propagated TV sets as key parts of the Soviet urban lifestyle.

Thereby, television fulfilled a dual function within the new Soviet consumer culture. On the one hand, television became the symbol of the steadily improving good Soviet life because it represented technological progress. On the other hand, the party elite finally discovered in the late 1960s that Khrushchev had shown keen sense to hook into the Moscow kitchen debate with Nixon not only to demonstrate technological equality but also to compete on pop-cultural terms. People's enthusiasm about television and their ambition to own a TV set suggested the new entanglements of the public and private spheres, as well as the regime's opportunity to reach out to people in their living rooms.[74] Consequently, the regime not only strongly invested in the expansion of television networks and the production of TV sets, but also aimed to guide people's use of the new medium. Advice literature and popular scientific journals characterized the TV set as necessary item in a modern apartment. Thus, the new lifestyle became closely related to the technical-scientific progress from which the New Soviet Man should benefit as rational consumer. As early as in 1956, journals like the youth periodical *Znanie—sila* (Knowledge—power) propagated the image of a happy family gathering in front of the KVN in the evening. All illustrations of an article entitled "The screen of Life" (*Ėkran zhizni*) emphasized the message of the article about the significance of television for Soviet society. The author explicitly stated that the "screen of life"—as television was commonly referred to—increasingly moved into the everyday life of Soviet people. At the same time, he claimed to see the Soviet audience within a global context, as TV satellites seemed to realize the "dream of mankind to see all."[75]

In the early 1960s, the expert discourse quickly connected questions of TV consumption with new design concepts. Several companions on home furnishings and housekeeping demonstrated that the new style of interior and living should generally be characterized by using technical devices. The new home design was defined by deliberate modesty, minimalism, and functionality in contrast to the flamboyant petty bourgeois lifestyle.[76] Aesthetical aspects played a secondary role during the Khrushchev period before they returned in the Brezhnev era.[77] Advice literature, which primarily addressed women with photographs and drawings, perpetuated traditional gender roles. Despite the omnipresent rhetoric of emancipation and gender equality, women usually took the responsibility for furnishings and caring for all household devices. In the mid-1960s, already-adolescent youths learnt at school that apart from a cabinet, sofa, armchair, table, and chairs "radio or TV" belonged to the "highly necessary furnishings" of one-room apartment.[78] The authors had probably not claimed the simultaneous purchase of radio and television as standard yet, because they addressed young people who at first normally moved into apartments with little space. The editors of a companion published one year later discussed the purchase, positioning, and maintenance of a TV set before mentioning the radio and directly after instructing the reader how to use the refrigerator. The advice on where to place the TV set was very precise: The TV set should be positioned on a little table without any underlay somewhat lower than the viewers' viewing height. The table should be placed at 20 to 40 cm to the wall. From time to time the owner should free the device from dust.[79]

Consumers who established their new way of life in their own apartment also encountered these rationalistic claims that aimed to normalize people's privacy in periodicals, TV programs, and movies.[80] Particularly, movies made clear to the audience how the modern Soviet way of life should look, as the abovementioned American correspondent Levine depicted. At an early stage, Villen A. Azarov's highly popular comedy *Vzroslye deti* (Grown-up children) of 1961 represented the rising importance of the TV set. Starring prominent actors, the movie covered generation conflicts resulting from the new norms of design and lifestyle. The adult daughter gets married and moves with her husband for lack of an apartment of their own into the parental home. Thus, the redesign of the two-room apartment and the quarrels about taste, design, comfort, and functionality constitute the background of the plot. The audience traces how the recently retired father (Aleksei Gribov) strides through the bright apartment where he would spend much more time in the future. Obviously, the home does not comply at all with the straight layout of the apartment presented in the typical manuals on home furnishings of the 1960s underlining modesty and minimalism. However, this imaginary apartment, the viewer inspects tracking the father, is clearly arranged around the position of the radio and TV set. The shelf unit with the TV set is on one side of the room, sofa and radio on the opposite. The movie showcased that the daily use of the still-new medium television was officially considered to be part of Soviet cultured life (*kul'turnyi byt*) as early as in 1961: Television should be of equal rank with reading books and the long-established radio.

The movie visualized the parental home characterized by coziness, a bit of plush and playful details but, nevertheless, modernly equipped with technical media devices. The contrast of the parents' lifestyle to the suggestions of the booming advice literature becomes even more obvious when the young couple starts to redesign their room. The result perfectly corresponded to the unfussy design typically presented in the advice literature.

Vzroslye deti was far from being the only movie to pick this issue of the new Soviet lifestyle and the role television played for all generations. Other movies of the early 1960s also represented this connection of old and new ways of life and made fully clear that precisely the TV set addressed not only children, youths, and young people but older generations as well.[81]

The advice literature presented the new way of life in newly built houses, thus rather addressing young people who had the opportunity to purchase new furnishings. Movies, however, were able to depict the parallel existence of old and new lifestyles. A typical furnishing showcased in the advice literature consisted of a modern wall unit, which made effective use of the space up to the ceiling, of armchairs designed with clear lines and unruffled elegance. The TV set appeared to be much trendier than the somewhat old-fashioned models with their wooden cases presented in the article "The Screen of Life" of 1956 and in many movies of the early 1960s. The modern TV set should be placed between the viewers with whom it potentially forms a triangle. Thus, the TV set seemed to become a communicator and part of the viewers' conversation including the TV program in their daily life. This arrangement at least suggested a more dynamic and interactive communication than the design of the apartments presented in the movies. They represent the traditional design, which forced the viewers to take place side by side and vis-à-vis the TV set. Both drafts of lifestyle, however, are characterized by the common, cross-generational practice of watching TV. How consumers appreciated the advice literature—and if they received them at all—is hardly comprehensible, as well as we can't trace the contemporary perception of the cinema and TV audience.[82] Some of the respondents, however, remember that they loved these films because of the way they covered questions of love, family, and friendship. But nobody explicitly remembered them because they might have obviously represented the new Soviet way of life. On the contrary, they conveyed any educating objectives more subtly. Only when explicitly asked by the interviewer, they reflected upon this way of presenting consumer practices in fictional pieces. Then some interviewees retrospectively perceived that consumption, the supply with consumer goods, the way of life, and so on were negotiated in the background of many movies. Samara engineer Rudol'f N. put it this way:

> It [i.e., the topic of trade and supply] existed, but it existed again by what? . . . Well, in some movies where they presented some TsUM, I think, in Moscow or Children's paradise (*Detskii mir*), and where, hence, people queued for some nonsense. . . . This happened in comedies. Well, you know, I mean light entertainment. All these movies were subject to rigid censorship. You don't show anything superfluous. There were such movies, but sort of playful, hence, you know, serious problems, serious problems—that appeared much later. . . . That began when they laid the manager of the store 'Eliseevskii'[83] open to very harsh attack, they shot him.[84]

Obviously, the "playful" way Rudol'f N. considered movies to address consumer and lifestyle issues was very effective. Subliminally picturing the new Soviet lifestyle, dress codes, behavior, and so on meant to pick out the positive aspects and make these issues present without exposing its problems. On the contrary, the embedding of lifestyle issues into the televised entertainment culture provided ground for affirmative emotional bonding. The fact that most of the respondents did not address the way consumer issues constituted an obvious background plot rather proves that the televised items and practices had already been normalized. Nobody wondered or felt outright disgusted about the way they were pictured. In this sense, fifty-nine-year-old historian Yuri Vasil'evich stated that consumer issues were present and pointed to the movie *Legkaia Zhizn'* (An Easy Life) of 1964. The popular movie covered the career of a young chemist. He prefers to stay in Moscow taking orders for his speculative "private enterprise" in order to make a comfortable living instead of pursuing scientific goals far away like his friend. Yuri Vasil'evich found this an "undoubtedly real" topic. He also stressed that television was "certainly" the source from which one could learn about fashion:

> Well, where did we learn about fashion? Certainly there. About fashion, about that you look for the announcer, the way he dressed, that announcer. Of course, the woman, she changed her hairstyle, her dress. Where from did we get to know about these jeans, what else? All from there, from television. But if they presented something purposefully—I don't remember.[85]

Apart from the examples of the advice literature and movies, we find other images that deliberately visualized the central role television was suggested to play in family life. This applies for photographs that pictured the achievements of blue and white collars' new everyday life. Most of them were official propaganda materials promoting the new lifestyle in newspapers, journals, and booklets, so that they probably reached a larger audience. Many of them were taken by professional photographers who aimed to present these sceneries as daily life. Some of them were included in contemporary photo exhibitions or advertising. This kind of photograph adhered to a rather transnational design as one finds very similar pictures of West German or Swiss families in front of the screen.[86] East and West European images were partly prepared for media reports about the new way of life and had one important aspect in common: They represented the contemporary discourse about the "normal" but modern way of life, which included watching TV on both sides of the Iron Curtain. The rising distribution of TV sets symbolized the increasing prosperity of all social groups in industrialized countries.

Because of the multiethnic population, as well as the size of the country and the great differences between urban and village life, the Soviet discourse was, however, more complex and ambivalent than West or East Middle European narratives of modernization. Figure 1.2[87] illustrates that the TV set should also be part of the new way of life at the non-Russian periphery. Therefore, the third channel started to operate via the regional Uzbek TV station as early

Figure 1.2 Team Leaders of a Construction Plant Madalin Mamadaliev (Left) and Abitbai Tokhtabaev Watch Television and Drink Tea with Their Families. *Source*: Photographed by V. S. Tarasevich, Uzbekistan, Syrdar'ia region, 1957–1959, courtesy of RGAKFD.

as in 1956. Concluding from the photos, the watching practices as such might not have differed much between Soviet center and periphery.

The photographer of Figure 1.2 pictured two Uzbek workers' families watching TV and drinking tea together. The picture was taken in the late 1950s and the TV set was deliberately presented as a symbol of the Soviet Imperial modernization project that aimed to implement the new Soviet way of life in the periphery. It is striking that the two men had benefited from the economic modernization of the periphery and had obviously reached a decent living now. The showcasing of Central Asian workers' daily life slightly differs from that of Russian ones in so far, as it emphasized the socializing effect of television bringing together two families. This picture also underlines the way these Uzbek families merged watching television with their traditional practices of meeting, chatting, and drinking tea. Interestingly, they wear popular clothes and headgear, which perhaps suggests that their socioeconomic advancement was one of technical progress and adaptation in the first place. Whereas the chairs and the sofa in the background have an Uzbek flair, the TV set is arranged like those in the Moscow apartments on a small table in the corner. Different, however, is the artistic signature of the photographer who let his protagonists face the camera with a clear message: They smiled happy with the TV set prominently in the background.

In contrast to the photo that pictures Uzbek life, the other images presented families and friends focused on watching TV. These viewers were sitting or even standing rather close to the screen suggesting that they adapted themselves to the position of TV set instead of the other way around. The way they sat down did not seem to constitute very convenient positions to watch television. Some were seated directly in front of the screen (Figure 1.3) or rather inconveniently on a sofa positioned besides instead of toward the TV set (Figure 1.4).

Probably, these photos partly documented the process of arranging the living room around the TV set. Some respondents, who referred to the mid-1960s onward, and thus to a somewhat later period than pictured on these selected photographs, explicitly mentioned that they relocated the sofa or the armchair to get a more convenient seating position to watch television. Seventy-three-year-old teacher Zinaida Mikhailovna from Tol'iatti recollected that "the room was organized so that it was convenient to watch TV."[88]

One could easily imagine that the actual living rooms differed strongly from the design the advice books proposed as the arrangement of furnishings and TV sets presented on the photographs. Similar to the apartments staged in the movies, the furnishings of the Moscow machine operator (Figure 1.3[89]), of the Moscow female meat factory worker (Figure 1.4) or of the kolkhoz couple

Figure 1.3 Family of a Machine Operator of the Depot im. Il'icha Zapadnoi zheleznoi dorogi in Front of a KVN–49. *Source*: Photographed by C. Tartakovskii, Moscow, December 1952, courtesy of RGAKFD.

Figure 1.4 Kostina A. V., Female Worker of a Meat Factory in Front of the Television. *Source*: Photographed by D. Pirogov, Moscow, April 1963, courtesy of RGAKFD.

(Figure 1.5) did not comply with the aesthetics the advice literature claimed: One TV set stood diagonally in front of a kitchen sideboard, the others in the corner of the room. Many respondents would not have wondered about these arrangements, as they often recalled that their TV set was placed on a small table in the corner near the window. This may have often been the best solution to space problems, but it was also due to very pragmatic considerations, as the viewers had to connect the TV set to an antenna. Thus, it was the easiest way to reach the roof antenna via cable through the window as it is seen in Figure 1.4.

The living rooms of the new urban apartments normally comprised less than 20 square meters. It is easy to imagine that there was only little scope to place the TV set. At least, people had to somehow creatively arrange the cabinet, the armchair, and the coffee table to give their living room an individual touch. Those who managed all these challenges to purchase a TV set and to

Figure 1.5 Collective Farmworker of the Collective Farm Frunze V. L. Atroshkin with His Wife in Their House in the Village Zabrod'e, Belarus. *Source*: Photographed by I. Krasutskii, October 1966, courtesy of RGAKFD.

find a convenient place for it at home were proud of it, as the sixty-one-year-old chemist and university lecturer Aleksandr Nikolaevich remembered. He depicted that in 1961 his family was one of the first among their friends who purchased a TV set and put it "like an icon at the most representative place of the apartment."[90]

Further, television was an important instrument to converge center and periphery, as well as urban and village leisure practices. Figure 1.5 illustrates the modernization of kolkhoz life in a Belarusian village in Mogilev region. This photo presented the comparatively spacious and at the same time scarcely furnished living room of the couple that might not have been fully representative of living rooms in village houses. Apart from the fact that less villagers owned a TV set of their own, the memories did not basically differ from urbanites' narratives about the efforts to find a convenient place to watch.

A typical way to personalize the TV set as an item of furniture was to use it as shelf, to put a little lamp, candleholders, or figures on it as is shown in Figure 1.3. The female worker of a Moscow meat factory (Figure 1.4[91]) and the collective farmer (Figure 1.5[92]) placed the TV set on a small cupboard with a shelf for books beneath. Connecting old and new media, they represented themselves as cultured people.

The attempt to embellish the TV set as furniture sometimes dramatically failed. Liudmila (born in 1951), a nurse then living in Tol'iatti, reported that she once placed a vase with fresh flowers on the *Rekord*. Unfortunately, the flower vase somehow dropped, and the water gushed over the TV set. Contrary to all advice on how to clean the device, she decided to switch on the TV hoping that the heat of the picture tube would dry it. Coming home later, her husband was appalled by her carelessness, but Liudmila was right: Luckily, the TV set survived the mishap. It comes as no surprise that this TV set captured a special place in the family's memory. The *Rekord*, Liudmila enthused, was a "very good TV set, we loved it!"

Many owners of TV sets exercised a lot of care to find a convenient place for the TV set in their apartment after they often had already invested a lot of time and money to get one. Looking back, they revealed—resembling the reports of foreign correspondents—that television normally considerably changed the way the apartments constituted communication spaces. Charlotte Curtis, *New York Times* correspondent reporting from Moscow, quoted a "well-to-do" Moscow colleague. Accordingly, the newspaper journalist stated in the mid-1960s: "Everybody I know gathers in the kitchen, because the television and the refrigerator are there."[93] Although most of the respondents reported that they placed the TV set in their living room, not few moved their dining place and watched sport, movies, entertainment, and music broadcasts while eating with the family.

The memories about the way television changed the furnishings and the daily life particularly reveal how much many former Soviet citizens charged the TV set as a seemingly impersonal item with affirmative emotions and opened for the regime a way into their homes.[94] Concluding from the interviews, many viewers really enjoyed owning a TV set and accepted this new intersection of the private and public spheres. Nearly all, who explicitly talked about the purchase of their first TV set, depicted it as an exciting family event. Some respondents like Elena from St. Petersburg remembered it very well merely because they could henceforth watch TV "at any time."[95] Thus, people let the television become an ambivalent force of privatization and communication with the regime. Soviet consumers approved the way the regime promoted the new home-centered Soviet lifestyle and the nuclear family. However, they also appreciated the opportunity to use the TV set according to their own ways.[96] Henceforth, the TV set linked the private and public spheres materially and discursively. Television considerably withdrew the perception and the making sense of the authoritative discourse from the public into people's homes. Only this physical shift opened an intermediate space for the possible hybridization of official meanings, when viewers started to discuss the programs without being immediately controlled in how to understand authoritative messages.

NOTES

1. https://www.youtube.com/watch?v=z6RLCw1OZFw, 0:20–1:32 min., (accessed October 15, 2015). Greg Castillo, *The Cold War on the Home Front. The Soft Power of Midcentury Design* (Minneapolis, MN, London: University of Minnesota Press, 2010), X–XIII, XXII, 140, 160–169; William Safire, "The Cold War's Hot Kitchen," *New York Times*, July 23, 2009 (accessed October 15, 2015), http://www.nytimes.com/2009/07/24/opinion/24safire.html?_r=0; for a transcript see http://www.TeachingAmericanHistory.org/library/index.asp?documentprint=176, (accessed October 15, 2015).

2. https://www.youtube.com/watch?v=z6RLCw1OZFw, 1:33–1:45 min., (accessed October 15, 2015).

3. Susan E. Reid, "The Khrushchev Kitchen: Domesticating the Scientific-Technological Revolution," *Journal of Contemporary History* 40 (2005): 289–316; idem, "Khrushchev Modern. Agency and modernization in the Soviet Home," *Cahiers du Monde russe* 47, 1–2 (2006): 227–268; idem, "Cold War in the Kitchen," 211–252; idem, "Communist Comfort," 465–498; idem, "The Meaning of Home," 145–170.

4. Lynn Spigel, "Media Homes: Then and Now," *International Journal of Cultural Studies* 4, (2001): 385–411; idem, *Welcome to the Dreamhouse: Popular Media and Postwar Suburbs* (Durham, NC: Duke University Press, 2001).

5. On advice programs on Soviet television, see chapter 5 of this book.

6. GARF, f. 6903, op. 1, d. 499, 1956: Stenogramma soveshchaniia u nachal'nika upravleniia po voprosam razvitiia televideniia ot 1 fevralia 1956 goda, l. 92.

7. GARF, f. 6903, op. 31, d. 2, 1956: Stenogramma konferentsii telezritelei zavoda "Optika" g. Krasnogorska l. 7.

8. GARF, f. 6903, op. 1, d. 499, 1956, ll. 84–88.

9. GARF, f. 6903, op. 3, d. 306, 1965: Referat "Zritel' – o televidenii, televidenie – zriteliu," podgot. NMO dlia seminara v g. Aleksandrii (ARE), l. 2.

10. GARF, f. 6903, op. 48 ch. 1, d. 21, 1971: "Chto govorilos' o televidenii i radioveshchanii na XX, XXI, XXII, XXIII s''ezdach KPSS." Spravka, l. 5.

11. GARF, f. 6903, op. 1, d. 499, 1956, l. 91.

12. GARF, f. 6903, op. 1, d. 578, 1959: Pis'ma v TsK KPSS i SM SSSR o rabote Komiteta, t. 1, l. 107; Iurovskii, *Televidenie*, 106. The expansion of airtime progressed slightly ahead of other Eastern bloc states. GDR television, for example, started in 1953 with fourteen hours per week. In 1960, it aired fifty-eight, and in 1962 sixty-five hours per week. Cf. for more details, Knut Hickethier, *Geschichte des deutschen Fernsehens* (Stuttgart: J. B. Metzler, 1998).

13. GARF, f. 6903, op. 31, d. 2, 1956, l. 3. The estimates for 1957 accounted for about one million TV sets and more than four million viewers. Cf. Aleksandr Ia. Iurovskii, Televidenie – Poiski i resheniia. Ocherki istorii i teorii sovetskoi telezhurnalistiki (Moscow: Iskusstvo 1983), 93.

14. GARF, f. 6903, op. 48 ch. 1, d. 21, 1971, l. 13.

15. The first nine regional TV stations broadcasting from 1955 onward were located in Riga, Baku, Khar'kov, Omsk, Tomsk, Tallinn, Minsk, Sverdlovsk, and Vladivostok. Soon, also stations in Yerevan, Novosibirsk, Kazan, Saratov, Molotov (Perm), and Rostov-on-Don were founded. See GARF, f. 6903, op. 31, d. 2, 1956, l. 6; Iurovskii, *Televidenie*, 41–43.

16. GARF, f. 6903, op. 1, d. 499, 1956: Stenogramma soveshchaniia u nachal'nika upravleniia po voprosam razvitiia televideniia ot 1 fevralia 1956 goda, ll. 2–3; ibid., op. 3, d. 306, 1965, ll. 2–3.

17. GARF, f. 6903, op. 1, d. 499, 1956, ll. 8–9; ibid., op. 48 ch. 1, d. 21, 1971, l. 9.

18. GARF, f. 6903, op. 3, d. 130, 1959: Protokol i stenogrammy konferentsii radioslushatelei i telezritelei g. Permi, Sverdlovska, l. 5.

19. Viktor Shepel', "I sokratilis' rasstoianiia," *Pravda*, April 13, 1969, 3; GARF, f. 6903, op. 48 ch. 1, d. 21, 1971, l. 13.

20. GARF, f. 6903, op. 31, d. 14, 1959: Stenogramma zasedaniia Obshchestvennogo soveta telezritelei pri Tsentral'noi studii televideniia, ll. 13–15.

21. GARF, f. 6903, op. 28, d. 31, 1967, t. 3, ll. 37–8; see also Roth-Ey, *Moscow*, 185–192.

22. GARF, f. 6903, op. 3, d. 306, 1965, ll. 2–3.

23. Iurovskii, *Televidenie*, 108–109.

24. The East German television launched a second channel only in 1969. Cf. Susanne Vollberg, "'Wiederholungssender,' 'Russenprogramm' oder alternatives Massenprogramm? Zur Konzeption und Realisation des zweiten Programms des DDR-Fernsehens," in *Das Ende der Langeweile? Programmgeschichte des*

DDR-Fernsehens, ed. Claudia Dittmar and Susanne Vollberg (Leipzig: Leipziger Universitätsverlag, 2003), 169–170.

25. Harrison E. Salisbury, "What Russians See on TV," *The New York Times Magazine*, July 11, 1954, 14.

26. Salisbury, "Russians," 14.

27. Marguerite Higgins, *Red Plush and Black Bread* (Garden City, NY: Doubleday & Company, Inc., 1955), 38–41. See, for example, the image behind page 128: "One of Zagorsk's new apartment houses loaded with television antennae."

28. The interview with Yuri Vasil'evich was conducted by Elena Zhidkova in Samara region in July 2010. Yuri was born in Kuibyshev (Samara) in 1951,and graduated from university. He was member of the CPSU and deputy of the district soviet.

29. Iurovskii, *Televidenie*, 100–101.

30. GARF, f. 6903, op. 1, d. 869, 1965: Stenogramma zasedaniia sektscii televideniia konferentsii "Televidenie i kritika" g. Moskva, l. 120.

31. GARF, f. 6903, op. 1, d. 542, 1958: Pis'ma v TsK KPSS i SM SSSR o rabote Komiteta, l. 30.

32. GARF, f. 6903, op. 1, d. 499, 1956: Stenogramma soveshchaniia u nachal'nika upravleniia po voprosam razvitiia televideniia ot 1 fevralia 1956 goda, ll. 20–22.

33. GARF, f. 6903, op. 3, d. 130, 1959, l. 7.

34. GARF, f. 6903, op. 1, d. 499, 1956: Stenogramma soveshchaniia u nachal'nika upravleniia po voprosam razvitiia televideniia ot 1 fevralia 1956 goda, ll. 17–20.

35. The data are compiled from: *Narodnoe khoziaistvo SSSR za 60 let. Iubileinyi statisticheskii ezhegodnik* (Moscow: Statistika, 1977), 513; *Narodnoe khoziaistvo SSSR, 1922–1972* (Moscow: Statistika, 1972), 588; *Narodnoe khoziaistvo SSSR v 1988 g. Statisticheskii ezhegodnik* (Moscow: Statistika, 1989), 119; Leonid A. Gordon, Eduard V. Klopov and Leon A. Onikov, *Cherty sotsialisticheskogo obraza zhizni: Byt gorodskikh rabochikh vchera, segodnia, zavtra* (Moscow: Izdatel'stvo Znanie 1977), 156; Thomas Lindenberger, "Geteilte Welt, geteilter Himmel? Der Kalte Krieg und die Massenmedien in gesellschaftsgeschichtlicher Perspektive," in *Zwischen Pop und Propaganda. Radio in der DDR*, ed. Klaus Arnold and Christoph Classen (Berlin: Christoph Links, 2004), 37; Stephan Merl, "Staat und Konsum in der Zentralverwaltungswirtschaft. Rußland und die ostmitteleuropäischen Länder," in *Europäische Konsumgeschichte, Zur Gesellschafts- und Kulturgeschichte des Konsums (18.-20. Jahrhundert)*, ed. Hannes Siegrist, Hartmut Kaelble and Jürgen Kocka (Frankfurt a.M., New York: Campus, 1997), 227.

36. Interview with Igor, born in 1949 near Moscow. He finished secondary education. The interview was conducted in March 2011; *Obzor pisem telezritelei za oktiabr' 1981 god* (Moscow: Gosudarstvennyi komitet po televideniiu i radio, 1982), 19; *Obzor pisem telezritelei za ianvar' 1983 g*. (Moscow: Gosudarstvennyi komitet po televideniiu i radio, 1983), 17; *Obzor pisem telezritelei za sentiabr' 1983 g*. (Moscow: Gosudarstvennyi komitet po televideniiu i radio, 1983), 11.

37. Fedor Razzakov, *Gibel' sovetskogo TV. Tainy televideniia: Ot Stalina do Gorbacheva, 1930–1991* (Moscow: Ėksmo, 2009), 12.

38. Razzakov, *Gibel'*, 78.

39. *Narodnoe khoziaistvo SSSR, 1922–1972*, 314.

40. Vladimir F. Goriachev, "K voprosu o tipologii detskoi auditorii," in *Televidenie i deti*, ed. Gosteleradio, Tsentr nauchnogo programmirovaniia and Nauchno-issledovatel'skii institut obshchikh problem vospitaniia APN SSSR (Moscow: Iskusstvo, 1976), 53.

41. Robert G. Kaiser, *Russia. The People & The Power* (New York: Atheneum, 1976), 52–53.

42. According to official statistics more than 38 million apartments were built, into which almost 141 million people moved between 1953 and 1970. See *Narodnoe khoziaistvo za 60 let*, 492–498.

43. For similar developments in Italy, see for example Cecilia Penati, "'Remembering Our First TV Set.' Personal Memories as a Source for Television Audience History," *VIEW. Journal of European Television History and Culture* 2, 3 (2013): 4–12; for a somewhat delayed domestication of television and a rather reluctant attitude toward the new medium in the Netherlands see Liesbet van Zoonen and Jan Wieten, "'It Wasn't Exactly a Miracle. The Arrival of Television in Dutch Family Life,'" *Media Culture & Society*, 16, 4 (1994): 641–659.

44. There are numerous websites documenting the history of TV sets. See for example http://www.rw6ase.narod.ru/00/tw1/rubin102.html, (accessed August 18, 2015) or http://www.fernsehmuseum.de/grundig-revue-1957.html, (accessed March 20, 2015).

45. The interview with Natal'ia Petrovna was conducted by Oksana Zaporozhets in Samara in October 2010. Natal'ia was born in Samara region in 1946, finished secondary school, and managed a trade company.

46. Lev N. Kogan, "Sblizhenie sotsial'nykh grupp v sfere sotsialisticheskoi kul'tury," *Sotsiologicheskie issledovaniia* 2 (1977): 38.

47. The data of the tables are selected from Tsentr ėkonomicheskogo analiza i ėkspertizy: http://www.ceae.ru/SSSR.htm, (accessed January 5, 2017).

48. The interview with Zinaida Mikhailovna was conducted by Elena Zhidkova in Tol'iatti/Samara region in July 2010. Zinaida was born in Riazan' region in 1937, graduated from Moscow pharmaceutical institute, moved to Arkhangelsk and Tol'iatti, worked as teacher.

49. Interview with Georgii (St. Petersburg, November 2010). The interview with Nadezhda Ivanovna was conducted by Alexander Ermakov in Usol'e-Sibirskoe/Irkutsk region in August 2010.

50. Interview with Natal'ia Petrovna (Samara, October 2010).

51. Interview with Zinaida Mikhailovna (Tol'iatti/Samara region, July 2010).

52. Michael Milenkovitch, *The View from Red Square: A Critique of Cartoons from* Pravda *and* Izvestiia, *1947–1964* (New York, Buenos Aires: Hobbs, Dorman, 1966), 3.

53. The interview with Oleg Igor'evich was conducted by Oksana Zaporozhets in Samara in November 2010. Oleg was born in a village in Kuibyshev (Samara) region in 1958. He graduated from university.

54. The interview with Liudmila was conducted by Ol'ga Zaporozhets in Samara in September 2010. Liudmila has secondary education.

55. The interview with Valerii Nikolaevich was conducted by Alexander Ermakov in Usol'e-Sibirskoe/Irkutsk region in August 2010. Valerii was born in 1958 and received secondary education.

56. Boris Firsov, "Srednego zritelia net," *Zhurnalist* 12, (1967): 43.

57. *Tovarnyi slovar'*, vol. 7: Plavki-svekla, ed. I (Moskva: Gosudarstvennoe izdatel'stvo torgovoj literatury, 1959), after column 640.

58. A typical sales department of TV sets in late 1950s is depicted in Natal'ia B. Lebina and Aleksandr N. Chistikov, *Obyvatel' i reformy. Kartiny povsednevnoi zhizni gorozhan v gody NEPa i khrushchevskogo desiatiletiia* (St. Petersburg: Dmitrii Bulanin, 2003), 168.

59. The interview with Masha was conducted by Elena Bogdanova in St. Petersburg in September 2010. She was born in Leningrad in 1946 and graduated from university. Masha worked as lab technician in a sociological institute in Leningrad.

60. The Interview with Antonina was conducted in St. Petersburg by Elena Bogdanova in December 2011. Antonina was born in 1947; she graduated from a technical university and worked as engineer.

61. Interview with Natal'ia Petrovna (Samara, October 2010).

62. Cf. the contemporary description of NBC's Moscow correspondent Irving R. Levine, *Main Street, U.S.S.R.* (Garden City, NY: Doubleday & Company, Inc., 1959), 151.

63. Bönker, "Depoliticalisation of the Private Life?" 210–220.

64. Alfred Jr. Evans, "The Decline of Developed Socialism? Some Trends in Recent Soviet Ideology," *Soviet Studies* 38, 1 (1986): 7–8; Grushin et al., *Zeit*, 64 (quote).

65. Grushin et al., *Zeit*, 68–69, 75.

66. Günther Anders, "Die Antiquiertheit des Menschen (1956)," in *Texte zur Theorie und Geschichte des Fernsehens*, ed. Michael Grisko (Stuttgart: Reclam, 2009), 108

67. GARF, f. 6903, op. 1, d. 869, 1965, l. 93.

68. GARF, f. 6903, op. 1, d. 869, 1965, l. 119.

69. Interview with Georgii (St. Petersburg, November 2010).

70. The interview with Katia was conducted by Elena Bogdanova in St. Petersburg in July 2010. Katia was born in 1955 and graduated from the faculty of biology Leningrad State University. She worked for years as a translator in Soviet times and as an editor of St. Petersburg 5-ii Kanal from 2004 to 2009.

71. The interview with Vladimir was conducted by E. Ivanova in Irkutsk in October 2010. Vladimir was born in 1947 and graduated from university.

72. GARF, f. 6903, op. 3, d. 130, 1959, l. 9.

73. GARF, f. 6903, op. 31, d. 2, 1956, l. 27.

74. Roth-Ey, *Moscow*, 178–185.

75. Michail Arlazorov, "Ėkran zhizni," *Znanie – sila: ezhemesiachnyi nauchno-populiarnoi i nauchno-khudozhestvennyi zhurnal dlia molodezhi* 1 (1956): 41–42.

76. Ol'ga G. Baiar and Raisa N. Blashkevich, *Kvartira i ee ubranstvo* (Moscow: Gosudarstvennoe izdatel'stvo literatury po stroitel'stvu, arkhitekture i stroitel'nym materialam, 1962), 83.

77. Chernyshova, *Soviet Consumer Culture in the Brezhnev Era*, 164–165.

78. Liudmila I. Vorob'eva et al., eds. *Domovodstvo: VII i VIII klassy* (Minsk: Izdatel'stvo Narodnaia Asveta, 1964), 78.

79. Nikolai V. Efremov, *Domovodstvo* (Kiev: Izdatel'stvo Urozhai, 1965), 459.

80. Susan E. Reid, "Communist Comfort: Socialist Modernism and the Making of Cosy Homes in the Khrushchev Era," *Gender & History* 21, 3 (2009): 473–476.

81. See for example, Georgii Daneliia's *Ia shagaiu po Moskve* (Walking the streets of Moscow) of 1964, a prize-winning movie at Cannes Film Festival starring young Nikita Mikhalkov (Kolia), which similarly represents the ambivalence of the old and new: Kolia's new friend comes along and is introduced to his family. The scene takes place in the living room equipped with a piano, radio, and a TV set which was placed prominently.

82. Perhaps there are letters of moviegoers, but I have not traced them. At least, there are no letters of TV viewers as the audience mail of 1962, when *Vzroslye deti* was aired on TV, is not handed down. Cf. the inventory of GARF, f. R-6903, op. 10, 1952–1970 gg.: Otdely pisem.

83. Rudol'f N. pointed to the so-called Eliseevskoe delo, the notorious episode during the campaign against corruption starting in 1982 after Yuri Andropov's inauguration as general secretary. The director of the famous Eliseevskii store in Moscow, Yuri K. Sokolov, was accused of having taken bribes. He was arrested, sentenced to death, and then shot in December 1984. Cf. Luc Duhamel, *The KGB Campaign against Corruption in Moscow, 1982–1987* (Pittsburgh, PA: University of Pittsburgh Press, 2010).

84. The interview with Rudol'f N. was conducted in July 2010 in Samara by Elena Zhidkova. Rudol'f N. was born in Tashkent in 1940. He graduated from a technical institute und worked as engineer partly in Iran.

85. Interview with Yuri Vasil'evich (Samara region in July 2010).

86. For West German pictures, see for example Simon Müller, Familie vor dem Fernseher, 9.11.1960: http://www.wdr.de/bilder/mediendb/Fotostrecken/wdr5/themenwochen_und_specials/2012/Sommer_1962/Bilder_zugeschnitten/0724_film_fernseher_akg-images_990_m.jpg; http://www.welt.de/fernsehen/article13396067/Huberty-die-Hausfrauen-und-das-Eckenverhaeltnis.html: Bild 2; ein Beispiel für die Schweiz: http://www.mfk.ch/fernsehen.html; all accessed March 18, 2013.

87. RGAKFD, fotodokumenty, 1-4232 tsv: V. S. Tarasevich. Brigadiry stroitel'nogo kombinata Madalin Mamadaliev (sleva) i Abitbai Tokhtabaev s sem'iami smotriat televizor, p'iut chai, Uzbekistan, Syrdar'inskaia obl., 1957–1959.

88. Interview with Zinaida Mikhailovna (July 2010, Tol'iatti/Samara region).

89. RGAKFD, fotodokumenty, 0-212658 ch/b: S. Tartakovskii. Mashinist depo im. Il'icha Zapadnoi zheleznoi dorogi Korolev s sem'ei v novoi kvartire v vysotnom zdanii u Krasnykh vorot za prosmotrom televizora, Moskva, dekabr' 1952.

90. The interview with Aleksandr Nikolaevich was conducted by Anna Tikhomirova in Iaroslavl' in April 2010. Aleksandr was born in 1949 and holds a PhD from university.

91. RGAKFD, fotodokumenty, 1-104926 ch/b: D. Pirogov. Rabotnitsa miasokombinata A. V. Kostina u televizora v svoei novoi kvartire, Moskva, avril' 1963.

92. RGAKFD, fotodokumenty, 1-88353 ch/b: I. Krasutskii. Kolkhoznik kolkhoza im. Frunze V. L. Atroshkin s zhenoi v svoem dome v derevne Zabrod'e, Mogilevskaia oblast'/Belorussiia, oktiabr' 1966.

93. Charlotte Curtis, "The Way People Live," Anatomy of the Soviet Union, ed. Harrison E. Salisbury (London et al.: Nelson, 1967), 43.

94. Cf. generally about emotional bonding to material items: Catriona Kelly, "Making a Home on the Neva: Domestic Space, Memory, and Local Identity in Leningrad and St. Petersburg, 1957-present," *Laboratorium: Russian Review of Social Research* 3 (2011), 3: 53–96.

95. The interview with Elena was conducted by Elena Bogdanova in St. Petersburg in November 2010. Elena was born in Leningrad in 1961 and graduated from the faculty of economics of Leningrad university.

96. Cf. Reid, "The Meaning of Home," 145–170.

Chapter 2

Winning the Time, Hearts, and Minds of the Viewers

In 1956, the director of Tomsk television reported that the amount of "immoral acts and bringings to court" in a city district of Tomsk had recently dropped due to the rising number of TV sets. "As early as 7 p.m., the kids sit in front of the TV," he depicted the magic appeal of the new medium that seemed to influence social relations and public behavior.[1]

This chapter takes up the contemporary perception of watching television as pleasure-giving leisure habit and focuses on the way the medium changed people's daily life. It is interested in the way viewers used and perceived television as an information source and entertaining medium. The question of use and perception immediately touches upon the question to what extent did television trigger viewers' emotional commitment and affirmation toward the medium and its messages. Referring to Martha Nussbaum's mentioned concept of political emotions, this perspective examines the way television provided an emotional basis of compassion and affection that contributed to the maintenance of Soviet society. Are we right to assume that the regime enforced emotional bonds to the Soviet way of life by engaging viewers into the new televised Soviet popular culture? And how did the viewers assess TV contents and the medium's credibility?

In the first part of this chapter, I trace how television's rise to the most popular leisure activity after the mid-1960s affected everyday life and interpersonal relations. The question is in what way gender, social, educational, cultural, and economic aspects determined watching television. The second part highlights the emotionalization of the viewers. The third and fourth sections investigate from different angles how the viewers made sense of the political meaning and the credibility of TV contents watching Soviet television on the one hand and listening to foreign radio stations on the other.

WATCHING TELEVISION AND EVERYDAY LIFE

Already by the late 1950s, residents of Moscow, Leningrad, and some regional capitals claimed that watching TV had become daily routine.[2] In fact, it had at least started to modify daily routine not only for the better off, who benefited from consumer privileges, but gradually for blue-collar workers as well. Complaints about the unreliable announcements of the television schedules illustrate that almost from television's early days all social groups watched the programs. Editors of Central Television recorded a typical statement of a discontent viewer on a factory meeting in 1956: "I have to drive home from my job to Moscow. You run, you fly like a crazy man not to miss the start of the program! But you arrive at home and start screeching in front of the TV, because the broadcast has been delayed. Nobody knows why."[3]

From the beginning, the television editors tried to exploit viewers' enthusiasm in order to gain official support for their medium. At the same factory meeting with viewers near Moscow in 1956, a TV editor claimed that television had already won "the appreciation of the masses." Using ideological phrases, he pathetically contended that television "had been established in the life of working people."[4] He surely exaggerated the development and coverage of television at this time, but the interesting point is that he considered television to be a changing force regarding leisure practices and the relation of public and private communication.

The obvious changes in leisure practices and private life quickly aroused the interest of the only recently revived Soviet sociology. Sociologists started to collect extensive data and to explore daily life and therefore almost necessarily paid considerable attention to watching TV. Renowned Soviet media sociologist Boris Firsov observed an "expansion" of television during the 1960s.[5] His dictum referred to contemporary empirical studies that found out that Soviet people were spending increasing time watching TV from 1960 on. Soviet citizens of the 1960s were regular media users. In 1967, Central Television's Scientific-Methodological Department (*Nauchno-metodicheskii otdel*: NMO) established that 89 percent of the Soviet citizens were reading newspapers, 82 percent were listening to the radio, and 85 percent were watching TV. Most of these media users were even daily consumers: 69 percent read at least one newspaper every day, 50 percent listened to the radio, and 45 percent watched TV.[6]

This was also due to the fact that Soviet people generally gained slightly more leisure time after the late 1950s.[7] An often-cited study focusing on Pskov, an old and growing city located in the North-West of the RSFSR, shows that in particular women's leisure time increased toward the 1980s.[8] Working men hardly gained leisure time during the period from 1966 (34.3

hours) to 1986 (34.5 hours). Working women, however, benefited from the fact that the workload for housework decreased from 28.7 to 22.0 hours, whereas their leisure time increased from 21.0 hours per week in 1966 to 26.5 hours in 1986. Nevertheless, men still had more than 8 hours more free time than women. Housewives spent 41.2 hours per week in 1966 to care for housework, a workload that decreased to 36.9 hours in 1986. Statistically, they had the considerably increased quantity of 51.0 hours of leisure time per week in 1986 compared to 44.1 hours in 1966.[9]

These statistics make clear that from the beginning watching television was a highly gendered leisure practice. Moreover, it was severely dependent on the professional and educational background of the viewers. Soviet sociological studies show that women watched TV more often while doing housework, whereas men often watched TV without engaging in any distracting activities. According to the study on Pskov, employed men watched TV 5.9 hours per week in 1965, whereas employed women only watched 3.7 hours per week.[10] According to another research project on workers in big cities of the European Soviet Union, the respondents indicated that they watched TV and listened to the radio between 3.9 and 6.2 hours per week, respectively. Generally, both sexes reported spending about half of their leisure time sitting in front of the TV set.[11] Particularly, watching TV became a more and more absorbing practice for urban employed men.[12]

Another survey claiming representativeness found that blue-collar workers in the Ukrainian SSR and in the Russian city of Kostroma watched between seven and ten hours per week, respectively, in the late 1960s.[13] Boris Grushin demonstrated, in a study that claimed representativeness for the whole Soviet Union, that TV consumption had on average become the most important leisure activity besides meeting friends in the mid-1960s.[14] In the mid-1980s, Soviet citizens only spent more time sleeping and working than watching TV.[15] The long-term study on Pskov proved the generally rising share of TV consumption in people's life. In 1986, urban employed men watched 14.5 hours and women 10.6 hours per week.[16]

Whereas men in 1965 had invested 7 hours more in housework than in watching TV, they watched more than twice as much compared to their household activities in 1986. Corresponding to traditional gender roles, women, however, were still much more occupied with housework (16 hours) than with sitting in front of the screen. Both interviews and contemporary empirical studies document that men clearly benefited from the patriarchal family structures, having much more time for media consumption compared to working women. Generally, men watched longer television and were more focused on the medium than women. Men also spent more time reading newspapers and listening to the radio.[17] Thus, there could be no talk of equal opportunities at home regarding leisure and housework.

This higher daily burden is convincingly reflected by female respondents who vividly described their physical and mental exertion. Some of them strongly emphasized the high workload and the troubles of Soviet everyday life. Liubov' Aleksandrovna, who was born in Leninsk-Kuznetskii, a city in Kemerovo region, in 1950 and worked as accountant and ship's cook, typically recounted: "Well, there was practically no leisure time. We have worked very much in Russia, we had a dacha, where we had to go for tillage. When I did not work with my husband on the ship, I looked after the two kids and the dacha. That was why I didn't have much leisure time."[18]

Sixty-four-year-old Masha similarly stated that she simply had too little time to spend in front of the screen: "Little. I have much time only during the last few years. Television was always last in those days. When everything else was done, then you could watch TV. Well, then my daughter was born, and I had to care for all the household—sewing, something sewing up, mending, I thought of activities that you could do in front of the screen."

Gendered differences often characterized the concrete viewing habits. This was even true to a great extent regardless of the viewers' educational background. In particular, women performed multiple tasks simultaneously and connected their TV consumption with other activities. Many female respondents reported that they often dealt with housework while the TV was switched on. Sixty-eight-year-old Liudmila from Leningrad bears concise but impressive witness to how Soviet women included watching TV into their daily housework: "You come home, switch on the TV set, run back and forth, and watch the program with one eye."[19]

For women, television gained considerable significance as a background medium of everyday work. Sixty-four-year-old Natal'ia Petrovna remembered how she normally switched on the TV in the morning and listened to it rather than watched the program while making herself up.[20] Like many other women, she stressed that men and women differed in their preferences and were principally not interested in the same programs. That was why she had a TV set of her own placed in the kitchen. This enabled her to watch while she was preparing meals—a typical watching habit many women favored, at least, if they could afford to buy a second TV set and put the old one into the kitchen. Although it did not seem to have been an exclusively female habit to switch on the TV set as soon as one got home and to keep it running in the background, more women claimed to have done so.

In contrast, male respondents recollected much more often than women that they either intently watched television or often spent time with friends in front of the television. They chatted, played board games or cards, and simply had the TV set switched on in the background, as Valerii Nikolaevich, a tractor driver and crane operator from a village in Irkutsk region recounted.[21]

Not surprisingly, the respondents remembered very different incidents regarding family quarrels about which program should be watched. Central Television's letter department steadily reported that viewers were complaining about annoying family discussions in cases when sports and movies or sports and children's programs were broadcast simultaneously.[22]

Some families could afford to own several TV sets during the 1970s, so that they placed the new color TV in the living room, the old black-and-white set in the kitchen, and some even another one in the bedroom. Some women grinningly recounted that their husbands had to watch soccer on the black-and-white TV set in the kitchen.

It is not only the interviews that document how program choices were gendered, as this issue was already contemporarily discussed. Cartoonists were among the first observers to depict the rising importance of TV in family life. As early as 1960, the popular journal *Ogonek* picked out the potential conflicts television brought about. A typical caricature made fun of a husband watching soccer, whereas the woman is enjoying a classical concert. Although television was actually still far away from being the key medium at this time, the cartoonist contended that television had such high importance as a pleasure-giving leisure habit that the quarreling couple made a radical decision: They sawed the TV set in two pieces insisting that both had the right to watch what they wanted to.[23] Obviously, television should be considered as a pleasure-giving medium addressing both sexes equally, but the time men and women spent watching greatly differed, so that cartoonist repeatedly mocked this gender gap (see Figure 2.3).[24]

Not surprisingly, watching television also depended on educational background. The better educated the people, the smaller was the scale of respondents who reported about their TV consumption. The fact that the time budget studies relied on self-reported behavior could have distorting effects on the share of TV consumption in the everyday life of the better-educated groups, because quite a few deliberately displayed a critical stance toward TV and preserved anti-TV snobbery.

Already by the mid-1960s, some studies established that male and female blue-collar workers watched more than male and female white-collar workers.[25] In contrast to this seemingly obvious trend, the findings of *Gosteleradio*'s NMO were more diverse in view of the impact of education on media consumption. They, for example, grouped their respondents into owner of a TV set and those without one of their own and took further into account the amount of work days, that is, five or six. Men who had secondary education, were working five days a week, and owned a TV set were the most eager TV viewers in 1967. All men who did not own a TV set of their own but who had secondary or higher education and were working five days a week watched more TV than those with only primary education. The picture of female viewing habits was even more complicated. Higher-educated women without a TV set of their own and working six days a week watched, for example, much more than women with primary or secondary education.[26] These data suggest that in the 1960s more of the better educated tended to own a TV set, which might be due to the still extravagant prices and the difficulties to buy one.

Also, the relative time expenditure for reading and watching TV changed after the late 1960s in general and with regard to education: Men and women read more than watched TV in 1965, while both sexes invested three times as much in watching TV in 1986.[27] Further, especially less-educated men favored watching TV over reading the daily press, journals, or books from the 1970s onward.[28] These findings seem, however, to be more self-evident with regard to the relation of reading and watching than they ultimately are. Even if the quantitative relation changed and reading newspapers remained socially determined, it did not generally lose importance. Already contemporarily, Boris Firsov found out that most of his Leningrad respondents were still reading newspapers in the mid-1960s. In fact, 65 percent of his respondents read at least three periodicals.[29]

Another important factor that influenced TV consumption and reading newspapers was geographical milieu. Although fewer people owned TV sets in the countryside, villagers watched even more television than did their urban counterparts because, Soviet sociologists believed, they lacked possibilities to enjoy other leisure activities. Contemporary studies found that, after the 1960s, villagers spent more time watching TV than urbanites.[30]

Another study on rural viewers found out that in 1971 more than 40 percent of the people who also regularly read literature discussed the programs after watching. The survey, however, was only interested in the relation between watching television and reading. Its aim was to prove that TV did not have any negative influence on reading practices and that reading TV users were more critical than nonreading ones.[31]

Contemporary sociologists considered the villagers' habits of media use as an important issue to study but also as a sensitive field in which, theoretically, the progress of the Soviet way of life should become obvious. Not surprisingly, empirical studies found out that prerevolutionary leisure practices had been overcome. Accordingly, church festivals were said to have lost meaning. The main result was that villagers' leisure practices had gradually converged with those of urbanites without reaching the quantity or diversity of the latter. It was, however, the increasing TV consumption that gave the impression of gradual alignment. Some studies even proved a considerably higher weekly viewing period of villagers compared to urban workers as early as in the late 1960s. To be sure, some sociologists contested these findings and criticized the heuristic presumptions that were derived from studies about urban population. The applied categories, therefore, were criticized as ignoring specific rural leisure practices like gardening or animal keeping. Nevertheless, sociologists generally agreed that television's importance for recreation of villagers had also considerably increased after the mid-1960s.[32]

These trends corresponded to the subjective remembrances of former villagers among the respondents who appreciated the socializing effect of television. Like urbanites, villagers gladly remembered how they talked about the programs and discussed them in the evening or the next day at the workplace. Thus, communicating about TV became increasingly involving with the rising number of TV sets and the growing appeal of shows, movies, and series during the 1970s.

The interviews further document that the vast majority of the higher-educated respondents under no circumstances abandoned reading newspapers. Many convincingly recollected that newspapers continuously played an important role as an information source and somewhat characterized their lifestyle. Former engineer Antonina, for example, impressively recounted:

> Newspapers, journals, we subscribed to a great number of them. I remember them, heaped up, for many years . . . *Nauka i zhizn'* (Science and life), *Tekhnika* (Technology)—those were the main journals. We subscribed to *Literaturnaia gazeta*, *Pravda*, we certainly subscribed to, we subscribed to them all. *Komsomolka* and *Pravda*—that's no life, if you don't subscribe to it, everywhere and always! Then we subscribed to *Nauka i zhizn'*, *Khimiia i zhizn'* (Chemistry and life), necessarily some of the thick literature journals . . . *Iunost'* (Youth) I subscribed to, I subscribed to many journals for my daughters, so that very many periodicals went through our mailbox.[33]

Only a few of the less educated admitted that they rather read fewer books and somewhat fewer newspapers. However, some of them said that they never read periodicals much. Thus, Soviet people generally consumed more media with only a slight shift between newspapers, journals, radio, and TV. Further, most of the respondents documented the importance of listening to the radio. Regardless of their educational background, the majority reported that the radio was still switched on as much as before. It remained a "daily companion," as sixty-two-year-old Igor' put it. Liudmila from St. Petersburg, who had finished secondary school, claimed that she "generally never switched off the radio in the kitchen" and that she basically obtained news and information from radio, but also from television. She combined listening to music on the radio with getting news, but at the same time recognized that television aired ideological programs she did not like to watch.

Similarly, seventy-three-year-old Zinaida Mikhailovna left no doubt about the high importance of Soviet radio: "Yes, we listened to the radio. There was a radio in any apartment, a radio point, and we listened to the radio."

Asked which differences occurred between watching TV and listening to the radio, she like many others described the spatial division of many apartments in terms of media consumption: "Well, basically we listened to the radio in the kitchen, when you prepared the meal there, when you eat there etc., there was the radio point, one switched on the radio. When you left for the living room, you watched TV."[34]

Many narratives reveal that media users seemed to have been more selective about watching television than about listening to the radio because of television's much broader range of program genres. Regarding the TV programs, it was important to distinguish entertainment, documentary, and news broadcasts, if one looked for pleasure-giving moments. The attitude of higher-educated people toward television was not homogenous but ranged from reservation to compliance. In comparison to television, the assessment of radio was relatively unambiguous. Especially for higher-educated media users, the radio was, as Zinaida Mikhailovna asserted, characterized by higher topicality compared to television until the early 1990s. That was why one

> listened more to the radio in order to obtain information, information about what is going on in the world, what is going on in the country, well, wherever. . . . Well, in the first period it took the place of television, because television didn't exist. Afterwards the television started to emerge, but people still listened to the radio, because not all owned a TV set, not all . . . , as one says, watched, in particular, if you were at the workplace. Nevertheless, you switched on the radio.[35]

Judging from the interviews, the relationship between radio and television was not strictly competitive but rather, from the point of view of most media users, complementary. This explanation applied for urban as well as village users, as soon as the supply of TV sets improved. The assertion of fifty-nine-year-old Mikhail, a meteorologist from Leningrad, that television replaced radio to a great extent was an exception rather than the rule for both urban and rural areas.[36]

Apart from gender, education and geographical background, sociologists applied different age groupings. However, the empirical breakdown of watching TV according to age groups remained rather vague. The studies on Pskov represented people's excitement about television as nearly independent of age. Only the interest in informative news and political programs increased with viewers' age.[37]

Overall, contemporary studies on television viewing underlined similar developments to those established in the interview narratives. The audience, contemporary observers and party officials had a clear opinion regarding the rise of television to become the uncontested key medium in the early 1970s. The available statistical data on media use are not that clear-cut but evidence a steadily increasing amount of time people spent in front of the screen: Still in 1966, less than 40 percent of urban working people watched TV daily. Sociologists assumed that by the end of the 1970s, 98 percent of factory workers watched daily. In 1982, 97 percent of the urban, 80 percent of the village and 93 percent of the whole population were considered to switch on the television daily. In the mid-1980s, 82 percent of the interviewed Muscovites stated that television had become not only the most important entertainment medium but also the most appreciated information source.[38]

These high daily viewing rates complied with the ideological charge placed on Soviet sociology to prove the superiority of the communist societal concept and of the concrete details of the Soviet lifestyle. Sociologists' main interest, therefore, was in the collective of people, rather than in the individual. This caveat heuristically limited their findings. The potential outcome was further determined by the fact that the increasing leisure time was considered as a logical step on the way toward communism. The change of leisure time was ideologically sketched out in the sense that the working people should make their way to cultural improvement.[39] According to Brezhnev's formulation on a party congress, leisure time should not primarily be a personal concern, but should be "used on behalf of the general development of the person, his abilities and thereby of the further enhancement of the material and intellectual potential of the whole society."[40] In this interpretation, leisure could not be considered to be private and unpolitical, because the individual should be responsible for the collective even at home in his or her living room.

Against this background, television's increasing significance as a leisure habit was highly welcomed and promoted with the shift of the party's attitudes toward the medium. This change toward political protection very much complied with the self-representation of TV producers: although many party leaders had been reluctant to praise television during the 1960s, in the 1970s, they granted people an unwritten "right to relax in front of the television after a day's work."[41] Moreover, in the mid-1970s, the chairman of the State Committee for Television and Radio, Sergei Lapin, confirmed *Gosteleradio*'s claim on TV's leading societal impact: In the *Bolshaia Sovetskaia Ėntsiklopediia*, he asserted that "TV broadcasting" helped to "organize people's leisure time."[42] Soviet media scholar Nikolai Shonin explained in an international journal in 1977 that TV genres should be especially "designed for recreation, for people resting after a day of hard work under a tight schedule dictated by the scientific and technical revolution and the need to fulfil the economic assignments of the State plan."[43] From the ideological point of view, hard work for the benefit of the Soviet society and watching TV became two sides of the same coin.

THE EMOTIONALIZATION OF THE VIEWERS

Despite television's obvious presence in many people's life, seventy-three-year-old teacher Zinaida Mikhailovna was not sure that television actually changed daily life. Her account, however, contradicts her own explicit assessment:

> and whether it changed life? No, I wouldn't say so. Of course, we watched it, of course, once we were at home, somehow tied to the home, you needed to watch some program, some, in general . . . yes, and in general, it would be switched on, as you might say, you come home from work, switch it on and then it's just on, whether you watch or don't watch—it's all the same, it's switched on, because—as I say—there was a painful lack of program guides. All the more so since we were at work and didn't hear what would be on TV today. Therefore . . . if there were interesting programs, so . . . we watched . . . I don't remember in which year *Semnadtsat' mgnovenii vesny* (Seventeen Moments of Spring) was aired? Seventy, already seventy somehow . . . well, ice hockey games were also aired, I remember, our team versus the Swedes—we did not sleep at night . . . we watched that, we drank valerian drops, when ours lost . . . Well, and television certainly provoked and carried a certain interest in life. But I don't want to say that it was the only means of communication, connecting to the world on so on . . . , namely we read a lot of newspapers, journals and so on.[44]

Zinaida Mikhailovna perfectly highlighted that television changed daily life a lot. It rendered home a place to perceive the world in a new and vivid way and to communicate about these impressions with family and friends. Apart from programs providing opportunities to "perceive the world," Zinaida Mikhailovna—like many other respondents—appreciated broadcasts of major sports events and entertainment series very much. Her narrative reveals the way viewers typically established affirmative bonds to the medium because it brought the popular national and international sports events directly to their homes. Although several women talked about watching sports broadcasts, Zinaida's enthusiasm about hockey and soccer events was hardly representative of female viewers. Male respondents remembered such television events much more frequently in a passionate way. However, both men and women emphasized the community-building impact when friends were sitting together in front of the screen cheering the Soviet hockey team or Olympic athletes. Especially during the 1960s, at a time when watching with friends was still common, football and hockey matches were frequent occasions to gather in front of the screen. Irina N., a former accountant born in 1953, recounted that she regularly met with ten to twelve people to watch TV in those times. Sports broadcasts were particularly popular: "The World Ice Hockey Championship—it was unambiguous that we watched. Soccer games, too. In other words, we knew everything, we knew the soccer clubs, we practically knew all our soccer players, we knew very much."[45]

That both Zinaida Mikhailovna and Irina warmly remembered international hockey matches let us trace how sports broadcast against the backdrop of the Cold War provided an emotional basis to which the governments on both sides of the Iron Curtain could link political narratives and concepts. Sport was a diplomatic means of political confrontation on the Cold War stage and part of political propaganda demonstrating the respective supremacy of capitalism or socialism.[46] The interviewees' statements also coincide both with the official discourse and with the description *Washington Post*'s correspondent Robert G. Kaiser gave to his Western audience in the mid-1970s. The attraction of sports that many respondents perpetuate seemed to have been an undisputed fact from different points of view: Kaiser reported that sports was "a modern substitute for religion, spirits and folk wisdom" and had become the "new opiate of the masses, in the Soviet Union as in the West. There are millions of Russians who think of little else outside their private lives." He even quoted the *Komsomol'skaia Pravda* of 1975 to emphasize his observation. The newspaper of the Soviet youth organization claimed the prolific collaboration of television and sports, as "the sports fan is a sign of the times. (. . .) Sports broadcasts are the most popular shows on television."[47] Clearly, the regime could link to the emotional and political dimension of sports in the domestic context. Focusing on soccer fans, Manfred Zeller has worked out the shaping role television played in the building of emotional communities. Television presented soccer to a broad range of viewers who otherwise probably would not have been able or even interested in physically attending a match at a stadium. Further, television addressed the collective of fans in a rather different way from the radio and printed press, as it invited viewers judge on their own what was happening on the soccer field, instead of solely relying on the radio commentator's report. The medium also increased the potential of soccer clubs to attract fan communities who were now sitting, emotionalized, in front of the screen, and were feeling at least virtually connected to those who cheered the same club. Thus, sport broadcasts on TV also linked the public and the private sphere in a specific way. As easy as it was to connect emotionally across a distance to other fan groups when sitting in front of the screen, watching television at home also made it possible to express these personal affiliations outside the private realm. Television even legitimized soccer fans to speak about the last match beyond the privacy of their home. Henceforth, soccer fans heatedly discussed the referee's performance or the public behavior of other fan communities in letters to the medium.[48] These discussions and emotional affiliations represented the voice of TV viewers and sport fans in the context of authoritative discourse. Although their expressions could but certainly must not mean to undermine authoritative discourse, the viewers' voices became audible and impacted on cultural negotiations because television addressed the sport fans as possible emotional and "interpretative communities" (Roger Chartier). This is at least what the interviewees' narratives about watching sporting events on television, whether alone or with family, friends, or colleagues, strongly suggest. International sport events easily allowed for a patriotic staging of Soviet athletes so that television could frame the viewers' emotionalization with a narrative of shared identity and belonging. Many interviewees and contemporary letter writers were very proud of sporting successes in international competitions. In the context of the Cold War, televised sport successes helped viewers to identify with "being Soviet": The athletes were "ours," as Zinaida Mikhailovna underlined in a very representative way. Hence, many respondents thought like Vladimir Ivanovich, who confirmed that "sports, hockey, that was sacred, even more provided that ours—you don't remember [he referred to the interviewer's age]—then they simply shined [he spoke about the high number of Soviet medal winners and triumphs in team championships]."[49]

The question, however, to what extent the viewers linked any sports successes, if not directly to the Soviet regime as such, but at least to the Soviet system of promotion and selection of athletes, opens a perspective of a certain range of attitudes, as well as of hybridizations of official meanings. Many respondents still retrospectively reveal their deep patriotic sentiments in cheering the Soviet athletes—an attitude that was at first sight perfectly complying with the official discourse. Their empathy, however, seemed predominantly to refer to a basic understanding of "being Soviet" and belonging to the Soviet society. At least, many of the respondents constructed their narrative in this way and dissociated themselves from any political implications. Retrospectively, some respondents like Zinaida Mikhailovna explicitly affirmed that they had not converted their enthusiasm about the athletes into identification with the Soviet regime as such. Zinaida insisted that she had, rather, distinguished between the "system" and "ours"—that is, the Soviet people. This categorization let us hear a voice undermining the way authoritative discourse tried to define the unity of the party and the people. Again, this hybridization of official meanings became possible because television not only transgressed the boundaries between public and private but also because it presented an ambivalent and dynamic picture of the sport event if it was live broadcast. The Soviet athlete might have failed, the viewer might have switched off the sound and commentary.

However, a contradicting voice or the tendency of hybridization must not necessarily mean opposition to the Soviet system as such. This is strongly suggested by some respondents' statements about international sport events.

Reflecting on the question if the Cold War competition and the confrontation with the capitalist system impacted on the sports events, Zinaida considered:

> Well, it was simply the Olympics. So, I remember there was a hockey match, our people played the final match just versus the Americans and were defeated by the American college team. And then practically all were shocked, someone was crying, others did not cry [. . .]. . . . Well, if our people just performed very well, we were somehow proud . . . but I do not say that we were proud of the system [she stressed the word]. Well, somehow our people . . . but only pride in the performance of our outstanding cross-country skiers, figure skaters, speed skaters.[50]

This feeling of pride became a strong cohesive factor that seemed to have bonded many viewers together. Historian and party member Yuri Vasil'evich, born in 1951, warmly remembered how one got excited about the top performances of the Soviet athletes: "Then you strongly kept track of everything, suffered, worked yourself up: 'However, again our people set the records, someone made a jump there in some competition, yes, our people, look!' You know, well, this was really a pleasant feeling. You know, we were proud of our motherland, that ours made a jump there."[51]

Several respondents referred to the "pleasant feeling" of pride Yuri Vasil'evich highlighted. They were proud of the country because of the successful athletes representing it and somehow avoided to assess the communist regime or to relate their pride to it. Nevertheless, patriotism was in any case meant to legitimize the own affirmation of the Soviet community. Thus, it even retrospectively sustained a continuing bonding to the glorious past of Soviet sports, as one of the youngest interviewees illustrates. The forty-two-year-old Ruslan from Samara emphasized that "all cheered the Soviet Union at that time, that's Russia now. Then you cheered the Soviet Union to win more gold medals and simply, simply cheered your people."[52]

The way the respondents still evoke their Soviet patriotism based on a concept of "ours" versus "them" reveals that the resultant imagined community gained its strength not least from the idea of being different from foreign states. Thus, the respondents' perception still sounded like the official Soviet idiom of being surrounded by allegedly hostile foreign countries—a wording that has, interestingly enough, gained importance in recent years again.[53]

However, the answer to the question, to what extent the viewers attributed an explicit political meaning to the televised sports successes and consequently related their commitment to the Soviet system as an alternative draft to the capitalist West, is less ambivalent regarding the contemporary letter writers. Whereas the respondents rather detached their idea of being Soviet from the political implications of having been convinced of the Soviet project, the letter writers predominantly reproduced authoritative discourse.

Although one must consider that the editors of the letters department quite likely published precisely those letters that exemplarily matched the official idiom, many letters did this in an impressive way. In 1980, the year when the Olympic Winter Games took place in Lake Placid and the Soviet Union was the host of the Summer Games, the incoming letters amounted to 8.333, reaching not only a quantitative but also an emotional peak. The viewer I. Dashkevich from Gor'kii (Nizhny Novgorod) region perfectly illustrated the acceptance of a Soviet identity and his emotional involvement in face of the Western Cold War propaganda:

> The Soviet athletes have to live in a real prison, in an overcrowded accommodation without any basic conveniences. Furthermore, the American authority launched an unbridled anti-Soviet campaign. However, our champions even in this case did not tremble, demonstrated the best features of the Soviet character: courage, endurance, determination to succeed, true collectivism. That was the genuine main triumph in Lake Placid. The best Soviet athletes did rebut the lies of the reactionary propaganda with their brilliant performances, they caught the American viewers to regard the representatives of the Soviet Union completely differently.[54]

Letter writers further reproduced authoritative discourse by being delighted that "the remarkable Soviet athletes duly defended the honor of their country with every new Olympic triumph, thanks to which the hymn of the Soviet Union sounded once again."[55] Regarding the boycott of several Western countries of the Summer Olympics, letter writers claimed to possess the moral superiority, as they accused the Carter administration of exploiting the Games for political reasons, whereas the Soviet Union was striving for peace, friendship, and international understanding.[56]

Another effect of watching sports on television was that the programs raised talking points for the next days just as the regime intended to. In 1975, *Komsomol'skaia Pravda* recognized that the "most heated arguments during lunch breaks are about sports."[57] This official picture came close to the retrospectives of the interviewees. According to them, the broadcastings typically provided ground for the fans to discuss the performance of the commentators, the sportsmen, and the referees. The respondents described that they often debated the sports events with family, friends, and colleagues on the job the next day. Television, thus, impacted on the interpersonal communication in the way that it caused a certain social pressure to be up-to-date and to be able to join in the small talks. In the context of authoritative discourse, sports programs obviously perfectly met the regime's demands to involve the viewers and let them discuss issues triggered by the programs. The audience mail makes clear that the viewers engaged in these discussions. Viewers praised the presenters who commentated with expertise and in a balanced way. Generally, the audience demanded a professionalization of the sports journalists and increasingly called for commentators who proficiently covered the competitions and matches.[58] Zinaida Mikhailovna was, however, one of the few who explicitly talked about watching television on the job: She emphasized that they had several TV sets in the medical school where she worked, and that they watched sports events together at the workplace.

Considering these points, the cohesive power of televised sports events seems to be rather obvious. At least, my available sources strongly suggest that television reinforced the building of an imagined Soviet community. Media scholars offer a fascinating explanation for the bonding effect of sports programs. Sports programs, in general, and highlights like Olympic games and championships, more particular, were outstanding examples of what Knut Hickethier calls "moderated emotions" regarding the GDR audience: Television presented those sporting events that offered the viewers not only a world of successes, but also exciting entertainment that distracted them from daily life. The audience could sympathize with "their" athletes experiencing their successes and defeats, thus engaging in a rather continuous process of cheering as television regularly started to broadcast sports events.[59] These intensive "para-social relationships" to sports heroes had been inconceivable without television before. Presenting a whole variety of heroes, television constituted the officially accepted frame of reference to which viewers could link their everyday life, as they observed, sympathized with, and judged these media actors. Media scholars consider these para-social relationships as the essence of entertainment, that is, a way to feel comfortable and to dock one's life to the broader world. At the same time, television itself benefited from the construction of heroes, as many viewers consciously consumed the programs to relate themselves to these heroes.[60] Again, *Washington Post* correspondent Kaiser highlighted the parallel developments on both sides of the Iron Curtain: "Both fans and athletes are virtually indistinguishable from their western counterparts. The crowds lose their hearts to star performers, wait for hours buy tickets for big games, deal with scalpers when they must, argue vehemently on behalf of their favorites."[61]

Even if Kaiser might have slightly exaggerated the star cult to trigger the sympathies of his audience for the ordinary Soviet citizen, his report matches the respondents' narratives and the interests of the contemporary television producers. For example, the editors of Central Television's sports desk considered this attraction of viewers a valuable advantage in their struggle for appreciation and money that should lastly ensure to expand the sports broadcasting. Interestingly enough, the editors argued in a similar vein compared to the modern media sciences regarding the audience's emotional involvement:

The editors claimed that a great number of fans would have lived through sports events: "They rejoice, get upset, make predictions, argue and feel themselves like participants. And television gives them this sense of delight about victory and of chagrin in the case of a defeat."[62]

The sports broadcasts accounted for a notable 10 percent of Central Television's whole airtime, compared to 40 to 50 percent made up of artistic programs during the 1970s. Although many viewers and interviewees were enthusiastic about sports broadcasts, the mailbag of the sports desk nevertheless constantly ranked among those with smaller numbers of incoming letters. The editors of the sports desk reported that their audience mail was decreasing after the late 1960s first and then stabilized in a range between ca. 6,000 and slightly more than 8,000 in 1980, the year of the Olympic Games in Moscow. They contended that this development essentially demonstrated that sports fans had become more satisfied than before, because Central Television's performance had steadily improved.[63] Based on the interviews, many viewers would perhaps have agreed with this interpretation, at least, because of the fortunate interplay between the television broadcasting sports triumphs.

The win-win situation for sports on television, however, applied not only for international competitions but also for national sports events. Both the audience mail and the interviews highlight that exciting broadcasts of national sports events contributed to the building of often competing fan communities, whereas international sports events guaranteed the synchronization and societal embedding of viewers' emotional bonding to "being Soviet." Especially in the context of the Cold War competition, international sports events became a major asset to emotionalize and involve the audience on a national scale. Therefore, sports broadcasts became one of the prominent television genres that—from the regime's point of view—guaranteed the recurring and partly calculated emotionalization of the audience.

This may explain why higher-educated Tat'iana Iakovlevna from Irkutsk (born in 1964) thought that television only broadcast those disciplines in which the Soviets were successful and which people loved to watch: "In fact, ours did not always triumph, there were disciplines in which ours were always winning, and those disciplines, in which ours were not involved. They simply did not broadcast them, just as if they did not exist."[64]

But did the television broadcasts of sporting successes generally invoke a feeling of pride among viewers of the interethnic Soviet Union that captured fragmenting dynamics? How did it contribute to build a supranational "Soviet people"? As all my respondents and presumably a considerable number of the letter writers who affirmed their love for the country were ethnic Russians or, at least, did not claim another ethnicity, we can only cautiously speculate about television's community-building impact on the empire's periphery.

From their perspective of the center, several respondents depicted their pride revealing an imperial idea of the Soviet people, which they constructed as being tantamount to ethnic Russians. However, the extent to which people belonging to non-Russian ethnicities complied to a transethnic Soviet identity while watching such sports programs on television probably ranged from open hostility to affirmation of "being Soviet," just as researchers have observed in other cases.[65] That many non-Russian people accepted Soviet values and norms and connected their ethnic affiliation with an emotional belonging to the Soviet Union as their homeland in a perhaps contradictory way suggests one of Maike Lehmann's respondents. In her study about nation and socialism in postwar Armenia, Lehmann quotes a former apparatchik, who ideal-typically described himself as having been a patriot of Armenia as well as of the Soviet Union. He distinguished between "the country," that is, the Soviet Union, and "my nation," that is, Armenia. For a former party member, the statement as such may not come as a surprise, but, interestingly enough, he aimed to prove his double patriotism by assuring that he cried in front of the TV set when "we lost to the Czechs in ice hockey."[66] Although this incident does not say much about ordinary Armenians' attitudes watching Central Television, we have further evidence from the Baltics: Hardly surprisingly, a majority of Estonian Soviets refused to watch the programs from Moscow, Leningrad, and the local Estonian TV station and switched over to Finnish television instead. There are only estimations, but Finnish television seemed to have steadily attracted considerably more Estonian viewers than Soviet Central Television during the 1970s and 1980s. In an oral history project, Estonian viewers claimed to have cheered the Finnish athletes and teams in sports events.[67] Looking back, the Finnish broadcast on the Los Angeles Olympics of 1984 became a symbol of the rejection of Moscow Central Television and generally of the Estonians' hostility toward the Russian-Soviet occupiers. Contrary to many other Soviet viewers, who felt morally superior because of the Soviet boycott of the Olympic Games this time, Estonians watched the Finnish broadcasts all the more. Watching was even more appealing as these Olympics were officially "forbidden games," as one Estonian viewer remembered. He claimed that he "had a feeling of superiority over this Soviet country, and its rulers."[68]

These dispersed indications suggest that watching television might have been a catalyst—in one case for bonding people together, in the other case for people's rejection of Moscow's authority. If the Soviet people was a community based on familiarity with the Russian culture and language, television might have been the adequate medium to promote multiethnic inclusion.[69] Historical media studies on transnational audiences have established that television proved be a powerful promoter of foreign languages or a catalyst for national conflicts related to language issues. This applied for cross-border watching in the Cold War as well as for a multilingual context like the former Yugoslavia.[70] There is much to suggest that Moscow Central Television also figured as facilitator of the Russian language and of the utopian idea of a supranational Soviet identity. Thus, television probably contributed to the hybridization of national ideas with the Soviet claim to homogenize the empire. It is, however, consigned to further research to highlight these urgent questions of how television concretely impacted on the reinvention and legitimization of the Soviet project on the peripheries not only via sports broadcastings but also through other entertaining genres.

Getting back to Zinaida Mikhailovna, her narrative lets us trace additional aspects that dominated the viewers' perception of the television programs apart from watching sports. Many, probably most viewers merged different attitudes toward TV contents. These ranged from compliance, through indifference, to critical reservation, often depending on the genre. Zinaida Mikhailovna emphasized that most viewers typically used television as a pleasure-giving medium. She depicted how amusing entertainment programs triggered cheering emotional assessments that had an affirmative effect on people's perception of "being Soviet" and living the Soviet way of life. This entertainment-oriented use of television implied a highly selective and, to a certain extent, ignorant viewer, who carefully chose the programs she/he actually turned on. Television's capacity to give pleasure and relaxation and to trigger affirmative commitment was not least due to those programs that attracted viewers of all age groups and let families gather together in front of the screen.

Unlike Zinaida Mikhailovna, most women were more interested in watching movies, series, plays, and concerts. These women approved that the increasing television programs legitimized high expenditures of money for a new TV set. Spending much money for a technical device was reasonable as soon as it provided pleasure and recreation. After the TV set had finally appeared in the living room, most people switched on entertainment programs, game shows, or movies. Former engineer Antonina recollected that television mainly broadcast movies in her youth in the 1950s and 1960s. That was why her family did not use it as an information source: "Really, television was none [i.e., an information source]." Antonina appreciated watching Soviet movies on television, although she and her parents were also passionate readers. Reading and watching TV was a combination of media practices that many respondents, not only women, favored. Antonina also underlined that she loved to discuss the movies and series watched on TV and that it was common habit among her friends and colleagues: "Permanently. In those years [the 1970s], when I still worked in the administration of civil aviation, we permanently debated all movies! Because there were only few—*TASS upolnomochen zaiavit* (TASS Is Authorized to Declare), *Semnadtsat' mgnovenii vesny* (Seventeen Moments of

Spring), such series were on—you always discussed them in the collective."[71] One could easily imagine how collectively suffering with the heroes of these movies and series while talking about them the next day at work created emotional relations among the followers.

The narrative of Svetlana Vladimirovna was rather typical of a woman talking about family watching habits. This sixty-three-year-old woman was raised on a state farm in Samara region. She recounted that she herself, her husband, and her daughter each had a favorite program in the 1970s. However, to watch together some highly popular programs such as *Goluboi ogonek* (Little Blue Flame) and *Ot vsei dushi* (With All My Heart) widely became a much-beloved family tradition. Svetlana Vladimirovna recounted that normally there were no quarrels about the TV program. This was, however, part of her self-representation as an always-busy working woman, caring mother, and careful housewife. Svetlana Vladimirovna said that she—in contrast to her husband—normally "had no time" to watch TV. She would have often left her place in front of the TV in the living room, because she had to prepare meals, mend clothes, and so on.[72]

Apart from this kind of gendered narrative that many women presented about former watching habits, many respondents depicted in a quite similar way how watching certain television programs together became a family tradition. *Ot vsei dushi* was one of the programs that was most often mentioned when the respondents came to remember their favorite programs. Their memories make this show quite representative of a retrospective discourse of "good mood" in Soviet times. Thus, the respondents ex post validated the strategy of Central Television to establish an emotionalizing show. The emotionalization was based on a programming concept that focused on talk, conversation, and a popular host. The TV editors felt vindicated to count on a strategy that emphasized the personality and let "the person on the TV screen subtly become 'theirs,' a friend, almost a family member," as the letter report of 1971 indicates. The editors claimed that if hosts like the writer Sergei S. Smirnov, the director Vladimir A. Shneiderov, or the announcer Valentina M. Leont'eva "have won the sympathies of the viewers, one addresses them, one personally shares impressions, thoughts, speaks one's mind." Therefore, they were sure that the viewers hoped "to get an answer at best, certainly, on the screen." For their own benefit, the staff of the letter department explicitly appreciated the viewers and their feedback, because they praised it to help improving the broadcasting.[73]

Ot vsei dushi matched all these requirements, as it was hosted by highly popular Valentina Leont'eva, who presented ordinary Soviet citizens on screen.[74] Leont'eva, who was still warmly remembered by many respondents, let the audience engage in often melodramatic stories of these people. Indeed, hosts like Leont'eva and her apparently ordinary guests in the show became—like athletes—objects of viewers' para-social projections. Leont'eva did so as an outstanding individual, whereas her guests represented the collective of ordinary people. Ol'ga Nikolaevna, who was born in Irkutsk in 1964, emphasized that she watched the show because Leont'eva "actually presented it from the bottom of her heart." Ol'ga Nikolaevna was attracted by Leont'eva's "unconventional compassion" with which she told the personal destinies of ordinary people.[75]

Thus, Central Television was able to imbue the normal Soviet life with official authenticity by staging it on the screen. The dramaturgic mélange of an engaging public personality and the normalcy of private persons like you and me rather successfully gave the viewers the impression of "finding the way to the human being," as Liudmila Vasil'evna, a sixty-four-year-old teacher of Russian language and literature from Rostov-on-Don observed for "practically all [programs] in Brezhnev times, yes." Looking back, she even generalized this characteristic for the Brezhnev era and affirmed that "these programs appeared especially in the era of stagnation."[76]

Talking about watching television with other family members, the respondents suggest that television might trigger their affirmative commitment and created more than transitional emotional communities. Television offered occasions to spend time with the children, a leisure activity that the respondents usually remembered rather warmly, because the children's programs were highly popular and accordingly appreciated. Higher-educated fifty-nine-year-old Liudmila from Samara agreed with many other interviewees that the whole family watched the programs during the New Year holiday season together. Normally, she watched *Spokoinoi nochi, malyshi!* (Good Night, Little Ones!) and children's movies together with her son. She attributed considerable relevance to this habit by stressing that she generally did not spend much time in front of the screen. Like many other female respondents, she explained this with reference to her time-consuming care for the household and her job. Not least, the way to work and back home did not leave her much time for watching TV, nor even enjoying time together with her son on other occasions: "Accordingly, I had only little leisure time."[77]

The incoming mail already contemporaneously reflected that television could give family life a new meaning and impacted on people's routines. Letter writers repeatedly complained about the airtime of children's programs that did not comply with family routines or overlapped with the care time in kindergartens and schools.[78]

Apart from proper children's programs, sports broadcasts often occurred as popular family events. Several respondents—male as well as female—remembered with pleasure how they watched soccer together with their father: "I remember when we watched soccer together with father, this was an event, we watched soccer together, we shouted, we cheered, the ball was on the screen. Generally, what a simple joint pastime! That's what I remember very well."[79] Higher-educated Tat'iana Iakovlevna from Irkutsk was still emotionally involved when she depicted these memories of watching television during her childhood. Again, many respondents illustrated how sports on television bonded the family together sitting in front of the screen.

Only very few of the interviewees completely prohibited their children from watching TV. It was Antonina who asserted: "Television was simply forbidden in our family."[80] According to her narrative, her family generally watched "exclusively selectively, we have preserved this method until today. Only grandma watches series in her room, but also selectively. . . . Our favorite program is 'What? Where? When?', *KVN*—that's for Aleshki (i.e., her son, she spoke about the present time and referred to the show *Klub veselykh i nakhodchivykh* [Club of the Merry and Resourceful])."

The respondents' stories fill the quite-reserved narratives of contemporary sociologists with passion and emotion. Many respondents would have agreed with the academic conclusion that watching TV became an important leisure habit that considerably impacted on family life.[81] Many also still retrospectively confirm that not only sports programs and shows like Valentina Leont'eva's *From the Bottom of My Heart,* but literary movies and documentaries too, became daily companions and sources of connecting oneself to the world around. Writing a letter to Central Television, the viewer S. V. Tiutchev from Severodvinsk warmly depicted his experience watching the documentary series *Derevenskie povesti* (Village Stories): "The feeling was as if I was sitting at the open window and seeing real life! May you do well in mastering further similar issues of this special civil meaning! The most successful aspect of the film was: Truth."[82]

Many viewers were wowed that television seemingly conveyed an immediate picture of life. They tended to perceive it as true and authentic. Thus, television let them feel emotionally involved with and committed to the televised heroes, as Iu. M Konishin, a worker from Moscow demonstrated. He wrote a letter about the documentary series *Village Stories*, because he was "thrilled by compassion and love for the people. And by anxiety for them. [The film] was filled with an amazing feeling of truth."[83] Quite a number of viewers contemporaneously reproduced and engaged in the discourse of good mood and affirmative commitment illustrated by another example from the mail about *Village Stories*: "There has not been a documentary of such an emotional impact for a long time. The heroes speak affectingly, spontaneously, it seems as if it was shot with a hidden camera: the beautiful, rich soil of Vologda, as if you smell the wild grasses."[84]

THE GOOD MOOD AND THE QUESTION OF CREDIBILITY

The way Soviet citizens used the newly acquired privacy was not easy to trace for contemporary observers and, thus, aroused the interest and fantasy of foreign correspondents. Many foreign journalists favored the story of the new consumerist lifestyle that depoliticized the Soviet private life. American journalist Harrison Salisbury painted the picture of Soviet citizens who found a personal retreat watching TV at home during their leisure time. Salisbury reported to the Western world that

> the new Soviet man was not a gentle utopian knight, motivated by a desire to better his brother's lot, but a hard-working, harassed citizen with simple wants: a decent two-room flat, a bit of leisure away from the dirty, noise factory, a chance to bring up the kids . . . to a little better life, and maybe a car. . . . At age fifty, it was more and more apparent that the Bolshevik Revolution was middle-aged, a bit wheezy, inclined to sit back in an easy chair, turn on the TV and watch a good night program. No speeches, please. No party exhortations![85]

Salisbury conveyed that the Soviets had not only been caught up with modest consumption desires but were also satisfied by watching entertainment programs at home. Watching television at home enabled the viewer to ignore any political programs. Television, thus, brought about a new choice within the private sphere that challenged the regime's traditional strategies of political communication and let the television producers rethink their strategies to reach out to the viewers. The perception of a Westerner interestingly complied with a contemporary cartoon (Figure 2.1) published in 1965. The cartoonist pointed to the obvious new gap between the viewer's consumerist habitus focused on entertainment and rest in front of the TV screen and the Soviet regime's concern of politically educating and enlightening the viewer: A quite uninspired-looking speaker, perhaps a propagandist or political observer, monotonously reads from a thick stack of papers, while the viewer has fallen asleep pretty much bored by the speaker's performance. This TV set has, however, a technical highlight, as it is equipped with an alarm clock to wake up the viewer as soon as the talk would be over. The cartoon is titled "TV viewer-innovator" and depicted the TV producer's fear of the autonomous viewer who might ignore propaganda broadcasts because of their boring performance.

The status of ideological messages within the private sphere, the question as to what extent television should simply entertain the audience, and the problem of how to inspire people for political education in the dawning era of television remained an important issue for cartoonists during the 1960s.

Figure 2.1 "TV Viewer-Innovator." *Source*: Cartoon by E. Gurov. *Krokodil*, 16 (June 10, 1965), 7.

Figure 2.2 was published in late 1968 and suggested how any hollowness of political lectures could be compensated by the engaging force of television—at least for a male audience and as long as it broadcast hockey. The openness with which the cartoonist depicted the instrumentalization of television by the Soviet regime at the expense of any serious ideological education after the Prague spring is striking.

The caricatures *Krokodil* published in the 1960s and 1970s show that the obvious tensions between a program design giving priority to political education in contrast to one promoting pop-cultural entertainment perhaps did not lose significance, but the ideological doggedness of the debates at least lessened around 1970. Henceforth, the regime's expectations of how the viewers should use the medium and which TV programs they should appreciate continued to rest on the idea of television as political educator. However, the idea that broadcasting a rising number of entertaining programs might be the more efficient way to win people's hearts and minds gained considerable strength. From this time on, the cartoonist also reflected this approach and paid much

Figure 2.2 "We Are Trying to Prevent Anyone from Leaving the Lectures. . . ." *Source*: Cartoon by E. Migunov. *Krokodil*, 33 (November 1968), 12.

more attention to television's influence on the private life, on gender and generational conflicts stirred by the medium. On the one hand, the cartoonist kept on exposing typical gender roles according to which the husband spent his time relaxing in front of the TV set, whereas the spouse was responsible for the housekeeping (Figure 2.3). On the other hand, the cartoonists not seldom depicted social relations and communication among family and friends substantially disturbed by television during the 1970s. As television was supposed to heavily impact on private communication, people were repeatedly mocked as having no issues to talk about without watching TV or because the TV set was considered to be the heart of family life.[86]

The caricatures in *Krokodil* very well represent the official strategy of the 1970s not only to present television as a symbol of the Soviet lifestyle but also to give it a communitarizing meaning: No private conversation without watching television. The message is, however, unambiguous, as the cartoonist

Figure 2.3 "Take a Look, How She Spins!—You Should Better Look At How I Turn Around!" *Source: Krokodil*, 6 (February 1971), front page.

suggests that spending leisure time enjoyably depended on the television and not on the books in the background or on personal communication.

The contemporary audience letters and the narratives of the interviewees impressively reflect the communicative changes that television brought about. For example, the locksmith Sergei Gorbunov ideal-typically described the high significance of watching TV for his family in a letter to Central Television in 1960: "With great pleasure our whole family watches Moscow television programs on our excellent TV set *Rubin–102* nearly every evening. Television has a high impact on our development, it encourages us and renders us support for work. We often discuss at home what we have watched on TV."[87]

However, the cartoonists mocked both sides of the TV program—boring propaganda broadcasts and engaging entertainment shows. And indeed, contemporary viewer letters also reflected the cartoonists' irony about boredom on screen and seriously criticized political programs. In this context, many viewers complained about the insufficient televisualization of the covered topics. This constituted a specific problem for programs on political issues that viewers criticized for focusing in an all-too-biased way on speech. Contemporary materials make evident that boredom was a quite common feeling among viewers watching propagandistic, non-news programs like *Leninskii universitet millionov* (Leninist University of Millions).[88] Its producers had obvious problems taking advantage of television's specific virtues to attract viewers visually: "The purpose of television is to display, not to tell. Currently, it comes down to the contrary, i.e. to dull reproduction of text without using the screen. It should not replace the radio, one should better employ the possibilities of television."[89] This viewer, who criticized the visual shortcomings of the program, participated in an anonymous survey of Central Television's audience research. He pointed to a central problem of authorized discourse, namely how to inspire communication with the audience about ideological issues and stage exchange and participation. The relation of television and political propaganda might be told as a mere story of missed opportunities, if some viewers had not made noteworthy seemingly serious suggestions for improvement as well. We find such supportive statements in letters and in anonymous surveys. With regard to *Leninist University of Millions,* viewers proposed to stage it in "a livelier way like a round-table-discussion" or suggested "programs should coincide with the issues of party training." A viewer who was not a professional propagandist or otherwise affiliated to the party training system applauded that the "programs broaden one's horizon, make you keep up with events."

However, viewers who were propagandists or party members normally praised propaganda shows. They reproduced phrases of authoritative discourse like a viewer who asserted that "the materials of '*Leninskii universitet millionov*' are presented in an interesting and clear way" and that "the programs are made up on a high ideological-political level."[90]

Apart from this minority of viewers, political propaganda on television remained a delicate problem for the television editors. Making sense of the viewers' comments, TV editors had to admit that political programs attracted a very specific and rather small share of the audience. The audience research concluded from its results that propaganda on television should be "uttermost concrete, systematic, consistent" and avoid any "multi-topicality and the stringing together of commonplaces."[91] It was the sobering realization of the journalists' failure to convert the medium's capabilities of communication that let the audience research staff think about improvements to the genre. In particular, they wanted political programs to benefit from television's specific seriality. Thus, one idea was to reorganize the program according to a principle that resembled the idea Raymond Williams called "flow."[92] The first step, therefore, meant to find a better broadcasting slot for *Leninskii universitet millionov* that would embed the program between entertaining broadcasts.[93]

Although the television editors definitely intended to gain an audience constantly watching ideological programs, only a small share of viewers felt seriously attracted by this kind of political messages. The lacking popularity of ideological programs involved conflicts, not only between the audience research and the editors of the Propaganda Desk over the right strategies to stage propaganda, but also between the Programming Directorate and the Propaganda Desk over the broadcasting slot. In the last instance, the Programming Directorate blocked any aspirations of the Propaganda Desk to air pure ideological programs on prime time.[94] Indeed, prime-time slots remained reserved for entertaining programs explicitly addressing the viewers' hearts and souls: They aimed to create a "good mood." It was the head of *Gosteleradio*, Sergei Lapin, who underlined this motto of prime time programming in an article in the professional journal *Zhurnalist* (*The Journalist*) published in 1972.[95]

The emotional appeal of the viewers became a leading guideline not only for the editorial desks producing entertainment programs, but for all editorial desks of Central Television. It was the task of the audience research to explore how the viewers perceived the programs and to investigate if the editors of documentary non-news programs chose the right strategies to create a good mood among the TV audience. Creating a good mood proved to be a complex task for propagandistic and educational programs, as the audience research of the early 1970s repeatedly criticized the insufficient emotional staging of "ordinary" people on the screen. The documentary programs in question did not adequately present the "normal" party delegates—that is, ordinary workers and collective farmers—to the XXIV party congress. The criticism of the audience researchers clearly reveals their idea of engaging political TV programs, as it complained about the unsatisfactory "emotional mood" of the reviewed programs. In this case, the audience researcher criticized the failure of the television journalists to present the ordinary delegates as remarkable individuals who engaged in the improvement of the Soviet society: They should have showcased the particular personality (*lichnost'*) of different delegates, which would have meant emphasizing their uniqueness—a strategy that from the point of view of the audience researchers would have made them "heroes of the screen" who would have wowed the audience.[96] The audience researchers ascribed the insufficient medial staging to a basic lack of knowledge about the "psychology of interpersonal relationships and the psychology of viewers perception." Hence, these programs were made "without the required empathy and familiarity with the topic of conversation."[97]

Further, the audience researchers requested more liveliness on the screen. The interviewees should not read their answers from a piece of paper, as such dull performances neither engaged the viewers' imagination nor let them perceive the message. On the contrary, "if the word 'enthusiasm' is pronounced without any enthusiasm, it compels the viewer to think of insincerity."[98] This was an extremely devastating review, as television was expected to create credibility and win the viewers' trust. Thus, the audience research linked the credibility of television programs not least to the competent and authoritative personality of the TV host, his commitment, and his manner of speaking. From the audience researchers' point of view, the exemplary way to let the viewers believe in what people talked about on the screen was demonstrated by the highly popular writer Sergei S. Smirnov. In 1965, he had emerged with the new feature *Podvig* (Heroic Deed), a TV almanac on the defense of the Brest fortress in 1941, which immediately reached a huge audience. His success was based on the strategy to present ordinary Soviet biographies on television in an authentic and empathetic way. In the early 1970s, Smirnov hosted a new program called *The Search* (*Poisk*), which again benefited from his concern to "investigate the human soul." The viewer was said to "believe Smirnov's interlocutors because their emotional state in the moment of talking deeply complies with the content of the story." The researchers perceived the "deliberate conversation to capture the viewer who thanks the television for the acquaintance with a remarkable person."[99]

From the retrospective point of view of many respondents, the issue of ideology, propaganda, and the impact of television in the indoctrination of the viewers remained ambiguous. Many respondents—compliant as well as those of critical reservation—denied having watched any obvious propaganda programs. Many even claimed to have never watched the evening news because of the programs' bias and unanimously glorifying stance of allegedly Soviet achievements. These respondents argued that they preferred programs that would have allowed for critical analyses of the Soviet system. Some respondents like higher-educated fifty-seven-year-old Ivan, who had earned a doctoral degree in philosophy, overtly dismissed the glorification of Soviet achievements, although he was a member of the Communist Party. However, he did not become a member because of his ideological conviction but because of his position at the university. He refused Soviet television as "source of information, which meant that I couldn't obtain anything from it." Regarding the question of credibility, he spoke about a "parallel world, precisely a television world. It was probably perceived even more dramatically, somehow."[100]

This stunning perception that television created a parallel reality is noteworthy and pertained to the most skeptical statements regarding the credibility of the medium. It is even more remarkable, as Ivan seriously loved to watch sports and series like *Seventeen Moments of Spring* and the miniseries *Ad''iutant ego prevoskhoditel'stva* (The Adjutant of His Excellency) of 1969. Thus, he himself drafted a sharp binary construction of television as a condemned source of information on the one hand and a popular entertaining medium on the other hand: "But so when the series were launched, yes, then television became cult."[101]

This strategy of separating television into different spheres of entertainment and information was no exception, but only a few formulated this concept so explicitly while reflecting on their own media use as Ivan did. Some respondents claimed to have principally ignored political programs, others devalued the importance of the propaganda genre for their TV consumption. Again, the respondents demonstrate a variety of attitudes toward propaganda and ideology on the TV screen. The interviews make clear that the extent of their personal ideological compliance was an important factor in the viewers' content assessment. A basic belief in the Soviet system surely increased the probability of watching propagandistic and news programs or the broadcasting of party congresses but did not automatically imply the appreciation of such political programs. The interviewees discussed the assessment and perception of these broadcasts by looking back on their credibility, plausibility, and appeal.

Several respondents made clear that political broadcasts, news, and documentaries did not basically fail to reach the audience and to impact on the viewers. Mikhail G., a sixty-two-year-old engineer from Samara, criticized the exclusivity of the Communist Party and the ruling elites but did not denounce the whole Soviet system and the way the television represented politics on the screen. Like many respondents, Mikhail G. remembered very warmly sports broadcasts and several entertainment programs like the youth comedy game show *KVN* (*Klub veselykh i nakhodchivykh*/Club of the Merry and Resourceful), the variety show and holiday program *Goluboi ogonek* (Little Blue Flame), Valentina Leont'eva's show *From the Bottom of My Heart*, as well as *A nu-ka devushki!* (Let's Go, Girls!) and *A nu-ka parni!* (Let's Go, Guys!). Regarding television as an information source, he found that

> there were also very interesting political broadcasts, by the way. There were those that were hosted by correspondents, there, for example Borovik. [. . .] Well, there were many, many. Well, Bovin, probably, they were all young. Zabrodin, Zubarev . . . their names faded away. He was also present for a long time. Well, the very interesting ones were not the news programs like *Vremia* (Time) and so on but the political commentators. They were, by the way, quite objective.

Thinking of political programs, Mikhail G. mainly referred to those that covered foreign policy issues. He especially appreciated the role of the political commentators and foreign correspondents. Because of their presentations, he was sure that television covered events like the Prague Spring or the rise of Polish Solidarność: "Well, yes, certainly, 'Solidarność,' and all that. No, if you say that we did not know anything, that everything was hidden from us—then watch television. Very good. [. . .] It was covered, yes. Yes, Yes, it was good. I think, yes."

As another striking evidence for his perception of rather accurate and credible television coverage, Mikhail G. pointed to the socioeconomic deficiencies and evils Valentin S. Zorin reported from the United States. Zorin was one of the most renowned Soviet political commentators and hosted several political programs on Central Television. It does not come as a surprise that he impressively demonstrated the rottenness of capitalism and the severe dichotomy of the Western World: the shiny wealth of the ruling classes on the one hand and the misery of the working people in the slum areas.[102]

Mikhail G.'s narrative illustrates both the relativity of evaluation criteria and the persuasiveness of TV pictures. As he did not basically abandon the traditional Soviet assessment schemes, he still retrospectively picked out other unfavorable phenomena to prove the basic credibility of Zorin's coverage of the decaying American society. Thus, he found Zorin's accounts validated by other TV broadcasts that highlighted the unambiguous racism and oppression of black people. In this case, it was the assassination of Martin Luther King in 1968, Bob Beamon's long jump world record, and the demonstration of black power by Olympic champion Tommie Smith in Mexico a few weeks later that made a plausible argument to support Zorin's criticism.[103]

Whereas TV documentaries offered several meanings and perspectives based on image and sound, the Communist Party congresses were difficult to televise. Accordingly, most respondents voiced an unambiguous attitude toward the broadcasts like seventy-three-year-old Aleksandra Apollonovna. Explicitly asked if she remembered any programs that evoked a direct negative association, she pointed to

> all those party congresses, endless plenary sessions, there, do you understand, this vain chatter . . . Nobody never listened to them, I always switched off the television at the sight of this. [. . .] They were aired fairly often because they loved to brainwash the people. But the people were no idiots. How can you brainwash the people, if the shelves in the shops were empty, nothing, one forgot about meat, one forgot about sausage . . . everything from under-the-counter, everything, whom could they brainwash, if all understood that this was throughout the country and certainly out in the sticks.[104]

Aleksandra Apollonovna, a language teacher, turned out to have been a highly critical and reserved Soviet viewer, who rejected any open propaganda programs on television. Unlike her, sixty-six-year-old Waldemar from Novosibirsk did not generally ignore the broadcasts of the Communist Party congresses. On the contrary, he found them "sometimes interesting to listen to."[105]

Others like the former Leningrad residents Katia and Georgii, both higher educated, similarly depicted these exemplary ways of assessing television.

Like seventy-three-year-old Aleksandra Apollonovna, fifty-five-year-old Katia presented herself as a critical and quite-reserved viewer. She reported that she had been an enthusiastic viewer in her childhood and youth days but had almost abandoned watching TV from the mid-1970s onward when she went to university. She claimed to have watched the news only "because of a mere zoological interest" and overall rejected television as an information source henceforth.[106]

Fifty-seven-year-old Georgii was an affirmative and confident Soviet TV viewer who offered to speak frankly because he had "nothing to hide" in the interview. Georgii originated from an apparently intelligent conformist family and became a member of the Communist Party.[107] In contrast to Katia, he underlined that he had "sincerely believed to live in the best country," adding that back then he thought that "everything was fine" in the Soviet Union. Georgii impressively and unambiguously depicted the affirmative effect of several propagandistic broadcasts that contributed to his confidence in the Soviet system: "I believed that everything was true."

Georgii was one of several respondents who ascribed a high credibility to the information given in the Soviet mass media so that he affirmed his strong belief in the television programs he watched, as well as in the newspaper articles and reports he read. He explicitly referred to programs that highlighted the achievements of the Soviet Union and presented the sociopolitical situation of the capitalist countries in a poor light by contrast with the Soviet Union. To what extent Georgii's compliant assessment of the lost social and political system was influenced by nostalgic considerations about his personal socioeconomic situation is hard to judge. Georgii's narrative, however, strongly illustrates that television offered interpretation frames for the Soviet life and the significance of the Soviet model in the Cold War competition that many people perceived as familiar and reliable.

Like Georgii, several respondents explicitly assessed the question to what extent they believed in the information presented in the news and documentaries. Again, the interviews reveal a certain range of confidence and credibility among the former viewers but are often influenced by the ex post knowledge about censorship. Waldemar and his sixty-five-year-old spouse Zoia depicted a high level of credibility with only slight doubts. Zoia figured that "one believed in 80 percent, let's say, yes, but somewhere inside of you, somewhere one perhaps did not agree or thought that it was not correctly explained."[108] Others responded concisely that they believed in the given information on Soviet television, like fifty-four-year-old Irina Alekseevna: "Yes. Well sort of truthfulness, sort of . . . well we quasi believed in what they presented. One thought that since they show it, it means that it exists."[109] Like Irina Alekseevna, other respondents also, who finished secondary school, tended to frankly depict their former confidence in the Soviet media contents. Fifty-five-year-old Ol'ga, a German-Russian woman from Karaganda, affirmed to have completely believed in the information presented on television because "newspapers and television were in unison."[110] In a similar sense, several others like Zoia pointed to the perception that the Soviet world of information were rather homogenous: "Well, as it [the television] was after all in the hands of the party, it was certainly like: 'one step to the right, one step to the left—execution' . . . certainly, its character was more political . . . there was censorship, please excuse me, within the context."

As mentioned above, the interviews strongly document that many, especially almost all higher-educated respondents, continued reading newspapers. Many pointed to the fact that newspapers continuously played an important role as an information source, because they perceived the press to be more credible than television. However, most respondents watched television to catch up quickly on the news, as forty-nine-year-old Elena attested. She found that television "probably" was the main source of information but specified more precisely: "Although, I say that if we speak about any political information, then probably still newspapers. But if we simply talk about news, then probably the television."[111]

Television obviously benefited not only from its capacity to underline information with image and sound but also from its greater up-to-dateness, as fifty-seven-year-old Anna B., a kindergarten teacher, characterized. She "partly" believed in the information given on television, although she claimed to have perceived their propagandistic narratives. This awareness did not necessarily affect the credibility of the medium, as Anna B. affirmatively answered the question if television might have had more persuasion: "But they—the information—were, practically identical everywhere, there was simply more information on television, let's put it that way . . . Or television had more possibilities, that's my opinion . . . and television, in any case, always had more possibilities . . . I think that still . . . I think, yes."[112]

Hardly anybody questioned the credibility and persuasiveness of television as such so explicitly as seventy-three-year-old Aleksandra Apollonovna from Tyumen did. She "certainly" was always critical about the information given on television and cynically generalized her attitude: "The people generally had an entirely negative attitude because they did not campaign for plenary sessions but for empty shelves in the shops. . . . This was the best agitation. . . . When you go to the shop and get nothing . . . no meat, no chicken, no egg, generally nothing, there was some strange sour cream, sometimes not even sour cream."[113]

Whereas Aleksandra Apollonovna generated her doubts by realizing a cognitive gap between her everyday life as a provincial consumer and the representation of Soviet achievements and material abundance on television, many other respondents explained that their confidence toward the reliability of the information was due to their education and the way ideology had impacted on all their daily perceptions. Several respondents invoked a high impact of Soviet propaganda on the everyday life. Regarding the credibility and persuasiveness of Soviet television, they painted a rather stereotypical picture of the individual media user. Waldemar explained that "one was educated the way that everything what is presented on television, what is told on the radio, that all that is the truth and all that shall be. . . . That's all. . . . As late as now, we have become 'wise,' but at that time, what they said, that was what we did."[114]

Sixty-two-year-old Mikhail G. argued in the same vein and claimed a special importance of the television for an ordinary Soviet family. He suggested that "the television was more than a trivial thing. Then, it was this sort of indispensable attribute. But we must also consider this peculiarity of the Soviet man to believe whatever is written, told on the screen or on a piece of paper, so generally one could not live without the television as it was a necessary attribute."[115]

On the one hand, the respondents themselves portray rather confident and affirmative Soviet TV viewers who were easily persuaded by propagandist messages of the mass media. On the other hand, they depict a certain range of attitudes toward propaganda and ideology. Many, like Aleksandra Apollonovna or Elena, criticized or at least identified the propagandistic construction of information, news, and argumentation. As mentioned above, many perceived two different spheres of communication—the field of information and news that was thoroughly ideologized and that of entertaining programs that allowed for "unpolitical" interpretations. In this sense, fifty-seven-year-old a kindergarten teacher Anna B. described the evening news program *Vremia* (Time) as

> highly ideologized, let's put it that way, and this ideology did not only consist in the way how they presented the news and what was said but also in the fact that this program gave this opportunity to present these congresses to all people . . . it highlighted some parts—statements and let them appear in the program and, at the same time, all these political broadcasts of the political meetings often were covered by this program, let's put it that way . . . I think that this was the most politically ideologized program.[116]

Anna B. was also clear about the idea that ideology did not permeate all television programs and claimed that "there were sufficient programs" without ideological messages. To evidence her observation, she referred to programs about animals, geography, and museums. She also mentioned art, cinema, and ballet but, while thinking of the question, she doubted a little whether these were indeed free from ideology: "This is a controversial question, let's put it that way."

In a similar way, seventy-one-year-old Konstantin, a historian, affirmed that the Soviet mass media were generally ideologized but that he could not say that he explicitly thought about it. Looking back, he thought that he was confident in the system, believed in most of the information and perceived it as "truth." Only today did Konstantin find it obvious that many facts had been concealed. He empathically pointed to the impact of belief and the relativity of "truth" and "telling lies" depending on the speaker's and the listener's position. Although his best friend continuously listened to foreign radio broadcasts already during the 1960s and 1970s and conveyed the information to him, Konstantin had been reluctant to deconstruct Soviet ideology. Only during perestroika had he started to refuse the ideologization of information and news.[117]

Like him, many other respondents did not principally condemn ideological frames but only few even appreciated the way Soviet TV had been "ideologized" when they compared it to the current media situation around 2010. Oleg Igor'evich depicted himself retrospectively as a "patriot of my country," who "was able to filter" as soon as he listened to foreign radio stations. He consumed them without being "fanatical." However, he conceded that he contemporarily did not realize the extent of ideologization. Oleg established that he only retrospectively understood the way ideology impacted on media contents. This recognition did not provide ground for rejecting ideology. On the contrary, he contended that ideology "ought to be," when he, for example, referred to the Vietnam War and the Soviet strategy to support the developing countries. He was still a child but claimed to have understood the events with the help of adult family members. His contemporary perception about the capitalist world was hostile because of media reports:

> Well, then I had a rough idea about what "communist" appeared to be; regarding the capitalist system they only reported about horror. Well, it might be that as a result there is something good in the capitalist world. Now we know it. But then, capitalism appeared to be bad all the time. . . . Well, and our ideology was to support the weak, developing countries to establish our socialist, communist system. That's all. Well, then, I probably liked it.

Against this personally experienced background of seemingly valuable political goals and biased information, he, nevertheless, considered it to be "bad" that "there is no ideology today," and even more: "I think that our ideology was right at that time. The system that they [the Soviet citizens] defended, protected and built." Statements like this frequently occur in the interviews and make clear that many interviewees even retrospectively do not dissociate themselves from the Soviet ideology but basically shared ideological values and ideas.

Many respondents depict how watching and perceiving television was influenced by the communication with family and friends about it. On the one hand, many interviewees highlighted how their parents impacted on the way they learned to use Soviet media during childhood and youth days. On the other hand, the way the respondents remembered the debates about the program of the previous evening documents that these conversations seriously supported the appropriation of media competences. Often the conversations directly framed the assessment of media contents and the way one should compare the media reports with their own daily experiences.

Not surprisingly, parents and grandparents were normally the first to direct children and young people in how to assess media contents. Particularly those respondents referred to their parents' guidance who considered themselves to have developed critical and reserved attitudes toward Soviet media. The reserved viewer Katia generally spoke in a somewhat contemptuous way about watching TV and responded to most questions rather taciturnly. She suggested it should be obvious that educated people rejected Soviet TV or, at least, used it highly selectively because of its propagandistic messages. Katia was very clear about the fact that her parents taught her how to consume Soviet mass media. She explicitly ascribed her critical view of television programs to her parents' education.[118]

Although we need to consider that some respondents might retrospectively credit themselves with a rather critical attitude toward media in Soviet times, Katia seemed to be an exception to the rule. The interviews do not indicate that private conversations about media consumption significantly triggered fundamentally critical views of the system. Even outright skeptical or reserved parents rather taught their children to balance information from different sources and carefully evaluate them.

At least, only a few respondents like seventy-one-year-old Konstantin and fifty-three-year-old Anatolii explicitly pointed to the fact that some of their immediate friends and family raised doubts about the system. Still, even they both retrospectively stressed that these doubts did not provide immediate ground to reject the Soviet system or its values and norms.[119]

Fifty-nine-year-old, higher-educated Mikhail from Leningrad also presented himself as a critical consumer of the Soviet mass media who was taught by his father:

> I tried to handle all information critically, by and large, depending on the way they presented it. All in all, my father taught me somehow, probably, I listened to him, most likely my father played an important role in this respect. And afterwards I became ascertained based on my own experiences of life that something is like this and not the other way, and where contradictions appeared. And newspapers, my father also taught me how to read Soviet newspapers between the

lines, because information was presented in such a way that you had to ponder over and think why it is this way and no other.[120]

These narratives about learning to interpret TV contents sounded rather similar. Higher-educated parents actively tried to provide models of interpretation for their children and aimed to teach them how to use the medium. Most of the higher-educated respondents were rather engaged with the question of how ideology generally impacted on television contents, how they would retrospectively assess its impact, and how ideology worked in comparison to the current post-Soviet media landscape. In this respect, the aspect of the credibility of television and generally trusting media contents mattered a lot.

Regarding the contemporary perception of ideology, sixty-nine-year-old Al'bina, an engineer from St. Petersburg, seemed to be representative of most of the higher-educated interviewees. She first underlined that newspapers constituted her basic source of information which she trusted in. Asked about ideology and the way she might have filtered media contents, she argued:

> Yes, I trusted. I understand now that, yes, we didn't notice, we thought that it must be that way. It was filtered. By what means? [. . .] First, I'm certainly a lazy person, but, in particular, it was such a funny moment. I started to be annoyed that they underlined all along only our accomplishments. Suddenly, I hear in this moment that we have, let's assume, an attitude towards ecology . . . but ecology was simply irrelevant. Something different, that actually the most wonderful zoo was somewhere, that there was such a relationship to nature. Really exactly from this point of view. But they screamed all the time, talked all the time about how we have the most remarkable, the most wonderful. Well, all the same, I probably still do not want to say that I'm such a curious or very critical person. For me it would have been easier and simpler, if something's wrong—forget about it! Life was so full of difficulties that absorbed and tortured me, that then there were such totally obvious discrepancies, but I didn't have it in me to care about them, I had my own problems—my relations with my parents, they affected me more than those public issues.[121]

Al'bina was rather caught in this ambivalence and explained that she had already started to doubt when she worked as political informer in the 1980s, because people spoke very openly about the economy at the meetings:

> They simply quite clearly spoke about the discrepancies in what is happening in our country. I was generally astonished about movies, you see, Aitmatov, his narratives, they did not correspond so much to reality, . . . I was generally astonished how such a thing could be published. I only then understood that there was some freedom . . . I adored Zoshchenko, I adored Averchenko. And Zoshchenko was forbidden. You understand, there all the time these things,

> they arose. For example, I have to this day this booklet which talks about how Khrushchev proclaimed communism for 1970 [*sic*]. And I also have this booklet from 1941, which I received in 1970, in which it talks about how one could read Zoshchenko, Achmatova, Pasternak, Ėrenburg differently, from a political perspective.

However, she convincingly explained how she got used to it and most of the time rather ignored these limitations of thinking. Comparing official presentations and personal experiences inevitably led to disturbances that people needed to make coherent. Looking back, the more reflective respondents tried to make sense of the ambivalences and explained them regarding the importance of ideology without essentially condemning ideological frameworks. This becomes especially clear if we look at the usage of foreign media sources.

COMPARING THE WORLD IN COLD WAR TIMES: LISTENING TO FOREIGN RADIO STATIONS

The rising distribution of TV sets—and of radios, books, and press products—paved the way for all social groups for a more individual, more private, and more flexible media use in terms of time, access, and interpretation. Not only reading newspapers, journals, and books, but also listening to the radio and watching TV became more informal leisure habits at home with family and friends. The technical improvements, better supply with media devices, and the shift of media consumption into the private sphere also enabled Soviet people to consume foreign media products. Owners of short-wave receivers could listen to Western radio services like *Radio Free Europe*, *BBC World*, or *Deutsche Welle*. Owners of TV sets living in the Soviet periphery were able to watch TV programs from the neighboring countries.[122] From the regime's perspective, these developments implied an increasing challenge to offer more entertainment and sophisticated news programs that were, nevertheless, ideologically compliant. The gain of the improving material side of TV consumption could be, however, new ways of emotional bonding to consumer goods and private life. It provided ground for viewers to write letters to Central Television in which they explicitly stated their gratitude for the rising living standard symbolized by the better supply of durable goods like TV sets.[123]

Listening to foreign radio might have potentially irritated the audience about the condition of Soviet living compared to the standard of living in capitalist countries as well as about political restrictions. The assessment of foreign media contents should be considered an indicator of the different social, cultural, and economic sorts of capital that impacted on media access and perceptions.

Only a few of the respondents were able, at least for some time, to receive foreign television programs during Soviet times. For most of them, the only possibility to obtain non-Soviet information was to tune in foreign radio stations. Indeed, a slight majority of the respondents had listened to foreign radio stations like *Radio Free Europe*, *BBC*, *Deutsche Welle*, or *Voice of America*. Most of these listeners lived in cities, so that we could suppose that listening foreign radio was a predominantly urban practice. Some villagers from Irkutsk region and Arkhangel'sk region or people from a small town in Nizhnii Novgorod region even indicated that they had not known about the existence of foreign radio stations.[124] This must have been not least due to an inadequate technical equipment, as many villagers still owned only wired receivers which prohibited that people could tune in any foreign broadcasters, if they owned radio sets at all.[125]

The foreign radio service constituted a thorn in the regime's side. In the official Soviet language, the broadcasts were "littering lies and slander on air" and represented nothing less than the "Imperialist aggression" toward the Soviet Union. Therefore, *Gosteleradio* was eager to explore how many people really listened to foreign radio programs. In this context, Soviet sources established a significant increase of the broadcasting time during the 1960s. In 1967, *Voice of America* aired around the clock. The same was true for *Radio Free Europe*, which broadcast in nineteen different languages present in the Soviet Union. One should add *BBC* and *Deutsche Welle*, both airing in Russian every day. In view of the Soviet perception that "counter-revolutionaries" captured radio, television, and press in the CSSR, the members of *Gosteleradio* saw a chance to underline the political importance of domestic propaganda in that difficult situation. They considered it even more necessary to counteract any information broadcast by Western radio, as "many Soviet people are listening to the broadcasts of foreign radio." Their explorations on Moscow and Riazan' revealed that approximately half of the owners of a radio set listened to *Voice of America*, *BBC*, and so on.[126]

In the 1950s and 1960s, party officials mostly perceived the foreign broadcasters as seriously compromising their claim on framing meanings and opinions. This was a common perception of all Communist regimes, as the foreign broadcasters covered all their territories. Although it was a balancing act to make the foreign broadcasters a subject of public discussion, political cartoons cautiously tried to reverse some of the messages and to use it for their own purpose. *Krokodil* printed a cartoon from the Chinese newspaper "Friendship" (Figure 2.4) mocking the Western attempts to fuel resentments between the two major Communist regimes in order to expose the hypocrisy of the Western radio broadcasters. In early 1957, the Soviet Union and China

Figure 2.4 "Voice of America" "Dear Radio Listeners, Don't You Hear the Sound of the Face Slaps? It's Moscow and Peking Arguing . . . It's the Genuine Truth . . ." *Source*: Cartoon by Mi Gu from the newspaper *Friendship. Krokodil*, 13 (May 10, 1957), 10.

were in a process of political rapprochement so that the cartoon did not contradict the official ideology of international friendship. Moreover, the topic was too remote for most of the *Krokodil* readers to prove its content based on daily experiences.[127]

At this point, the interests of Soviet media producers perfectly coincided with the propagandist strategies of the party and made the Western radio stations an important spur for improving the work of the Soviet media producers. Realizing the creative and entertaining potential of radio and television, the party's willingness grew to invest more money in the technical infrastructure, programming, and the salaries to attract creative professionals. In the end, the hope to prevent people listening to foreign radio stations pushed more attractive entertaining radio and television programs.[128]

The interviews give no hint that the official strategy to set up an increasingly entertaining radio and television program had a lasting effect on people listing to foreign radio. At least, nobody claimed to have abandoned listening to *Radio Free Europe* because of the intriguing TV program. Normally, the respondents used both information sources in parallel. The concrete use and assessment of the different media was highly dependent on the social environment and the respondent's own social position. Some of the respondents, like the rather content and confident viewers Oleg Igor'evich and Antolii Ivanovich, suggested that listening to *Radio Free Europe* or *Voice of America* "was simply fashionable." According to Oleg Igor'evich, nobody would have contemporarily attached importance to that.[129] Without being asked, fifty-seven-year-old Irina N., a former accountant, volunteered that she never listened to *Voice of America*, simply because she did not find it particularly interesting. According to her, one listened instead to Vysotsky everywhere and all around in the late 1960s.[130]

Irina N.'s assessment typically evidences that the opinions of the respondents about the political and cultural significance of foreign radio services were quite diverse. Many appreciated listening, to get alternative information, and to compare the information from official and Soviet media. Some even got up at night to tune in to the *BBC*, while others felt anxious about violating the law but listened anyway. Several respondents claimed to have remained reserved, some opposed it, others explicitly ignored it, and a few even claimed not to have been aware of its existence. In this sense, the respondents described a certain spectrum of listening habits, especially the way they handled the information, if they talked about the foreign broadcasts and discussed them with others.

Debates about foreign broadcasts were very delicate in Soviet times, as it was officially forbidden to listen to "Imperialist" radio stations. If the respondents contemporarily spoke at all about them, they mainly spoke within family circles and with very good friends. Higher-educated fifty-nine-year-old Yuri Vasil'evich from Samara was one of the few who claimed to have discussed somewhat more openly and who obviously did not:

> We discussed it. We talked over it, yes, *Voice of America* broadcast this and that. That was why we thought that all were informed. Well, certainly, not all were informed. If you talk with young people, yes, I say: Did you hear this or that?—They look at me with surprise. And it appeared that some fact that was clear to me . . . [he interrupted] And he says: I did not hear that, I don't need it, I do not listen to it.[131]

Seventy-six-year-old Roman R., who later became a renowned director and documentary filmmaker, recounted with evident pleasure how he listened with very good friends at the Komsomol committee to *Voice of America*. They locked themselves in a room within the committee and tried to tune into the foreign service in the early 1950s, although the quality was rather bad because of the strong jamming. Roman R. emphasized that the group had a strong consensus about the condition for collective listening to foreign radio even in Stalinist times. He thought that he and nobody else contemporarily perceived any "pernicious influence," but something might have started to "germ."[132]

Only some respondents reflected like Roman R. about any potential short- or long-term impacts of foreign radio services. The producers of *Voice of America* would have liked to hear how sixty-seven-year-old higher-educated Tamara assessed its effect looking back from today. She confirmed that she had never thought about this question before the interviewer drew her attention to it. Tamara herself only listened somewhat frightened and trembling to the voice sounding from afar that she even perceived as somehow enigmatic and terrifying. She justified this feeling by her Soviet upbringing and her knowledge about its prohibition. Nevertheless, she recollected that she talked with friends about the broadcasts and the fact that they compared them with the information given by Moscow Central radio. Interestingly, her current reflections let her retrospectively realize that those people who regularly listened to foreign services "always behaved more courageously and more impudently in their life, they were rather uninhibited. Well, now I say this, I never in my life thought about it and never in my life spoke that way."[133] Moreover, one could learn from her narration that since Soviet times she has at least unconsciously grouped people in those who listened to foreign services, perhaps believed them, and even did not conceal their hostile attitude toward the Soviet Union on the one hand and those who complied with Soviet values and rules on the other. The latter were people like her who would have never dared to say "I hate the Soviet authority," as a friend put it bluntly.

Many of those respondents who listened to foreign services, like the sixty-one-year-old locksmith Anatolii Ivanovich, attested that "you sit down in the evening, listening to *Voice of America*," in order to hear "what they say about us." Anatolii Ivanovich, even as a confident viewer, explained that "one compared who lied, who did not lie." He recollected that listening to foreign broadcasts was a common practice: "everyone who had a radio, listened to it," but it would not have been "wise to chat" about one's perceptions. While this respondent claimed to have already contemporarily rather secretly spoken about his listening to *Radio Free Europe* at that time, he was during the interview not eager to talk about what kind of programs he had tuned in, let alone to reflect upon its social and political significance.[134]

Mikhail from Leningrad, a higher-educated and rather reserved viewer, represented himself as a regular radio listener until television came up and, at least in his daily media consumption, replaced radio. Like Anatolii Ivanovich, he affirmed that many tried to listen to foreign radio, and all knew that it was forbidden. Mikhail, who claimed to have been critical toward Soviet media and an experienced newspaper reader, explained that he rather soon realized that *Voice of America* only broadcast "their anti-propaganda and so on, and they reversed one and the same question, one question to that side, the other to another, but that was not the truth about the question. The truth always lies somewhere in between." Thus, even media users who had a compliant attitude like Anatolii Ivanovich and someone like Mikhail, who claimed to have been a critical Soviet media user, agreed on the practice of comparing different information. Even critical listeners were not automatically inclined to simply believe in the interpretation frames the foreign radio offered. After a period of listening to *BBC* and *Voice of America*, Mikhail was sure that the only way was to analyze "and to compare the information from one side and from the other; at first, I believed '*BBC*' and '*Voice of America*,' later, I did not. I understood that they got their own way."[135]

Looking back, many respondents considered the listening to foreign radio as a rather common Soviet media practice. They recalled that contemporary perceptions tend to represent it as a private but normal and generally accepted leisure activity. The minor group of the respondents who never or only seldom listened to *Voice of America*, however, let us trace not only the different attitudes toward Soviet media and the disparate use of domestic and foreign mass media. The opinions about the extent and prevalence of listening to foreign radio in Soviet times are from today's view rather diverse and reveal the sociocultural diversity of the late Soviet society.

Whereas rather different media users like compliant Anatolii Ivanovich and reserved Mikhail both claimed to have not really believed Western broadcasts, seventy-three-year-old teacher Zinaida Mikhailovna represented a nuanced type of compliant media users. She again offered different attitudes toward media contents that based on the mixture of compliant, indifferent, and critical reserved perception. She basically asserted that she had only very rarely listened to *Voice of America*. This was partly due to technical problems and bad reception, but mostly resulted from her self-perception as "ordinary citizen" living in "a not very politicized family." On the one hand, they were "actually content with what the media officially communicated." In contrast to Anatolii Ivanovich, Mikhail, and many other respondents, Zinaida even thought that most Soviet citizens were not interested in listening to foreign radio, because they were "content with what they fed us, what they gave us." On the other hand, Zinaida's presentation ranged from rejection to a certain interest and cautious consumption of foreign radio. Once, she fundamentally refused any significance of information broadcast by foreign radio: "No, no, no, we did not search for answers there." However, she also admitted that they tuned into *Voice of America* on the rare occasion of being interested in the actual reasons for the emigration of Solzhenitsyn, Aksenov, or Rostropovich. But she insisted that they did not perceive the Western information "as complete truth" and lastly quoted her father to summarize her general attitude toward *Voice of America*. This assessment got closer again to that of Anatolii Ivanovich and Mikhail, as Zinaida's father used to say about the foreign radio: "That's poppycock, that's propaganda."[136]

Several respondents, who knew about the existence of foreign services, but denied having listened to them, referred like Zinaida Mikhailovna or sixty-three-year-old Tat'iana to their unpolitical attitudes and lack of interest in political issues. Sixty-year-old Natal'ia was working at a state optical institute in Leningrad and said that she never listened to any foreign radio station. But she heard her male colleagues at work chatting about foreign broadcasts but claimed for herself "not to love politics."[137] Tat'iana was raised in Novogorod region and moved as a young woman to Leningrad. She even denied having known about the existence of foreign services and insisted that she had never been interested in politics, football, or hockey.[138]

The narrative that links the lack of interest in politics to the rejection of foreign radio services could also be turned around. Seventy-nine-year-old Aleksandra from St. Petersburg argued that she regularly got up at night in order to listen to *BBC* and *Voice of America* because of her political awareness. Somewhat curiously, she believed that her interest was due to her mother's habit to read out loud the newspaper while being pregnant with her.[139]

In contrast to all these just-mentioned respondents, sixty-three-year-old Svetlana Vladimirovna from Samara was highly confident toward the Soviet system and totally uncritical toward Soviet media. Svetlana Vladimirovna was a member of the local Communist Party committee. Her affiliation still seemed to have shaped her attitude toward the Soviet Union during the time of the interview. Looking back, she assured that the foreign radio stations "were not able to re-orient me, I, nevertheless, believed that socialism is the best system for the common people."[140]

Compared with Svetlana Vladimirovna, seventy-one-year-old, higher-educated Nikolai argued even more rigorously about his attitude toward the "hostile" radio services, as he called *Voice of America* and *BBC*. He was member of the Communist Party by deep conviction, allegedly loved to read *Pravda*, and staunchly refused foreign radio. Nikolai was one of a minority who had a very clear and binary worldview based on his gratitude toward the Soviet state for receiving free education and a good job.[141]

There are other factors that help to explain in what respect the attitudes of media consumers may have differed concerning the political system and the way television was related to it. Again, Katia and Georgii, two earlier-mentioned higher-educated respondents from Leningrad, represent two different user types of young grown-ups in the 1970. Georgii became a member of the Communist Party and was a rather compliant and confident viewer who did not criticize the political system. In contrast to him, Katia appeared as a reserved, perhaps subversive, and sovereign viewer who talked about her "liberal Leningrad environment." Despite her father's party membership, she represented her family as "anti-Soviet." They regularly listened to *Radio Free Europe*, *BBC*, and *Voice of America*. Neither the own party membership (Georgii) nor that of the father (Katia) seemed to have ultimately determined the use of media in the respective family as many only strove for a membership because of their career. All party members subscribed, however, *Pravda* and often also *Izvestiia*. Some of these respondents explicitly stated like Katia that they had read newspapers "between the lines" without basically pursuing to demonstrate a nonconformist attitude toward the regime. Looking back, fifty-five-year-old Katia, however, was one of very few respondents who established a link between listening to foreign radio stations and any anti-Soviet attitudes. The others normally did not explicitly reflect upon this aspect, as the idea of being fundamentally "anti-Soviet" was, at least contemporarily, too far away for most of the respondents. As a rule, even the majority of conformist, confident viewers—as Georgii perfectly showed—was able to compare and balance Soviet and foreign media contents. They did not realize direct contradictions between them, but just separated the two worlds of information without categorically depreciating Soviet messages or deriving nonconformist, critical opinions from possible discrepancies. Some of the respondents depicted, however, that the programs and news they listened to on foreign radio stations had let them doubt more or to be more cautious about Soviet presentations. But these seeds would never have been so strong to prevent them from consuming Soviet entertaining or information programs.

Regarding those who listened to the foreign radio services, we can conclude that these radio programs did not effectively destabilize the Soviet society or initiated a profound hybridization of official meanings leading to oppositional interpretations. Even if the foreign radio stations attracted the attention of many Soviet citizens, they did not win the Soviet hearts and minds to that extent that it could broadly compromise the emotional link to the Soviet society. Rather, *Voice of America* became part of the more complex media landscape by stimulating its listeners to compare the information from both sides of the Iron Curtain. Soviet media consumers did not need to have an overwhelmingly compliant attitude toward the Soviet system to establish that "it was not possible to prove the information that *Voice of America* broadcast," as Vladimir Ivanovich did.[142] Soviet listeners, thus, felt that the "truth" was not self-evident and could be a relative category. Vladimir Ivanovich, who confirmed not to be particularly interested in politics, summarized:

> Yes, something happened there. But if this was the truth or not, you don't know. It did not simply soak in. Now it's possible, yes, you get thoughtful, but really at that time, no, you wouldn't believe it. And why go there and, excuse me, lick the lips, how good it is there and how bad everything is at our place, as they presented it to us? I was more interested in music.

Yet, we could learn from his narrative that even if the Soviet media users were not able to deconstruct the "reality" Soviet media presented, many of them, like Vladimir Ivanovich, saw the discrepancies between the media representation and their own experiences. For many Soviet listeners, the main conclusion, however, was that "truth" was a relative category and that the broadcast and use of information was a complex process. There could even be a dialog of different voices, but the capacity of the Soviet TV to frame issues and to involve viewers emotionally was a strong asset on the side of the Soviet regime.

NOTES

1. GARF, f. 6903, op. 1, d. 499, 1956, l. 20.
2. GARF, f. 6903, op. 31, d. 2, 1956, Stenogramma konferentsii telezritelei zavoda "Optika" g. Krasnogorska, l. 17.
3. GARF, f. 6903, op. 31, d. 2, 1956, l. 17 (quote); ibid., op. 3, d. 178, 1962, l. 15.
4. GARF, f. 6903, op. 31, d. 2, 1956, l. 3.
5. Firsov, *Televidenie*, 105.
6. GARF, f. 6903, op. 3, d. 404, 1967: Otchet o sotsiologicheskom issledovanii na temu "Ob effektivnosti informatsii i propagandy na vneshnepoliticheskie temi po radio i televideniiu" dlia naseleniia Sovetskogo soiuza, l. 11.
7. Vasilii D. Patrushev, "Osnovnye izmeneniia v sfere svobodnogo vremeni gorodskogo naseleniia v posledniuiu tret' XX veka," in Tat'iana M. Karakhanova, ed., *Biudzhet vremeni i peremeny v zhiznedeiatel'nosti gorodskikh zhitelei v 1965–1998 godakh* (Moscow: Izdatel'stvo Instituta sotsiologii RAN, 2001), 35.
8. Pskov accounted for 98,000 inhabitants in 1962, 126,000 in 1970 and 182,00 in 1982. Cf. http://www.mojgorod.ru/pskovsk_obl/pskov/index.html, (accessed October 20, 2015).
9. John P. Robinson, Vladimir G. Andreyenkov and Vasily D. Patrushev, *The Rhythm of Everyday Life. How Soviet and American Citizens Use Time* (Boulder, CO, San Francisco, CA, London: Westview Press, 1989), 91–92.
10. In 1967, NMO presented somewhat higher figures concerning the amount of leisure time. According to their findings for 1967, women who worked five days per week had 23.45 hours and men 37.30 hours of spare time. The share of TV consumption also slightly differed from the Pskov study: Women spent 15 percent of their leisure time or 3.34 hours watching TV, and men 23 percent or 8.36 hours. The data

depicted, however, the same trend. See GARF, f. 6903, op. 3, d. 417, 1967: Statisticheskie tablitsy o rezul'tatach issledovanii televizionnoi auditoria v 1967 godu, l. 2.

11. Most of the contemporary research focused on cities located in the RSFSR. See Boris T. Kolpakov, Vasilii D. Patrushev, *Biudzhety vremeni gorodskogo naseleniia* (Moscow: Statistika, 1971), 209, 212; Gordon and Onikov *Cherty sotsialisticheskogo obraza zhizni*, 59–66, 149–150.

12. Once more referring to Pskov: Patrushev, "Osnovnye izmeneniia," 37–39.

13. Leonid Gordon, and Ėduard Klopov, *Der Mensch in seiner Freizeit* (Moskau: Progreß, 1976), 14, 300, 309–313.

14. Boris Grushin et al., *Die freie Zeit als Problem. Soziologische Untersuchungen in Bulgarien, Polen, Ungarn und der Sowjetunion* (Berlin: Berlin Verlag, 1970), 48–49.

15. Boris Sapunov, "Televidenie v sisteme sotsialisticheskoi kul'tury," in *Televidenie vchera, segodnia, zavtra: '86*, ed. Ėduard Efimov (Moscow: Iskusstvo, 1986), 17.

16. Patrushev, "Osnovnye izmeneniia," 38.

17. Robinson, Andreyenkov and Patrushev, *Rhythm*, 93. In 1967, *Gosteleradio*'s NMO confirmed these findings: GARF, f. 6903, op. 3, d. 404, 1967, l. 13; GARF, f. 6903, op. 3, d. 417, 1967, l. 2.

18. The interview with Liubov' Aleksandrovna was conducted by E. Peters in Lilienthal (Germany) in August 2012.

19. The interview with Liudmila was conducted by Elena Bogdanova in St. Petersburg in November 2010. Liudmila was born in Penza region in 1942 and moved to Leningrad in 1962. She finished secondary school and worked as waitress and in a fish factory.

20. The interview with Natal'ia Petrovna was conducted by Oksana Zaporozhets in Samara in October 2010. Natal'ia was born in Samara region in 1946. She finished secondary school and managed a trade company afterward.

21. The interview with Valerii Nikolaevich was conducted by Alexander Ermakov in Usol'e-Sibirskoe, Irkutsk region in August 2010. Valerii was born in 1958 and has secondary education.

22. *Obzor pisem telezritelei za 1980 god* (Moscow: Gosudarstvennyi komitet SSSR po televideniiu i radio, 1981), 164.

23. *Ogonek* 34, 1960, 10. The caricature was entitled "No more quarrels."

24. See, e.g., *Krokodil* 31, 1978, 10.

25. GARF, f. 6903, op. 3, d. 289, 1965, Dokumenty o provedenii ezhednevnogo operativnogo oprosa zhitelej Moskvy i Mosk. obl. o peredachakh Vsesoiuznogo radio i TsT, t. 2, ll. 5–6.

26. GARF, f. 6903, op. 3, d. 417, 1967, l. 6.

27. Gordon, Klopov, and Onikov, *Cherty*, 151; Robinson, Andreyenkov and Patrushev, *Rhythm*, 93.

28. Lev N. Kogan, "Buch und audio-visuelle Medien," in *Massenkommunikation in der UdSSR. Sowjetische Beiträge zur empirischen Soziologie der Journalistik*, ed. Hansjürgen Koschwitz (Freiburg, München: Verlag Karl Alber, 1979), 158–163; Rosemarie Rogers, "Normative Aspects of Leisure Time Behavior in the Soviet

Union," *Sociology and Social Research* 58 (1974): 369–379; Gordon, Klopov, Onikov, *Cherty*, 151.

29. Boris Firsov, "Srednego zritelia net,“ *Zhurnalist* 12 (1967): 43.

30. Rozalina V. Ryvkina, *Obraz zhizni sel'skogo naseleniia. Metodologiia, metodika i rezul'taty izucheniia sotsial'no-ėkonomicheskikh aspektov zhiznedeiatel'nosti* (Novosibirsk: Izdatel'stvo »Nauka« Sibirskoe otdelenie, 1979), 215–216.

31. Kogan, "Buch und audio-visuelle Medien," 162.

32. Ryvkina, *Obraz zhizni*, 214–217.

33. The interview with Antonina was conducted in St. Petersburg by Elena Bogdanova in December 2011. Antonina was born in 1947, graduated from a technical university, and worked as engineer.

34. Interview with Zinaida Mikhailovna (Tol'iatti, Samara region, July 2010).

35. Interview with Zinaida Mikhailovna (Tol'iatti, Samara region, July 2010).

36. The interview with Mikhail was conducted by Elena Bogdanova in St. Petersburg in December 2011. Mikhail was born in Leningrad in 1952. He holds a doctoral degree in meteorology.

37. Robinson, Andreyenkov and Patrushev, *Rhythm*, 115–116.

38. The number for the whole population refers to 1986. Grušin et al., *Zeit*, 40; Mickiewicz, *Split Signals*, 3, 188, 204–205; Sapunov, "Televidenie," 17.

39. Leonid A. Gordon and Ėduard V. Klopow, *Der Mensch in seiner Freizeit* (Moskau: Verlag Progreß, 1976), 289–293; Elizabeth A. Weinberg, *The Development of Sociology in the Soviet Union* (London and Boston, MA: Routledge & Kegan Paul, 1974, 111–112; Edward Beliaev and Pavel Butorin, "The Institutionalization of Soviet Sociology. Its Social and Political Context," *Social Forces* 61 (1982): 418–435.

40. Quoted from Gordon, Klopow, *Mensch*, 4.

41. The phrase, probably apocryphal, was attributed to none other than Leonid Brezhnev. Cited from Roth-Ey, *Moscow*, 201.

42. Sergei Lapin, "Televizionnoe veshchanie," *Bolshaia Sovetskaia Ėntsiklopediia* (3rd ed. Moscow: Sovetskaia ėntsiklopedia, 1976), 378–380.

43. Nikolai Shonin, "Radio and Television as Art Forms in the Soviet Union," *Cultures* 4, 3 (1977): 91.

44. Interview with Zinaida Mikhailovna (Tol'iatti, Samara region, July 2010).

45. The interview with Irina N. was conducted by Elena Zhidkova in Samara in July 2010. Irina was born in Cherniakhovsk, Kaliningrad region, in 1953. Her family moved to Kuibyshev (Samara) region in 1957. She finished secondary school and worked as accountant.

46. There are several studies on sport events during the Cold War, see for example the recent volume of Philippe Vonnard, Nicola Sbetti and Grégory Quin eds., *Beyond Boycotts. Sport during the Cold War in Europe* (Berlin, Boston, MA: De Gruyter/Oldenbourg, 2018).

47. Kaiser, *Russia*, 287–288.

48. Zeller, Das sowjetische Fieber, 145–180.

49. The interview with Vladimir Ivanovich was conducted by Elena Bogdanova in St. Petersburg in April 2012. Vladimir Ivanovich was born in Leningrad in 1958 and has specialized secondary education.

50. Interview with Zinaida Mikhailovna (Tol'iatti, Samara region, July 2010).

51. Interview with Yuri Vasil'evich (Samara region in July 2010).

52. The interview with Ruslan was conducted by Oksana Zaporozhets in Samara in October 2010. Ruslan was born in Kuibyshev (Samara) in 1968. He has secondary education and works as driver at the time of the interview.

53. Vladimir V. Putin's regime still gains legitimization by a similar strategy of depicting Russia in "splendid isolation." The long continuities are evident, as this justification for patriotism or even chauvinism already rooted in the Stalinist concept of the Soviet nation and was transferred during the late Soviet Union. With regard to the Stalinist era see Terry Martin, *The Affirmative Action Empire: Nations and Nationalism in the Soviet Union, 1923–1939* (Ithaca, NY: Cornell University Press, 2001), 450.

54. Obzor pisem za 1980, 145.

55. Obzor pisem za 1980, 146.

56. Obzor pisem za 1980, 147.

57. Quoted from Kaiser, *Russia*, 288.

58. Viewers recurrently praised and criticized the performance of the sports commentators. Letter writers disliked biased reporting and formulaic and long-winded speech. However, the opinions upon the performances sometimes strongly diverted. *Obzor pisem telezritelei za 1971 god* (Moscow: Gosudarstvennyi Komitet SSSR po televideniiu i radioveshchaniiu, 1972), 154; *Obzor pisem za 1979*, 181–183.

59. Hickethier, "Das Fernsehen der DDR," 123–124; idem, "Die kulturelle Bedeutung medialer Emotionserzeugung," in *Audiovisuelle Emotionen. Emotionsdarstellung und Emotionsvermittlung durch audiovisuelle Medienangebote*, ed. Anne Bartsch, Jens Eder and Kathrin Fahlenbrach (Köln: Herbert von Halem Verlag, 2007), 104–122, 118 (quote); Meyen, *Denver Clan*, 82–84.

60. Donald Horton and R. Richard Wohl, who introduced the concept of parasocial interaction, did not have, however, movie stars or famous athletes in mind but rather announcers or quizmasters who became everyday companions for the audience. As the Communist Party and, along with the official line, the cultural workers were reluctant to establish any stars, starlets, and heroes, prominent athletes should also appear as ordinary people. Cf. Donald Horton and R. Richard Wohl, "Mass Communication and Para-Social Interaction. Observations on Intimacy at a Distance," *Psychiatry*, 19, 215–229; Peter Vorderer, "Unterhaltung durch Fernsehen: Welche Rolle spielen parasoziale Beziehungen zwischen Zuschauern und Fernsehakteuren? " in *Fernsehforschung in Deutschland. Themen—Akteure—Methoden*, vol. 2, ed. Walter Klingler, Gunnar Roters and Oliver Zöllner (Baden-Baden: Nomos, 1998), 689–707; Tilo Hartmann, "Mass Communication and Para-Social Interaction. Observations on Intimacy at a Distance von Donald Horton und R. Richard Wohl," in *Schlüsselwerke der Medienwirkungsforschung*, ed. Matthias Potthoff (Wiesbaden: Springer, 2016), 75–84.

61. Kaiser, *Russia*, 288.

62. *Obzor pisem za 1971*, 156.

63. *Obzor pisem za 1971*, 5, 154, 172; *Obzor pisem za 1975*, 3, 129, 144; *Obzor pisem za 1979*, 185; *Obzor pisem za 1980*, 169.

64. The interview with Tat'iana Iakovlevna was conducted by E. Ivanova in Irkutsk in October 2010. She was born in Irkutsk in 1964 and graduated from the faculty of law of Irkutsk university. She works in the state admistration.

65. Ronald Grigor Suny, "The Contradictions of Identity: Being Soviet and National in the USSR and After," in *Soviet and Post-Soviet Identities*, ed. Mark Bassin and Catriona Kelly (Cambridge: Cambridge University Press, 2012), 17–36, here 22–23.

66. Maike Lehmann, "Apricot Socialism: The National Past, the Soviet Project, and the Imagining of Community in Late Soviet Armenia," *Slavic Review* 74, 1 (2015): 9–31, 12 (quote).

67. Annika Lepp and Mervi Pantti, "Window to the West. Memories of Watching Finnish Television in Estonia During the Soviet Period," *VIEW. Journal of European Television History and Culture* 2, 3 (2013): 76–86, 76–77, 83.

68. Quoted from Lepp and Pantti, "Window to the West," 82.

69. See for a recent outline of research perspectives on interethnic relations after Stalin's death: Zbigniew Wojnowski, "The Soviet People: National and Supranational Identities in the USSR After 1945," *Nationalities Papers. The Journal of Nationalism and Ethnicity* 43, 1 (2015): 1–7. Further: Maike Lehmann, "The Local Reinvention of the Soviet Project. Nation and Socialism in the Republic of Armenia after 1945," *Jahrbücher für Geschichte Osteuropas* 59, 4 (2011): 481–508; Jeff Sahadeo, "Soviet 'Blacks' and Place Making in Leningrad and Moscow," *Slavic Review* 71, 2 (2012): 331–358.

70. Lucia Gaja Scuteri, "TV as a Linguistic Issue in Yugoslavian Slovenia: A Brief Chronology from the 1960s to the 1980s," in *Television Beyond and Across the Iron Curtain*, ed. Kirsten Bönker, Julia Obertreis and Sven Grampp (Cambridge: Cambridge Scholars Publishing, 2016), 196–226; Idrit Idrizi, "Magic Apparatus" and "Window to the Foreign World"? The Impact of Television and Foreign Broadcasts on Society and State-Society Relations in Socialist Albania," in *Television Beyond and Across the Iron Curtain*, ed. Kirsten Bönker, Julia Obertreis and Sven Grampp (Cambridge: Cambridge Scholars Publishing, 2016), 227–256; Dana Mustata, "Geographies of Power: The Case of Foreign Broadcasting in Dictatorial Romania," in *Airy Curtains: European Broadcasting during the Cold War*, ed. Alexander Badenoch, Andreas Fickers and Christian Henrich-Franke (Baden-Baden: Nomos-Verlag, 2013), 151–179; Annemarie Sorescu-Marinković, "Serbian Language Acquisition in Communist Romania," *Balcanica* XLI (2010): 7–31; idem, "The World through the TV Screen. Everyday Life under Communism on the Western Romanian Border," *Martor – The Museum of the Romanian Peasant Anthropology Review* 17 (2012): 173–187.

71. Interview with Antonina (St. Petersburg, December 2011).

72. The interview with Svetlana Vladimirovna was conducted by Elena Zhidkova in Samara in June 2010. She was born in Kuibyshev (Samara) in 1947, has specialized secondary education by completing a technical college, and was a member of CPSU.

73. *Obzor pisem za 1971*, 13.

74. Cf. Evans, "The Soviet Way of Life."

75. The interview with Ol'ga Nikolaevna was conducted by E. Ivanova in Irkutsk. Ol'ga Nikolaevna graduated from both a technical and an economic college. Afterward, she worked in a housing trust.

76. Interview with Liudmila Vasil'evna (Rostov-on-Don, October 2010).

77. Interview with Liudmila (Samara, September 2010).

78. Obzor pisem za 1980, 166.

79. Interview with Tat'iana Iakovlevna (Irkutsk, October 2010).

80. Interview with Antonina (St. Petersburg, December 2011).

81. Kolpakov, Patrushev, *Biudzhety*, 186.

82. *Obzor pisem telezritelei za 1982 god* (Moscow: Gosudarstvennyi komitet po televideniiu i radio, 1983), 111.

83. Obzor pisem za 1982, 110.

84. Obzor pisem za 1982, 111.

85. Harrison E. Salisbury, "Fifty Years that Shook the World." In *Anatomy of the Soviet Union*, ed. idem. (London et al.: Nelson, 1967), 28–29.

86. See for example the caricatures "The Sphere of grandmother's and mother's influence. Dad remains completely neutral" in *Krokodil*, 7 (March 1959), 6 or "You see! In fact, I warned you to invite any guests, until we will have repaired the TV!" in *Krokodil*, 33 (November 1975), 15.

87. GARF, f. 6903, op. 1, d. 667, 1960: Dokumenty o rabote obshchestvennogo soveta telezritelei pri tsentral'noi studii televideniia (spravki, predlozheniia), l. 28.

88. The thirty-minute program was aired on television from 1966 to 1985 once a week.

89. GARF, f. 6903, op. 48 ch. 1, d. 22, 1971: Svedeniia anketnogo oprosa auditorii teleperedachi "Leninskii universitet millionov," l. 17.

90. GARF, f. 6903, op. 48 ch. 1, d. 22, 1971, ll. 18, 20.

91. GARF, f. 6903, op. 48 ch. 1, d. 22, 1971, l. 23.

92. Raymond Williams, *Television. Technology and Cultural form* (2nd ed. London: Routledge, 1990), 71–111.

93. In 1970, they changed the days of broadcasting and the time slot, however surely not in the sense the Propaganda Desk would have preferred to: The Program Directorate moved it from Mondays and Tuesdays at 6:30 p.m. to Tuesdays—for a while—at 5:30 p.m. and Thursdays at 6:30 p.m. See GARF, f. 6903, op. 48 ch. 1, d. 55, 1971: Mironovich G. M. Rubriki obshchestvenno-politicheskikh peredach TsT v 1970, l. 3. *Pravda* even pointed to *Leninist University of Millions* in its television program in bold type: See, e.g., "Na ėkranakh televizorov," *Pravda*, May 25, 1971, 6; "Na ėkranakh televizorov," *Pravda*, May 27, 1971, 6.

94. Evans, *Between Truth and Time*, 73–81.

95. Sergei Lapin, "Tribuna liudei truda," *Zhurnalist* 5 (1972), 14–16.

96. GARF, f. 6903, op. 48, ch. 1, d. 48, 1971: Kuznecov G. V. Obzor peredach TsT SSSR s uchastiem delegatov XXIV s''ezda KPSS (31 marta–7 aprelia 1971), ll. 7–8. Georgii Kuznecov was one of the prominent journalists who bridged theory and practice, as he researched and taught at the MGU and hosted several popular TV programs such as *Dobryi vecher, Moskva!* (Good evening, Moscow!) and *Moi adres—Sovetskii Soiuz* (My address—the Soviet Union). See http://2005.novayagazeta.ru/nomer/2005/19n/n19n-s10.shtml (accessed January 15, 2016).

97. GARF, f. 6903, op. 48, ch. 1, d. 48, 1971, l. 9.

98. GARF, f. 6903, op. 48, ch. 1, d. 48, 1971, l. 9.

99. GARF, f. 6903, op. 48, ch. 1, d. 48, 1971, ll. 8–10, 13. On the success of *Podvig*, see also chapter 4.

100. The interview with Ivan was conducted by Elena Bogdanova in St. Petersburg in March 2012. Ivan was born in the city of Efremov, Tula region, in 1955. He studied at the workers' faculty in Leningrad and worked as an assistant professor before he switched over to a private enterprise offering business consulting in the 1990s.

101. Interview with Ivan (St. Petersburg, March 2012).

102. See for example Valentin S. Zorin's famous documentaries of the 1970s and 1980s *Protivorechivaia Amerika* or *Amerika semidesiatykh*, which were aired on Central Television. They are still available on youtube: https://www.youtube.com/watch?v=ax0HRidnIR8&list=PL1JMl3ASqksMFfurB1yhHy5WhSY8E9YDV, https://www.youtube.com/watch?v=nC_reaiwbIM&list=PL5qjMlQ_LZ0N7bU9YMDau3I6O2qdvOkOj (both accessed November 19, 2016).

103. The interview with Mikhail G. was conducted by Elena Zhidkova in Samara in July 2010. Mikhail G. was born in Samara region in 1948 and graduated from university. He worked as engineer of telecommunication.

104. The interview with Aleksandra Apollonovna was conducted by Elena Mingaleva by phone in November 2011. Aleksandra Apollonovna was born in Rostov-on-Don in 1938 and lives in Tyumen since 1967. She graduated from a philological faculty and worked as translator and language teacher for English and German. Her father was a White Army emigrant to Paris who returned to the Soviet Union and was shot in 1941. Her mother was an accountant.

105. The interview with Waldemar was conducted by Vadim Korolik in Germany in February 2012. Waldemar was born in Novosibirsk in 1946. He is of German origin, and his parents were deported to a NKVD sovkhoz near Novosibirsk, where his father worked as a transport worker. Waldemar completed a technical secondary school.

106. Interview with Katia (St. Petersburg, July 2010).

107. Interview with Georgii (St. Petersburg, November 2010). Georgii's father was a scientist and doctor specialized in sports medicine. The father was also a member of the Communist Party and worked a while for the KGB.

108. The interview with Zoia was conducted in Germany by Vadim Korolik in February 2012. Zoia was born in Novosibirsk in 1947. She is married to Waldemar and of German origin. She completed a technical secondary school.

109. The interview with Irina Alekseevna from Omsk was conducted by Elena Mingaleva by phone in March 2012. Irina Alekseevna was born in the village of Kochenevo/Novosibirsk region in 1958. She finished secondary school and worked as a cook.

110. The interview with Ol'ga was conducted in Germany by Julia Schmidt in March 2012. Ol'ga is of German origin and was born in Karaganda in 1957. She completed a technical secondary school.

111. Interview with Elena (St. Petersburg, October 2010).

112. The interview with Anna B. was conducted in Germany by Vadim Korolik in March 2012. Anna B. is of German origin and was born in Kazakhstan in 1955.

She finished a secondary school and went to a pedagogical institute that she did not complete. She lived in Kazakhstan and in Altai region for a while before the family moved back to Kazakhstan in 1963.

113. Interview with Aleksandra Apollonovna (Tyumen, November 2011).

114. Interview with Waldemar (Germany, February 2012).

115. Interview with Mikhail G. (Samara, July 2010).

116. Interview with Anna B. (Germany, March 2012).

117. The interview with Konstantin was conducted by Elena Bogdanova in St. Petersburg in January 2012. Konstantin was born in Kokchetav (Kokshetau)/Kazakhstan in 1941, to where his father was deported in 1937. Before the repression, his father was director of a workers' faculty and local head of censorship, and his mother worked as teacher for ancient history. Konstantin graduated from Leningrad university, faculty of history. He worked in several positions, for example in the agency for historical preservation, as researcher at the Academy of Sciences and at the university. He wanted to become a member of the Communist Party but was not accepted. During perestroika, he engaged in civic activism and became the co-chairman of the committee to save the prisoners of war in Afghanistan. Today he is a member of *Edinaia Rossiia*, and is involved in several societal activities as well as in his academic work.

118. Interview with Katia (St. Petersburg, July 2010).

119. The interview with Anatolii was conducted by Kirsten Bönker in Moscow in November 2011. Anatolii was born in 1957, and graduated from university.

120. Interview with Mikhail (St. Petersburg, December 2011).

121. The interview with Al'bina was conducted by Elena Bogdanova in St. Petersburg in January 2012. Al'bina was born in 1941, graduated from a technical university, and worked as engineer.

122. For example, Finnish Television in Estonia: Lepp and Pantti, "Window to the West."

123. *Obzor pisem telezritelei za ianvar' 1983 g.*, 17.

124. The interview with Liudmila was conducted by Elena Bogdanova in St. Petersburg in January 2012. She was born in a small village in Arkhangel'sk region in 1951 and moved to Leningrad in 1964. She graduated from the polytechnic and worked as metallurgical engineer. Further, Ol'ga Galanova conducted a table talk with some of her relatives in a small city in Nizhnii Novgorod region in August 2014.

125. Stephen Lovell, *Russia in the Microphone Age: A History of Soviet Radio, 1919–1970* (Oxford: Oxford University Press, 2015), 155–158.

126. GARF, f. 6903, op. 3, d. 404, 1967, ll. 2–3.

127. The Sino-Soviet rapprochement only reentered the state of crises at the end of 1957, see Lorenz M. Lüthi, "The People's Republic of China and the Warsaw Pact Organization, 1955–63," *Cold War History* 7, 4 (2007): 479–494.

128. RGANI, f. 5, op. 33, l. 26; Roth-Ey, *Moscow*, 174–175, 191.

129. Interview with Oleg Igor'evich (Samara, November 2010).

130. Interview with Irina N. (Samara, July 2010).

131. Interview with Yuri Vasil'evich (Samara region, July 2010).

132. The interview with Roman R. was conducted by Galina Orlova and Sof'ia Kontorovich in Rostov-on-Don in October 2010. Roman was born in 1934; he graduated from Moscow Library Institute, and was a director and a documentary filmmaker.

133. The interview with Tamara was conducted by Elena Bogdanova in St. Petersburg in December 2011. Tamara was born in Kazakhstan in 1944, and lived in the Far North before she moved to Leningrad. She graduated from the faculty of physics and mathematics of the Pedagogical University of Astana.

134. The interview with Anatolii Ivanovich was conducted by Elena Mingaleva via phone in December 2011. Anatolii was born in 1950. He lives in Dzerzhinsk, Nizhnii Novgorod region, and has specialized secondary education.

135. Interview with Mikhail (St. Petersburg, December 2011).

136. Interview with Zinaida Mikhailovna (Tol'iatti, Samara region, July 2010).

137. The interview with Natal'ia G. was conducted by Elena Bogdanova in St. Petersburg in February 2011. Natal'ia was born in Leningrad in 1951, and graduated from university.

138. The interview with Tat'iana was conducted by Elena Bogdanova in St. Petersburg in March 2012. Tat'iana was born in a village in Gor'kii (Nizhnii Novgorod) region in 1949 and moved as a child to Novgorod region and later to Leningrad in 1972. She worked in a woodworking factory and has secondary school education.

139. The interview with Aleksandra was conducted by Elena Bogdanova in St. Petersburg in November 2010. Aleksandr was born in Leningrad in 1931. He has a doctorate in biology.

140. The interview with Svetlana Vladimirovna was conducted by Elena Zhidkova in Samara in June 2010. She was born in Kuibyshev (Samara) in 1947, specialized in secondary school education/technical college, and was a member of CPSS.

141. The interview with Nikolai was conducted by Elena Bogdanova in St. Petersburg in January 2011. Nikolai was born in Briansk region in 1940 and moved to Leningrad in the 1960s. He graduated from university and was member of the Communist Party.

142. The interview with Vladimir Ivanovich was conducted by Elena Bogdanova in St. Petersburg in April 2012. Vladimir Ivanovich was born in Leningrad in 1958 and has specialized in secondary education.

Chapter 3

Negotiating the Boundaries of Popular Culture

Soviet Television, Leisure, and the Educational Mission

"Regarding television as the most far-reaching mass media, I regret the loss of certain substance (. . .), if you wish a fine aesthetic content."[1] In 2010, seventy-two-year-old Yuri from St. Petersburg started with these words to describe differences between the current Russian and the former Soviet television. Looking back, the retired engineer represented himself to have been a critical and rather distanced consumer of Soviet media. He was often annoyed by the Soviet television program, but, nevertheless, remembered some "aesthetic television events," as he put it, very positively. He was thinking of some—in his view—sophisticated films and live broadcasts of theater plays, ballet, opera, and concerts from Moscow and Leningrad. Television leveraged the increasing proliferation of highbrow performing arts: "In those places, where there was no theater at all, television suddenly gave people the opportunity to watch and to listen." Compared to these Soviet programs, the modern Russian television has been "Americanized" and just follows commercial interests.

Yuri's ambivalent characterizations of Soviet television raise questions not only about contemporary television uses. They let us also ask for television's social, cultural, and political meanings for the late Soviet society, as well as for the way television challenged the common acceptance of classical genres like ballet, opera, and drama as to be more sophisticated and highbrow. This chapter, therefore, explores how television modified the hierarchy of performing arts in the perception of contemporaries. How did it change the assessment of cultural professionals, creative artists, and journalists of the established hierarchy? How did the new medium change the debates on Soviet mass culture and the claim to educate the New Soviet man? To what extent did television shift the established boundaries between pop culture and entertainment on the one hand and the classical genres like ballet, opera, and drama on the other? To what extent did contemporaries classify television as a cultural facilitator which let "ordinary" citizens gain access to the so far dominating cultural canon? To what extent did the television broadcasts of classical performing arts impact on the new medium's standing?

This chapter focuses on contemporary elite debates about television and brings them together with the retrospective perspectives of former Soviet television viewers formulated in the interviews. The contemporary observers and representatives of the established arts dominated the important feuilleton and culture sections of print media and radio which set the tone of the discussions and especially reviewed the new medium. They were keenly and somehow unrelentingly engaged in drawing the line between legitimate and illegitimate pleasure, between lowbrow entertainment and highbrow arts. The debates aimed to define the social, political, and cultural significance of the new medium and its programs. Thus, this chapter discusses in particular the contemporary perceptions of television as a rising medium and is not interested in the artistic and aesthetic arrangements of the television programs. Also, it concentrates not on how television affected the audience but, following Elihu Katz, on how the viewers used the medium.[2] This shifts the perspective on viewers' motivation to watch TV, on their ideas of entertainment and pleasure. Entertainment and pleasure are cognitive effects that are not easy to grasp, as they depend on time, social, political, and cultural contexts and so on. They were closely related to those debates that aimed to define sociocultural boundaries.[3] As mentioned earlier, I apply a broad notion of entertainment and pleasure that includes laughter, relaxation, and distraction felt by viewers watching a film or a philharmonic concert, as well as historical documentaries, animal features, or Central Television's nightly news program.[4] Television entertained and gave pleasure to the viewers, because it offered sites and topics of conversations with friends. Television also represented pictures of lifestyle and society on which basis people could interpret their own home und life.[5] Thus, television turned out to become a medium of social reproduction and self-reassurance. The medium entailed interpretative communities in front of the screen whose members then engaged into the negotiations on sociocultural classifications.[6]

All this meant that television became a key element of the late Soviet lifestyle and popular culture. Both constituted important sites of struggle in the context of the Cold War bloc competition. On both sides of the Iron Curtain, artists, cultural and media professionals, scientists, and journalists debated the notions of mass and of highbrow culture highly controversially according to the respective sociopolitical position of the observer.[7]

Therefore, we take a closer look at how contemporaries reinterpreted concepts of popular culture and of the established hierarchy of performing arts considering the new technical capacities of television. The chapter first focuses on journalistic views, in order to illustrate the normative claims about the use and consumption of television and the cultural educational mission they envisioned for the new medium. The second section explores how the studies of Soviet social scientists on leisure practices reflected the contemporary cultural criticism. The third section demonstrates based on my interview sample in an exemplary way to what extent respondents perceived television as cultural facilitator and its impact on the established cultural canon. For this purpose, I chose narratives of interviewees who looking back represented themselves as critical and rather distanced media consumers. They either addressed this issue by themselves or had simply more to say about it than less educated or rather content viewers.

THE SOVIET TELEVISION AS CONVERGENCE ZONE OF POPULAR CULTURE AND PERFORMING ARTS

On January 1, 1955, Moscow Central Television Station started transmitting daily programming covering the fields of information, entertainment, and arts. In all these domains, the television producers needed to develop the new medium in order to compete with established hierarchies of the printed press, radio, and performing arts. Television not only became the key site of changes in political communication with the audience, because it experimented with new journalistic and artistic forms to update political rituals, to engage viewers, and to address the emotional dimension of Soviet lifestyle.[8] As television basically challenged established cultural classifications of popular and highbrow culture, television provoked intensive discussions. The enthusiasm about television, as well as the attitude of cultural criticism toward it, was as such a rather common phenomenon on both sides of the Iron Curtain. The specificity of the Soviet discourse was, however, that the somehow utopian idea of the new medium as a transforming force in social and political life did not fade away as quickly as it did probably in Western debates.[9]

In the beginning, the Communist Party underestimated the communicative potential of television. Therefore, the party rather late officially engaged for the first time in the course of television's development, with a decree in January 1960. The Soviet artistic intelligentsia and print journalists, however, put the relation of the new medium to the established performing arts on the feuilleton's agendas as early as the mid-1950s. They initiated a public debate not only about norms and values of popular culture, but also about the meaning of high culture for the Soviet society. This meant that the different claims of the audience, the television producers, the party, and, last but not least, the artistic intelligentsia politicized the negotiations about the use and the future development of the new medium right from the start. This was an obvious sign of television's transnational character bridging the gap between the political blocs.

On both sides of the Iron Curtain, the established educated elites met television and the upcoming entertainment culture with a highly patronizing attitude. To a certain extent, the different political frameworks—even within the Western political landscape—set, nevertheless, different priorities within the debates. West European discussions seemed to be more critical and fractured than the American one, but in France and West Germany, for example, left and conservative intellectuals agreed to devalue television as pop-cultural medium.[10]

Academic and artistic protagonists in East and West were struggling with the concept of "masses." Basically, they wondered to what extent television would constitute a new form of art. But they also asked for the relation between television and established arts and about the medium's reach. The question of television's impact on the audience was fundamental, as the new communities of viewers in front of the screen considerably grew in terms of numbers because of the technical possibilities.

The Soviet idea of "*massovost*" as the core of socialist culture implied the positive characteristics of comprehensiveness and widespread impact. The Soviet intelligentsia, however, deemed the concept of "*massovaia kul'tura*" (mass culture) to be merely the degenerate product of Western capitalism. From the intelligentsia's point of view, this sort of capitalist mass culture simply served to lull the "exploited working people" and disguise "societal antagonisms." A distinct socialist pop culture should therefore oppose the seemingly Americanized pop culture of the West. Socialist pop culture should revolutionize existing concepts of high culture and should blur traditional cultural boundaries. Regardless of educational levels and cultural training, all citizens should have access to highbrow cultural products.[11] That "our mass intellectual [*massovyi intelligent*]" got to know Russian classics like *Voina i mir* (War and peace) or *Anna Karenina* not through comics or poor movies, as the Americans allegedly did, would, according to the chief editor of *Novyi Mir* (New World) Aleksandr T. Tvardovskii, maintain an important difference between the Soviet and American media consumers. In his speech on the XXI congress of the CPSS in 1959, Tvardovskii presented an ambivalent but all-embracing concept of television that became rather typical for those artistic elites at this time who were willing to collaborate with the new medium: Although Tvardovskii supposed books in a quite traditional way to be the "highest and irreplaceable form of the intellectual development of man," reading would not prevent people from any "facilitated acquisition of artistic materials" as television would not exclude any interest in painting, theater, concerts in their own premises."[12]

"*Massovost*" seemed to be per se an inherent part of television. In the 1960s, its technical characteristics still implied opportunities, as well as challenges. Television was not only an aesthetic experimental field, but also technically new ground which attracted people with different professional backgrounds. Some who failed to make their way in other areas found a promising niche in the context of the new medium. Established journalists, directors, or dramaturges initially rather refrained from television, as it did not pay competitive salaries.[13] Kristin Roth-Ey and Christine E. Evans have convincingly pointed out that the television producers in the beginning imagined a special societal role for their medium within the de-Stalinizing society. Television should bring back truthfulness and sincerity and, thus, help to shape the New Man. The noble claim of the television staff and its search for "cultural authority"—to borrow Roth-Ey's term—became, however, rather contested, as soon as the party confronted them with its own concept of cultural education.[14] The situation got even more complicated as television producers realized as early as 1959 that Soviet viewers did not basically differ from "foreign" ones, and preferred watching TV as pleasure-giving leisure practice.[15] Nevertheless, television producers tried the balancing act and represented the medium as "intermediator between theater and cinema."[16]

Thus, from the beginning, television producers also aimed to culturally educate the audience, to mold viewers' taste, to trigger their preferences for highbrow arts and culture, and, last but not least, to promote their capability to distinguish "good" and "bad" arts.[17] They further tried to assure their position toward the performing arts by pointing to the medium's broad impact. Similarly to a statement of earlier-quoted interviewee Yuri, an editor of East German television emphasized television's mediating role: "Television actually prevents that whole generations have to wait until they might enjoy someday the opera (i.e., State Opera *Unter den Linden*) or the *Komische Oper* etc."[18]

In view of television's strong expansion after the mid-1950s, the position of praising its mediating power invited sharp criticism from theater and cinema artists. Representing many of his colleagues, the renowned theater director and actor Igor Il'inskii reprimanded the new medium in an article published in *Literaturnaia gazeta* in 1956. Although Il'inskii recognized a few positive characteristics of the new medium, he ultimately came to a scathing judgment of television's achievements. He considered favorable that there were a number of intriguing and smart programs. Further, he found obvious that the television producers paid attention to the diverse interests of TV viewers. Apart from that, he accused the television producers of wide-ranging dilettantism. Similarly to cinema and radio in their early days, television had not found its own specific characteristics yet, but "has taken possession of fields which do not belong to it." With this, he targeted those live broadcasts he thought to be inadequate representations of performing arts. He assessed the program to be eclectic and crude. In his view, much of the program was even "primitive." He did not like the way television tried to stage reality on the screen at all.

Il'inskii, moreover, pointed to the fact that television randomly broadcast movies and transmitted rather thoughtless concerts and plays live from the concert halls and theaters. He considered it "to be very annoying that the station tests its strengths only very reluctantly." In this, perspective television's indecisiveness seemed to be simply detrimental, as it were only the "genuine television programs that attracted the viewer, that is, what he cannot see in the theater or the cinema." Hence, "television should duplicate as little as possible. It should do everything for 'the first time,' and what it realizes should, however, reach the highest level as soon as it concerns millions." The logical consequence was to demand the professionalization of television staffers on the one hand. On the other, he advocated for the founding of a genuine TV theater with actors and directors who should specialize according to the medium's requirements.[19]

This criticism perfectly met television's efforts to "create own authentic television productions and programs."[20] As early as in the late 1950s, television producers obviously intended to reduce the share of live broadcasts from theaters and so on in favor of new pop-cultural formats. Also, television staffers agreed with Il'inskii, who outlined how the relation between television and performing arts should further develop. In his view, the competition with television let cinema and theater run the risk of losing viewers in favor of the new medium. Il'inskii found it rather paradoxical that Central Television broadcast the latest movies "not before 10 to 11 days after their release," whereas the audience of millions was deprived of the best theater plays. Instead, the television stations only aired those plays "the theaters did not want for their own purposes."[21]

Not only from today's perspective, it seems absurd that Central Television broadcast movies only shortly after their release. Already American radio and television correspondent Irving R. Levine wondered about this policy. He found the plausible explanation that the movie studios were state-run so that no private business interest impacted on the relation between cinema and television.[22]

Il'inskii's claim served on the one hand to emphasize his argument that television should broadcast only the best productions: "How could we present only the worst examples of arts to the people instead of the best?"[23] On the other hand, Il'inskii's demand to enable TV viewers to watch the latest movies at home was rather strategic to lead the discussion about which state agency should control the new medium in a certain direction. In accordance with viewers' demands, television considered movies to constitute a very important program share, so that Il'inskii's demand was rather congruent with Central Television's own ideas about broadcasting movies. The Ministry of Culture, to which Central Television was subordinated until 1957, was, however, not very eager to promote the new medium at the expense of cinema and theater. Contrary to the expectations of television, it decreed in 1958 that new artistic movies might be aired not earlier than twenty days and not later than two months after their release. Central Television, however, complained that the film studios made the movies only seldom available in accordance with the decree.[24] Only Central Television's program director R. N. Zdobnov realistically appraised the availability of new movies for Soviet viewers compared with other countries. This asset led him even to the typical Soviet assessment that "Soviet television is considered to be the best in the world."[25] Indeed, looking back, not a few viewers like fifty-nine-year-old historian Yuri Vasil'evich really appreciated being able to watch the latest movies on TV: "And then Kuibyshev TV was good: literally one month after it was aired at the cinemas, it was on TV."[26]

Apart from Soviet movies, television aired foreign productions. As early as in the mid-1950s, television proved to be a transnational medium that brought foreign images to Soviet living rooms. Letters and interviews reveal a certain spectrum of preferences with regard to movies. Many of the respondents explicitly mentioned that they loved to watch foreign, especially Italian, and French movies. Some viewers complained about the need to read subtitles and found "our Soviet movies much better."[27] Looking back, nearly all interviewees regardless of their educational level spoke rather critically about American productions, especially about those aired in post-Soviet times compared to Soviet movies. But even with regard to the Soviet period, some of the higher-educated interviewees, like the then sixty-four-year-old Masha, remembered: "French idiotic comedies were terribly annoying, but I very much loved French serious movies. For a very long time I didn't love American movies, for a very long time those they aired."[28] Whereas higher-educated viewers tended to differentiate among Western movies, comparing them to Soviet ones, the less educated tended to be rather indifferent and were not particularly interested in foreign productions. Then sixty-eight-year-old Liudmila detests modern American movies because they often stage violence. Asked if she was interested in foreign movies and the different lifestyle they might have presented in Soviet times, she simply denied any attraction.[29]

Further, the ministry did not even support television producers to meet their official task of propagating highbrow culture, although many viewers expressed in their letters to television right from the start that they would have preferred to watch more premieres.[30] Directors of many important theaters in Kiev, Leningrad, Moscow, and elsewhere often successfully protested in cases where television wanted to broadcast plays, and simply did not let cameras in. Television workers rather desperately asked for administrative help in order to do a better job, but did not have any profound lobby for years.[31] These struggles with theater and opera directors continued and did not really relax throughout the 1960s. Together with technical problems, which made it difficult to meet the aesthetic claims to live broadcasts of ballet and operas, television workers found themselves rather constrained to create television-specific performances. The basic requirement was, however, a TV studio spacious enough for a stage on which the great operas like Boris Godunov could be performed.[32] Nevertheless, Central Television started to broadcast seventy to eighty plays each year from the mid-1950s onward.[33]

Because of these quarrels, Il'inskii favored a reorganization of responsibilities. The administrative development endorsed his suggestions. Thus, the moment when Central Television was administratively assigned to a new State Committee for Radio Broadcasting and Television (*Gosteleradio*) gave a strong impetus for its further development. From then on, *Gosteleradio* was directly subordinated to the Council of Ministers, a fact ensuring better financial and infrastructural support and highly appraised by television workers.[34]

Il'inskii's statements not only highlighted that the questions of control, censorship, and aesthetic guidelines were highly contested, but generally documented the controversy about the new medium's place in the cultural hierarchy in the early days. His concept to present only "the best" performances revealed that he himself considered television basically to be a facilitator of precious and highbrow arts. In particular, television should not only make the performing arts available to the enormous TV audience, but also "popularize the work of theaters, actors, and directors." In his view, television would never take the place of the theater and cinema, but could, at best, attract new viewers for them or, at worst, discourage viewers to consume highbrow cultural products by bad broadcasting. As television was "the best stage in town," it was "an honorable right" to perform a theater play on the screen.[35]

Il'inskii's article presented the whole spectrum of the television discourse. It ranged from perceptions of a deplorable present state of the medium to utopian elevation of the medium. It also included the already-hopeless attempt to reinforce the position of the established performing arts toward television by seeking much more cooperation. The author seamlessly fit into the unbalanced contemporary discussion. Most of the Soviet viewers admired the medium because of its technical opportunities as "window to the world," whereas the print media hailed it for its potentials to culturally educate the New Man and to shape the new Soviet lifestyle. At the same time, we trace in comments such as Il'inski's that the artistic establishment, however, adopted a cultural critical attitude toward television.

Il'inski's descriptions about the early aesthetic and dramatic program strategies did not differ from those in the East and the West Germany: In the 1950s and 1960s, television initially constituted a new forum for the established performing arts on both sides of the Iron Curtain.[36] The TV program sections took up well-known forms and genres from theater, concert, variety and cabaret performances, circus, or cinema and adapted them to the requirements of television. In fact, in television's early days, live broadcasts from renowned theaters, operas, and concert halls let many viewers in East and West enjoy performances they otherwise would not have been accessible. Television substantially expanded the participation in the established arts not only in social but also in geographical respect. By offering a different sensual experience, television, at once, complemented and changed the perception of arts. The new way of consuming high culture at home also differentiated the segments of the public sphere (*Teilöffentlichkeiten*) in which people communicated about culture.

Comparing the Soviet debates with those in West Germany, for example, the similarities between the arguments and practices of television's promoters, as well as its critics, are striking. In television's early days, West German educational elites could not take pleasure in live broadcasts from *Deutsche Oper* in West Berlin nor from *Ohnsorg-Theater* Hamburg—one of the most famous popular theaters in Germany performing plays in dialect.[37]

The critical tone constantly characterized the discussion during the 1960s. On the one hand, often unidentifiable authors praised the technical opportunities of television and considered it "as medium of Communist education of the masses." They emphasized the fact that the Soviet people enthusiastically accepted the new medium. On the other hand, the official perception of television was rather tenacious. In 1960, the party's mouthpiece *Pravda* saw television still of equal rank with press and radio. The newspaper considered television as only one instrument among others to mobilize the people to meet the demands of the seven-year plan. In addition, *Pravda* even sharply criticized the deplorable, often boring program. This critique particularly referred to broadcastings covering literature and fine and performing arts. The *Pravda* journalist explicitly noticed that both groups of actors—television producers and art professionals—carried responsibility for the unsatisfactory situation. In his view, the television workers were not able to establish a fruitful working atmosphere to cooperate with the art professionals. The artists, in turn, preserved their "known resistance against broadcastings of theater plays, concerts, movies on television." In the same breath, *Pravda* cemented the fundamental distinction between high and popular culture. The party newspaper demanded that television should air more performances of popular theaters and of other "collectives of artistic self-activity."[38] This question about television's relationship toward folk art and artistic self-activity remained highly disputed throughout the 1960s. The discussion revealed the clear dividing line between the promoter of classical highbrow culture versus television as medium of the less educated.[39]

The debates about how to promote arts and literature via television gave revealing insight into common assumptions of the discussants. Apart from their critical attitude, Il'inskii, the *Pravda* journalist and other members of the intelligentsia implicitly agreed that the typical television viewer belonged to the less-educated social groups. Television producers themselves considered these people to focus more on "light" entertainment than on highbrow programs. From the beginning, many viewers gave evidence to this assumption and complained in their letters about too many operas on television.[40]

This, however, supported the argument of the television producers to offer the best conditions for shaping the audience's preferences. All the contemporary observers assumed that television was the best medium to reach out to an audience as large as possible. In contrast to their implicit assumption, they explicitly charged television to address all social groups. From the perspective of 1960, television should offer a forum to all genres of art. Looking back in an expert interview, renowned sociologist Boris M. Firsov, who managed Leningrad Television from 1962 until early 1966, explained that this claim was put into practice: "The sphere of artistic and cultural broadcasting was more or less leased out to television."[41] This contemporary claim did not mean, however, that any shift in established cultural boundaries had already been commonly accepted. Rather, it was still a balancing act. In 1960, contemporary viewers even warned that television should air more about the "cultural life of our society, the cultural beauty and the beauty of Soviet people's actual current life." Instead, "television often shows the miserable life of our people with clear substance of banality, some amoral, inadequate and sometimes implausible Soviet people."[42]

A discussion at a viewer conference in Kaluga illustrated the broad range of *Pravda's* editorials. In the 1950s and 1960s, Central Television regularly convened viewers in order to trigger direct feedbacks from the audience. The pensioner M. M. Luchkin referred in his statement on the Kaluga viewer conference to the just-quoted editorial. He interlinked his criticism of the television program with the claims *Pravda* had put forward. Luchkin made clear how much provincial viewers benefited from any broadcastings of plays and concerts from Moscow, as "who could watch good artists, could listen to good singers—you could only do this in Moscow." He explicitly wished for broadcastings from the Moscow Chekhov Art Theater (MChat). He emphatically distinguished this art discipline from "mass musical comedies," although he considered them, as well as dancing, worthy to be aired on screen. The viewer Luchkin, thus, still took the classical classification of high and mass culture for granted. Generally, he felt television should expand the artistic program. In particular, the medium should—according to his words—satisfy the increasing demand for high cultural productions.[43]

At a similar conference in 1962, the viewer Zlotnikova made an even clearer diagnosis about the importance of "contemporary fine arts" on television:

> The content of television programs is partly dictated by commercial considerations, but this is wrong. (. . .) Movies and programs are broadcast on television that do not constitute the best examples of art, but which simply are no art at all. TsT (i.e., Central Television) does not have a permanent theater of its own (. . .). I must say that our Moscow theaters do not permit their best plays to be broadcast due to financial reasons.[44]

Participants of such viewer conferences often pointed to the ideological phrase on the educational task of television toward all social groups. Any reference to socialist education reinforced the claim for better cooperation with the theater and for bridging traditional boundaries between cultural genres.[45]

These public forums for television viewer also made clear that the audience developed a whole spectrum of preferences, claims for entertainment, and media uses from the early days on. Some disdained broadcastings of football matches, whereas others would not like to miss them and still others were looking forward to enjoying Chaikovskii's "Swan Lake," a theater program with Leningrad actors or the popular comedian Arkadii Raikin in the evening.[46]

The internal reports of Central Television's Scientific-Methodological Department (*Nauchno-metodicheskii otdel*: NMO) revealed during the 1960s that Central Television considerably expanded its programs covering theater, movie, literature, and concert. Artistic movies and concerts of light music (*estrada*) appealed most to the steadily growing audience in the field of entertainment. Opera and operetta enjoyed considerably less popularity.[47]

However, this program policy was not least the logical consequence of the dilemma in which Soviet television producers of the early days found themselves constrained to resort to artistic and cultural programs. Former director of Leningrad Television, Boris Firsov, characterized the concept of focusing on artistic and cultural programs as circumvention scheme to avoid obviously "political" issues:

> One thing is clear that all political broadcasts were tabooed. It should only come down to mere information, what happens in the different fields of the party state leadership. No controversies, no discussions about serious programs, although it would have been possible to do something, this was impossible. Therefore, we were leaning to the side of television cultural broadcasting in Leningrad. We decided, that as far as the country's state of cultural isolation was rather lasting, it was necessary to remind people what they were deprived of—I mean Russian, Soviet and world cultural heritages.

This statement should not necessarily be read as a subversive strategy to remind people of the aspects limiting Soviet life and of the prize they pay for enduring the authoritarian regime. But it very well explains why Soviet television producers were so inclined to promote high culture and why it took a while until the program became more popular in the proper sense of the word with specific TV formats engaging the viewers. It turned out to be a cluster of pragmatic and deliberate decisions. Although there were few commercial pressures, television with its challenging technical infrastructure was expensive. Further, the daily schedule must be filled somehow. Broadcasts from theaters and so on were relatively easy to realize, and, moreover, they matched the ideas of many of the early television producers to air sophisticated and highbrow programs.[48] From a TV professional's perspective, Boris Firsov underlined that in view of the political constraints Leningrad Television rather depended on a cooperative relationship with the theaters. He highlighted that it proved to be a rather fruitful cooperation with mutual inspiration and exchange.[49] Although television could not pay competitive salaries, Firsov underlined that, at least, in Leningrad performing arts professionals welcomed the cooperation with television to take extra income. According to Firsov, they, moreover, gladly accepted to win fame and "take up a cultural dialog with the TV viewers."[50]

These few insights into contemporary elite representations illustrate how the boundaries between media and artistic genres had gradually started to change from television's early days on. Television quickly adjusted the pop-cultural canon for the mass audience by adapting parts of the established performing arts and of classical music. Its rising self-confidence became apparent on a conference with newspaper journalists in 1965. Although television workers conceded that the medium entailed certain social and aesthetic problems, they claimed to represent a "powerful medium of intellectual communication for the masses." Meanwhile, they raised the issue of television's place in the cultural hierarchy much more offensive stressing the specific cultural and artistic value of the medium. Television workers did not feel any need to take a humble stance any longer toward the established performing arts: "And you see that television talks about life more truthful than other arts."[51]

Based on this new self-perception, television producers even asked the printed press to improve and stabilize its reporting about television programs. Television, they argued, deserved much more analysis, propaganda, and theoretical comprehension. They considered necessary that journalists must specifically deal with the medium and understand its specificities. Reviews of broadcasts should not consist of merely replacing the words "theater, theatrical" by "television." The television editors complained that the newspapers of the capitals took considerably less notice of television than of theater and cinema.[52] Despite this new self-confident manner, the cultural stance of television remained contested throughout the 1960s and 1970s. Every newspaper reader could easily trace this debate on the pages of *Pravda*, *Literaturnaia gazeta*, or *Sovetskaia kul'tura*.

During the late 1960s, television very successfully developed the new format of art competitions to bring high cultural and popular cultural performances together. This was a particular way to involve viewers all over the country and let the participants, as well as the audience, join in this "fan community." In 1969, *Pravda* highly appreciated the "Music tournament of cities" (*Muzykal'nyi turnir gorodov*) that had been broadcast over a whole week in January 1968. The competing teams came from all Soviet republics and showcased peoples' friendship, as well as the cultural diversity of the Soviet Union. The journalist Mikhail Kapustin applauded the competitive but friendly atmosphere, in which the amateur performing arts took place, and the competent work of the jury. The jury was headed by the renowned composer Aram Khachaturian, a fact that should symbolize the merging of high and pop-cultural performances. Kapustin praised Central Television's positive impact on amateur performing arts, as the medium brought together local ensembles with prominent artists. The competition would be continued on the occasion of Lenin's 100th birthday in April 1970.[53]

Critical voices, such as that of the *Pravda* correspondent Viktor Shepel, based in Frunze (Bishkek), still officially complained about the artistic quality of the television programs. Shepel's article, published in 1969, formulated a double argument, as he, on the one hand, requested that *Gosteleradio* should be much more concerned about the professionalization of its staff, especially of the employees of the regional television stations. In his view, the television workers were still overtaxed by the specific characteristics of the medium with regard to its creative potentials. Thus, he pointed to the fact that television was still "in need of being supported in the creative fields by newspaper journalists, writers, artists of theaters and cinema. It's a pity that there is only little support." In addition to this artistic focus, Shepel's article could also be understood as a statement from the non-Russian periphery. It made clear to all *Pravda* readers that television was supposed to have a manifold unifying impact on journalistic standards, viewing habits, tastes, and artistic performances.[54]

THE "TELEVISION-REPRESSION": NEGOTIATIONS ABOUT MEANINGFUL LEISURE ACTIVITIES

The new medium not only raised artistic and aesthetic questions, but also challenged the traditional ideas about meaningful leisure activities. As we have seen in chapter 2, hardly any other consumer good or medium changed the leisure practices of so many people living in Eastern and Western industrialized countries. From an ideological viewpoint, working only five days a week along with more leisure time were considered as "lawful steps" on the way toward communism. "The cultural promotion of man" (*povyshenie kul'tury cheloveka*) through "cultured leisure activities" (*kul'turnyi dosug*) was considered an inherent part of this societal progress.[55] The ideological claim to lead a meaningful leisure time referred to several aspects of the new Soviet lifestyle. On the one hand, leisure activities in the field of reading, listening to music, and so on should comprise the classical Soviet cultural canon. Although the components of the canon had changed after the Russian Revolution of 1917, it basically included nineteenth-century classics of literature and music, as well as visits to theater, cinema, and concerts that would traditionally be attributed to high culture.[56] On the other hand, sitting in front of the television screen complicated any concepts of active recreation.[57] Therefore, television consumption quickly became involved in negotiations about meaningful leisure activities. In particular, protagonists of the revived discipline of empirical sociology were highly interested in researching leisure practices. Studying Soviet people's time budgets, they established that watching television became one of the most popular leisure activities during the 1960s.[58] This observation obviously challenged the ideological claims to rational leisure activities. As a result, the scope for interpretation the scientists had without violating the ideological compliance of their results became rather limited.[59] The sociologists needed to assess to what extent television impacted on and potentially replaced the traditional and officially approved leisure practices like visiting theater, cinema, and concerts, as well as reading books and press. They also needed to evaluate the change of leisure activities entailed by the increasing television consumption. Third, the researchers needed to find categories to explain people's preferences. The categories should cover questions of how TV viewers perceived the contents and how the parameters of profession, education, age, and gender influenced media consumption.

The choice of words sounded rather cumbersome but should suggest neutral assessments. Thus, reading figured as the "first main activity for leisure" and the performing arts of cinema and theater as the second. To television the sociologists referred as the "third main activity in the spare time with regard to cultural consumption (*tret'ii vazhneishii form dosugogo potrebleniia kul'tury*)."[60]

Soviet sociologists basically agreed that television had a lasting effect on everyday life. They found out that as early as in the mid-1960s the urban population devoted at least half of the time they spent in the context of "cultural life" to watching television. The sociologists considered television a propagator of "universal cultural and artistic-aesthetical information" as it enabled all people to conveniently enjoy theater plays and concerts at home. Thus, television considerably changed how people could be introduced to culture.[61]

The obvious intellectual and temporal impact television exerted on other art forms provoked *kulturkritische* assessments also among leisure researchers. As for their Western colleagues and Soviet art professionals, their choice of words often revealed that Soviet sociologists were rather reserved toward the new medium. Their critique referred openly to technical aspects of television and partly as well as to watching TV as a new social practice within families. However, West German left-wing intellectuals such as Theodor W. Adorno, Hans Magnus Enzensberger, and Austrian Günther Anders condemned television because of its allegedly lulling contents. Whereas they frankly supposed television to have a manipulative power that ultimately modified "reality," the Soviet sociologists could at best cautiously hint at these aspects. The German observers held television responsible for a loss of values, a depoliticizing "massification" (*Vermassung*), and the leveling of quality. The Soviet sociologists would have perhaps agreed with this assessment. Yet unlike the German observers, they did not dare to link the cultural critique to any social change.[62] The young Soviet media sociology rather tried to embed television into concepts of utopian society. For ideological reasons, however, they rejected the strong resentments of Western adherents to left "Critical theory" toward the idea of "masses" and of a "mass culture." "Manipulation" was the catchword of Western media criticism of the 1960s. Western critics were concerned how television as instrument of power influenced the powerless audience.[63]

The Soviet sociologists addressed more basic and perhaps innocuous problems and asked how television viewers used the medium. They searched for ways to adapt the viewers' television use to the official interpretations and norms of the Soviet cultural canon. Consequently, they needed to examine the relations between television and the other media, their aesthetical forms, and contents. Interestingly enough, the Soviet researchers refined their analytical methods and categories by engaging in the UNESCO project of time budget research. Henceforth, they cooperated with Western and East European colleagues across the Iron Curtain. The project pursued the target to collect broad data and to enable transnational comparisons. Indeed, the Soviet sociologists consulted the international data, compared concepts and methods. Moscow scientists Leonid Gordon, Eduard Klopov, and Leon Onikov emphasized the longer experience of American television. Its history showed that "the possibility of an excessive development supported the risk of a literal 'television-repression' towards other forms of cultural leisure activities." It does not come as a surprise that Gordon's, Klopov's, and Onikov's concern was to highlight categorical differences between American and

Soviet television. The principal differences between television contents and programs were rather obvious, which is why they particularly highlighted the importance of the "rational relation" between the duration of television consumption and other "forms of cultural introduction." It was exactly this relation that in their view determined the specificity of the Soviet lifestyle in contrast to Western ways of life.[64] Like other Soviet colleagues, these three researchers, however, did not say precisely what the concrete differences in content looked like. Their Moscow colleagues, sociologists Boris Kolpakov and Vasilii Patrushev, for example, worked with the same data records. They opted for an ideologically secure diction and merely pointed to the fact that mass media in capitalist countries were clearly commercial. This was why the capitalist media were "not subordinated to the task of the comprehensive development of the people's personality like under socialism." On the contrary, they were "devoted to the propaganda of bourgeois ideology."[65]

Kolpakov and Patrushev focused more on self-education and the general cultural and technical level. They found that the working Soviet population spent more time raising their cultural level than people living in capitalist societies.[66] Gordon, Klopov, and Onikov were, however, more interested in the impact of education on cultural consumption and the question of how television consumption was related to the enjoyment of traditional high culture. They emphasized that even Soviet television had undoubtedly repressed sometimes extremely valuable forms of cultural leisure activities. The time people now spent in front of the television was at the expense of leisure practices like "visits to the cinema or the 'Live theater'(*zhivoi teatr*)." In addition, the authors stressed the fact that members of families owning a television set spent less time reading than members of families without one. Nevertheless, the authors gave their analysis a fundamental, however, hardly convincing turn to explain that watching television was lastly only of secondary importance regarding the decrease of cultural leisure practices: The transformation of leisure practices would have been facilitated by idleness, housework, or even body care. Gordon, Klopov, and Onikov actually affirmed that Soviet television had not considerably expanded, because socialist culture was allegedly "systematically organized." As a result, such planning would prevent any competition between mass media and performing arts. On the contrary, they should be seen as a complementary system in which especially television even promoted traditional cultural activities.[67]

The studies of Soviet TV's Scientific Methodological Department (NMO) suggested that this claim, as Gordon, Klopov, and Onikov put forward, remained mere theory. The surveys of NMO documented a considerable change of practices from the early 1960s onward. They highlighted that the majority of the viewers preferred to watch movies and theater plays on television at home. Even though NMO had a vivid self-interest to stress the importance of television within the Soviet media landscape and the audience's enthusiasm toward the medium, the NMO researchers did not fuel the competition with cinema, theater, and literature. On the contrary, the television staff itself presented their medium as propagating forum for other media, genres, and arts. Television's "task was . . . to arrange cultural leisure activities for the working people, to propagate classical music, literature."[68] To underline this claim, they, for example, deliberately quoted interviewees who demanded more television programs covering literature in general or releases of books in particular.[69]

The argument that television stimulated interest in literature, theater, or new exhibitions included the new medium into a specific Soviet concept of the relation between high and popular culture. Social scientists considered both as two sides of one coin renewed and recently interwoven by television. As they obviously could not totally suppress their basic skepticism toward the new medium, they explicitly admitted that the "vigorous development of television" might result in a "televisual 'omnivorousness' (televizionnaia 'vsejadnost')." However, the sociologists also identified the remedy, namely education. In their view, it was the impact of education that basically distinguished Soviet television and its societal embedding from media consumption in capitalist societies. The considerably increased level of education would prevent Soviet people from spending their time thoughtlessly in front of the screen. Thus, the sociologists argued that the rise of television would not lead to intellectual impoverishment of everyday cultural life, but in contrast contribute to its diversity.[70]

Hence, the cultural assessment and social classification of television were inconsistent and ambivalent. On the one hand, the scientists claimed to evaluate their collected data in a neutral way. On the other hand, they did not hide that reading, visits to cinema, concerts, and theater plays were in the vanguard of the artistic-cultural hierarchy, whereas television figured as the preferred medium of the less educated.[71] In this way, sociologists considered television consumption as part of the "rational cultural leisure activities," but they reduced it to the question of quantitative use. Despite their derogatory attitude, they did not suggest what proportion of time watching television should account for. They chose the easy way and just considered the surveyed watching periods as ideologically compliant.[72]

Deliberately or not, Gordon, Klopov, and Onikov put forward lines of argumentation closely resembling those of Western observers and television lobbyists. The underlying thesis, that television was not the preferred medium of educated people, appeared to be consensus on both sides of the Iron Curtain. Just like Western lobbyists, the Soviet sociologists supposed that television had an integrating and harmonizing effect on society. Hence, the sociologists focused on the society and its collectives rather than on the individuals sitting in front of the screen. This heuristic approach, at least, ignored the fact that pursuant to the increasing supply of television sets fewer and fewer people watched in groups and viewing practices gradually individualized.[73]

The issue of how television consumption influenced the relation between the personality development and societal leisure culture belonged to the basic problems Soviet sociologists had to explain. The doyen of Soviet opinion and time budget research, Boris Grushin, also considered education as the decisive factor of meaningful cultural leisure activities.[74] Grushin regarded the "antagonism of masses and culture" as the central philosophical and sociological problem. Comparing "the broad masses" with the cultural elite—a group into which he subsumed intellectuals, artists, and top athletes—he found the elite to be very small. Grushin's analysis adhered to a clear hierarchy of leisure activities, differing levels of which complied with social groups that were relatively easy to distinguish. Grushin could not exactly determine whether the level of education rose, but overall he assessed it as insufficient. He argued that large sections of the population were incapable of acquiring "the treasures of culture." That meant that "for the time being, they are even offside with regard to such important activities as reading newspapers, journals and books, visiting museums, theaters and concerts."[75]

The fact that Grushin did not mention television in this context was probably due to the officially required critical attitude toward the new medium. He quoted German playwright Bert Brecht to evoke the urgent task of sociocultural homogenization: From "a narrow circle of experts" should emerge a great one. In contrast to academics, people who had not finished the intermediate education mainly enjoyed cinema, open-air concerts, and television.[76]

In Grushin's view, television threatened to prevent the envisaged sociocultural homogenization as soon as its consumption exceeded "the appropriate measure." As television sets quickly spread all over the country and symbolized technical progress, it became more and more difficult for the regime to define the limits of "correct" consumption. Hence, it does not come as a surprise that sociologists like Grushin warned about excessive television consumption. As early as in 1970, he noticed that in many families the phenomenon of what he called "visophilia" had become rampant. For an increasing number of people, television already preoccupied their leisure time. Television had, for this reason, turned from an important source of education, entertainment, and information into its opposite. The medium now led to "the mental and physical atrophy of man. This is not only explained by the fact that television is very time-consuming, which precludes other activities. It is even worse: It forms a type of human being who may only passively 'consume' culture."[77]

Passivity was diametrically opposed to the idea of meaningful, self-improving leisure activities. The criticism always focused on the media use. Sociologists and pedagogues discussed these issues not only regarding adults, but also to children, to whose education the experts paid special attention. As early as 1962, the pedagogue Ol'ga Bogdanova stated that children had become "constant television viewers." The parents, therefore, were absolutely obliged to watch the limits of children's television consumption. They had to teach them reasonable media use to enable children to benefit from television as source of education, information, and entertainment. In no case should children be allowed to watch every program. According to Bogdanova, the result of excessive television consumption was that children wasted their leisure time and quickly got tired. In principle, the possession of a television set posed the risk that children would do their homework only superficially in order to have more time to spend in front of the screen. Parents living with their children in a small one-room apartment were called to limit their own television consumption.[78]

However, already at this early juncture some of Bogdanova's colleagues underlined the value of the new medium for educational purposes at home and in school. Looking back, earlier-mentioned sociologist Boris M. Firsov explained that

> television should get into school, should not only reach the child at home, but also at school. And it should develop synchronously with the education process in school. It means, (there were) two children's televisions. More precise, (there were) even three, because from the moment the child goes to the shelf and chooses a book on its own . . . from this moment it has become the manager of its own television time.[79]

Apart from live broadcasts of theater plays and concerts, the observers considered television, so to speak, as a displacement or prolongation of cinema, as it could broadcast movies and documentary features completely independent of the spatial arrangement of cinema auditoriums.[80] The television producers quickly realized this specific potential of their medium for educational purposes in schools, as well as adult education. Therefore, Central Television pursued two strategies. It more strongly promoted its own Scientific-Methodological Department and took initiatives to engage in debates with sociologists, teachers, and pedagogues about the new medium from the late 1950s onward. The Central Television staff was very interested in positioning itself as an appropriate medium in the growing field of education and enlightenment. In contrast to sociologists like Boris Grushin, who remained skeptical concerning television's homogenizing capacity, Central Television's NMO optimistically praised the medium as serving "the interests of the whole society." NMO underlined that radio and television supported more than any other medium the convergence of social groups with regard to their "degree of knowledge, level of cultural development and public conscience, political and labor activities."[81] The fact, however, that until the late 1960s *Gostelera-dio* still mentioned television along with radio, witnessed that television was still struggling for its place in the hierarchy of media and culture. To improve its standing, Central Television launched a third channel explicitly dedicated to educational programs in March 1965. Firsov remarked about its founding that this was a key part of the differentiation of the programming according to the diversifying audience.[82]

Pravda prepared this development propagandistically from the early 1960s. In this context, the renowned writer and publicist Oleg Pisarzhevskii published a long article in 1964 with which he conceptualized television as excellent facilitator of the work of adult evening classes (*narodnye univer-sitety*). In his view, television and cinema should closely cooperate in the popular scientific field and thus bring the *narodnye universitety* to the broad TV audience.[83]

In addition to the wide field of adult education, not only Central television's executives but also the local Leningrad TV station focused on children and youths as specific viewers at a rather early stage. According to Boris Firsov, Leningrad Television got in contact with him as early as 1956, because he was secretary of the regional Leningrad Komsomol committee at this juncture. The TV station was looking for a person to set up the youth program. Looking back, Firsov interpreted this as a first step in journalist professionalization and—just like the establishment of Channel III—the next step in recognizing the heterogeneity of the audience.[84]

Central Television only somewhat later created a Youth Programming Desk (*Glavnaia redaktsiia programm dlia molodezhi*) in 1958 and started to set up particular programs for the younger age groups.[85] During the 1960s, the broadcasts started to range from cartoons, animal films, and concerts to documentaries aiming to impart knowledge. This proved that television increasingly regarded the promotion of its educational mission as adequate strategy to further improve its standing. From the television's point of view, it was logical to seek close collaboration with the experts. To involve them in the debates and to draw on their expertise would obviously strengthen television's scientific basis. Initially, the official stance toward children's television consumption remained, nevertheless, ambivalent. The experts' assessments tried to balance the arguments. Thus, they evaluated the pros and cons of the medium, for example, with regard to the sphere and the social environment in which young people consumed television programs. In contrast to the dangers that watching television was supposed to entail in the private sphere during the 1960s, the experts basically attributed a high value to television consumption outside the families and the private sphere. In particular, schools were tasked with teaching children reasonable television consumption by integrating programs for teaching purposes. As early as in 1959, the Soviet Ministry of Education launched a pilot project on the use of television in classes. In the initial phase, twenty-nine Moscow schools participated in the experiment and regularly presented television programs in literature classes. Together with creative workers of Central Television, teachers started to use programs to support pupils' creative works, such as essays, written accounts, and reviews. In Odessa, teachers and television workers arranged specific television programs "for the class" on different issues. The pedagogues highly appreciated the first results. As early as 1966, the Soviet Academy of Pedagogy organized a conference on the use of television in classes. The scientists predicted a bright future for television as educational medium; however, they emphasized that close cooperation with television workers, physicians, psychologists, and teachers would be required.[86] The cooperation of pedagogical experts and Central Television was based on the optimistic theory that the "'effect of presence' is amazing. Especially children sense the power of television—as a matter of fact, their belief in the electronic picture is unlimited."[87] Although the experts supposed a potentially high impact of television on the education and personality development of children, they warned, at the same time, that television could not replace the teacher and should only be applied to ensure improved teaching results. Therefore, teacher should be trained in the use of the medium in class.[88]

With regard to the actual media use of children and young people at home, pedagogues also increasingly appreciated television's potential for aesthetic education to the extent it improved and expanded the children's and youth's programs. In contrast to the very critical but characteristic stance of earlier-mentioned pedagogue Ol'ga Bogdanova, Voronezh pedagogue Zinaida Keilina represented the changing attitudes of this discipline toward the end of the 1960s. Keilina more carefully considered the pros and cons of television's role in domestic education and schools. She declined the idea that television merely condemned viewers to passiveness, as it inspired the audience to reflect on broadcasts. She highly appreciated programs and shows like the youth comedy game show *KVN* (*Klub veselykh i nakhodchivykh*/Club of the Merry and Resourceful), the variety show and holiday program *Goluboi ogonek* (Little Blue Flame), and *Gorizont* (Horizon) for schools.

Existing research has thoroughly explored the youth game show *KVN* that Central Television aired from 1961 to 1972. The findings make clear why it was highly consistent with the television broadcasting policy, as well as with societal interests during the 1960s that Keilina precisely instanced the genre of game shows for the use of television in schools. A youth game show like the extremely popular *KVN* represented best the way television producers of the 1960s sought to engage young viewers in negotiations of the modern consumerist Soviet lifestyle, as well as to promote and control them at the same time. It might come as a surprise that television opted for a pop-cultural format that was most likely suspected of propagating lowbrow capitalist consumerism. Television positioned itself in the debates about high- and lowbrow culture by establishing entertaining and, at once, educating shows. Onc key feature was that they deliberately showcased the faces of "ordinary" people on the screen. In case of *KVN* and *Goluboi Ogonek*, however, these "ordinary" representatives of the Soviet society mostly belonged to the growing-up intellectual elite or to the intelligentsia. Nevertheless, the shows aimed to provide societal role models. The participants of *KVN* were set off from schools, colleges, and labor collectives and drew on team building. They competed, for example, in sports like soccer and in joking, calculating, puzzling, singing, or playing chess. The show drew on humor and satire; it was closely connected to amateur theaters and became—apart from schools—an activity in youth camps and factory clubs. Most of its popularity derived from the explicit amateur status of the teams. The jury, which was first even seated on stage and communicated with the players, elected the winners and rewarded them with prizes. As the protagonists of these game shows sought to entertain the audience and thereby at least implicitly negotiated questions on competition, fair play, and the rules on the basis of which the public performances were judged, television made a very specific statement in the negotiations of cultural boundaries: The medium established a wide-ranging site to publicly debate central societal questions such as fairness, obeying rules, corruption, authority, and cultural tastes in terms of entertaining contributions and involving the audience.[89]

The crucial point was that the audience not seldom heatedly criticized the judgments of the *KVN* jury in viewer letters. The jury was allegedly modeled on the CP's Politburo as to what made any decision highly delicate with regard to the decision process and the approval of the audience. As the show missed any transparent measures for choosing the winners of the regular contests of satire and telling jokes or of amateur theater plays, viewers keenly debated the jury's decisions.[90] What topics and characters were presented on the screen could be seen as political comments on the shifting boundaries of high and popular culture. Television provided a new competitive format that directly addressed children and youths and strongly differed from the traditional teaching of literature and drama. Keilina found it very valuable for school education not least because of its competitive and engaging elements. That the intelligentsia protagonists of *KVN* were also able to entertain the audience said much about the increasing role of entertainment in the cultural hierarchy.

However, Keilina was a bit ambivalent, as she also pointed to the risk television entailed, if the parents did not guide their children carefully how to use the medium at home. Thus, children had to learn to limit their television consumption, and parents should not exploit television to keep their children at home and prevent them in the worst case from societal commitment. Keilina further criticized that some parents gave watching television priority to visiting cinema and theater. She established that theater plays, for example, convey "a flow of autonomous emotions and thoughts" basically differing from television. Television, therefore, "should not take the place of direct perception of reality."[91]

How television impacted the perception of reality and the question as to what extent viewers and especially adolescents should therefore refrain from watching remained contested among experts. In the mid-1970s, Boris T. Likhachev, renowned director of the Scientific-Research Institute of Art Education of the Academy of Pedagogy, found it "one of today's main tasks is to train pupils in the need to 'switch on or off' the TV set."[92]

Although one has to take into account that Likhachev participated on a joint conference of *Gosteleradio* and the Academy of Pedagogy and therefore very likely represented opinions tending to be in favor of television, his statement suggested that television's reputation in the cultural hierarchy had been considerably enhanced. As soon as children and the young had learned to consume television adequately, the mental and physical dangers of excessive television consumption would have been averted. In this way, Likhachev pushed the debate by urging to focus much more on the audience's intellectual and emotional needs, to align the programs to the viewer demands and the scientific exploration of reception.[93]

He considered the analysis of television reception and the scientific alignment of the programs highly necessary because he basically recognized "the immeasurable role of television for the aesthetic education of youth, the propaganda of contemporary art, art of the past. Television generously introduces its viewers to theater, cinema, painting, music." Likhachev pointed to the broad feedback these broadcasts normally triggered, as it showed that "television is in the field of the propaganda of arts the university of culture for many millions."[94] Moreover, according to Likhachev the viewer mail constituted a rich source to research the audience reception—a research he considered highly necessary in particular with regard to the young viewers.[95] This view perfectly matched the affirmative attitude of Central Television's producers toward the high significance of audience letters after Sergei Lapin's appointment as *Gosteleradio* chair in April 1970.[96]

Basically, the pedagogues' considerations bolstered the position of television with regard to its general educational mission and the regime's claim to provide appropriate programs for the aesthetic education of children and youths. The researchers underscored that the programs must definitely attain a certain ideological-artistic level (*ideino-khudozhestvennaia uroven'*). However, the criteria to measure this level and to determine concrete contents remained vague.[97]

This short outline of the pedagogical-sociological discourse from the late 1950s to the mid-1970s gives an insight into elite representations of Soviet television. It reveals that television challenged the established performing arts by its "*massovost*." This characteristic urged the experts and the television staff to conceptualize who actually sat in front of the screen and how to satisfy his preferences. The television responded by promoting the audience research and drawing up a model audience to which the programming should be oriented.[98]

Yet, "*massovost*" had another, previously unknown dimension in the context of the Soviet media landscape. The social, geographical, and cultural scope of its live broadcasts from theaters, operas, and concert halls changed the traditional concept of high culture and, in particular, interlinked it with the audience demand for more entertaining programs. According to Central Television's letter desk, as well as to its audience research, respondents of all social groups called "entertainment" and "rest" television's prime functions.[99]

The challenge the demand for entertainment constituted surely had a strong retrospective effect on television as soon as Soviet television producers realized the close relationship cultural production and cultural consumption. Faced with this kind of audience demand, the television workers—like Soviet radio staffers—became more sensitized for the actual popularity of the television program. Further, the spread of foreign radio broadcasting increased the television workers' awareness of viewers' agency.[100] After Khrushchev's removal from office, Central television programming policies professionalized. The exposure of televised performing arts to the mass audience tastes brought about the diversification of TV genres. This development was not least due to the fact that the basically optimistic assumptions of television's educational capacities crumbled when faced with "real" audience feedback. Already the adjustment of the television schedule in the late 1960s limited too-open propaganda programs or banished them to the margins of the schedule, placing more entertainment programs at prime time and weekends instead.[101] Thus, the overall response of television to cultural criticism and to trump the established performing arts relied on the one hand on the strategy to televise them. On the other, television professed its entertaining capacities, diversified its genres, and made popular culture encompassing parts of traditional high culture. This allowed to maintain its educational mission until perestroika without contradicting the audience demand for entertainment.

Critical voices among the urban intelligentsia and the better educated were still to be heard during the 1970s and 1980s, in particular during perestroika. They became part of popular and youth culture, as we could trace them in the song *Zhertva televideniia* (The Victims of Television) by singer-songwriter Vladimir Vysotsky, in *Pesnia o televidenii* (Song about television) by Aleksandr Gradskii, or in Rashid Nugmanov's experimental perestroika movie

Igla (The Needle) starring rock poet Viktor Tsoi. In 1990 Tsoi and his group *Kino* released a song with the meaningful title *Ia vykliuchaiu televizor, ia pishu tebe pis'mo* (I switch off the television, I write a letter to you).

Although these artistic contributions emerged from different periods and political contexts—Vysotsky published his song as early as in 1972, and Gradskii's and Nugmanov's pieces are from 1988—they had much in common. Their razor-sharp satires were directed against television's social, political, and cultural role in Soviet society: They presented television as a powerful tool of propaganda and brainwashing in order to politically control the passive and partly even stunned audience. Nugmanov paralleled television's alleged addictive and lulling quality with the drug abuse of one of the movie's main characters. The "needle" titling the movie is obviously a metaphor for *Ostankino*, the mighty Moscow television tower that was said to be "a thorn in the flesh" of the intelligentsia and the dissidents.[102]

Vysotsky's song of 1972 is perhaps the more intriguing one with regard to the contemporary context. In his typical sarcastic manner, he sang about the "Victims of Television" who preferred to stay at home and watch TV instead of accompanying the girlfriend to the cinema. Vysotsky attributed television a magical appeal, as the TV set seemed to be like a "door to the whole world," getting Angela Davis, Richard Nixon, and Georges Pompidou into the living room. Television thus seemed to those who would like to spend the spare time away from home just to be "a stupid box for an idiot (*glupyi iashchik dlia idiota*)."[103]

LOOKING BACK ON WATCHING TELEVISION AND THE CULTURE OF TELEVISION ENTERTAINMENT REVISITED

Looking back on their Soviet life, how do those people who sat in front of the "stupid box for an idiot" assess its entertaining and educational qualities? How do they judge the alleged lulling impact? To what extent might they contemporarily have shared the high claims toward television as a forum for high culture products or did they rather adhere to the critical attitudes of the academic elites? How did "ordinary" Soviet TV viewers perceive the shifting boundaries of popular and high culture and the role television played in this process?

Let us look at the narratives of the interviewees to gain a few insights in this complex field of contemporary assessments and retrospective comparisons. As mentioned earlier, listening to the respondents one quickly realizes that watching television did not only change leisure practices and considerably impact on the private life. The way Soviet people were now able to negotiate about issues on the private and public life via television consumption also set standards for those who grew up in Soviet times how to assess media contents with regard to its entertaining, artistic, aesthetic, and informative value. At the beginning of this chapter, I quoted seventy-two-year-old Yuri from St. Petersburg comparing aesthetic contents of Soviet and post-Soviet programs. His comments are quite typical for those who retrospectively present themselves as critical Soviet media users. I pick out this media-user type, because these respondents explicitly spoke about sociocultural changes television brought about. Nevertheless, also, those less critical and educated among the respondents appreciated the broadcasts from Moscow and Leningrad theaters and concert halls. Many remembered, like the former engineer Antonina, that television in her youth in the 1950s and 1960s mainly broadcast movies, which was why her family did not use it as information source: "Really, television was none (i.e., information source)."[104] Antonina, however, appreciated watching Soviet movies on television, although she and her parents were also passionate readers. A combination of media practices that many respondents favored.

Yuri was a rather particular interlocutor even compared to others who perceived themselves as critical users, as he turned out to be especially eloquent and interested in arts and culture. Yuri enjoyed the cultural programs Soviet television offered, but he also used it as source of information—as he still does today. Generally, he was very interested in American jazz music and joined a group of youths who imitated the dress style of young Americans in the late 1950s. He loved to listen to *Voice of America* from the mid-1950s onward. As his stepmother was friends with the director of the Leningrad Pushkinskii Theater, he was able to attend plays "until I was ready to drop."[105]

During the conversation, he referred to the very popular literary scholar, writer, and entertainer Iraklii Andronikov in order to describe the importance of the Leningrad Philharmonic Orchestra on the one hand and the Moscow theaters on the other. Not least because of his numerous television appearances, Andronikov was widely recognized as a renowned art expert and critic. He started his first Central Television show "*Iraklii Andronikov rasskazyvaet* (Iraklii Andronikov is talking)" in 1954, in which he reviewed literary works. Looking back, this kind of literary and artistic programs let Soviet television appear to Yuri

> surely not like ours (i.e., today's), it was not independent, but purer. I do not really talk about language. This could also be bad, as they, let's say, almost shot the poor, unfortunate announcers for a slip of the tongue, in cases they mispronounced some words. In more liberal times they simply lost their job, but there were such incidents. But the Russian language was flawless. What we

> hear sometimes today is terrible. . . . Of course, mass media have changed. Not to grumble and to be discontent—what is possible today, that was not possible in the Soviet Union, of course, including bad things. This is in the first place. In the second, certainly, let's say, regarding the television as the most far-reaching mass media, I regret the loss of certain substance (. . .), if you wish a fine aesthetic content. They aired artistic productions that, in contrast to today, in most cases also offered further information and food for the mind. . . . And, indeed, all, who prefer more substance and thoughtful things, could no longer count on television. One probably has to get used to it. Because when I travelled to the West for the first time, I turned to television there. I was overwhelmed then by the 20 channels, whereas we had only 3 or 4, I don't remember. But I then already saw that you also could actually watch nothing there. Although I did not understand the language, I understood that it was some crock of shit. And we very strongly and very well adopt this crock of shit with the Western style of mass media. . . . The mass media have changed. Some information that could be broadcast today was taboo in the Soviet Union.

Against the background of the current television program, Yuri retrospectively characterized Soviet television not only as pop-cultural entertainment medium, but also as reinforcing facilitator of the traditional performing arts. According to Yuri's description, only television enabled the traditional genres to gain a widespread impact all over the country. In this context, he conceded that Soviet television offered programs of aesthetical value:

> Television suddenly killed me because of its mere existence and the daily routine of life, of television life. They broadcast Hamlet, just like that. And suddenly we were caught by Hamlet. This struck me somehow. Then I got somehow annoyed at television, that you should not sort of combine [things]. No, of course, there were (programs of aesthetic value). Some television events were aesthetic. I don't remember. But some movies, some plays were aired. There was a time when they taped plays directly in the theater and broadcast them on television. They taped operas and broadcast them directly on television. There is nothing similar now. The thing is that this probably was the breakthrough. Because somewhere, where generally no theater existed, this meant that people suddenly also had the opportunity through television to watch and listen to something.

On the one hand, Yuri's comments could be understood as part of a cultural critical comparison of Soviet television with the current media landscape. Diachronically compared in the moment of the interview in 2010, Soviet television did much worse in Yuri's view because of the censored provision of information. On the other hand, Yuri was not the only respondent who observed a certain deterioration in the quality and accuracy of Russian language. Sixty-three-year-old Zina from St. Petersburg condensed her dismay about the changed Russian language spoken on television today in contrast to Soviet times into a few words: "It's very awful, how they speak and what kind of things they say. The language was much more accurate then [in Soviet times]."[106]

From this perspective, Soviet television appears to have preserved the quality not only of spoken Russian but also of aesthetics and entertainment. Yuri further explained that he now and then assessed the program according to its range of information and the level of entertaining. In Soviet times, he is said to have differentiated like many other Soviet media users between "propagandistic and less ideologized programs." He illustrated this dichotomous practice with an anecdote about the evening news show "*Vremia*" (Time).[107] With regard to the early days of *Vremia*, Yuri recalled that Leningrad regional channel always aired symphony concerts. As according to Yuri, "most people did not want to watch *Vremia*, because nobody wanted to kill time with propaganda, the evening news show educated the people to appreciate fine music."

Higher-educated Masha, who was born in Leningrad in 1946, spoke about the traditional arts on television in a similar detailed way like Yuri. At the time of the interview Masha was sixty-four years old and presented herself as "*chelovek knizhnyi*" (book person). Her mother spoke five languages and worked in the library of the Academy of Sciences (BAN). She regularly brought books home for reading. Masha also reported to have read the newspapers *Pravda* and *Leningradskaia Pravda* (Leningrad Truth) since her early age. Later she also read the journals *Novyi Mir* (New World) and *Iunost'* (Youth). Nevertheless, television attracted her curiosity as early as the mid-1950s. As her parents initially could not afford to buy a TV set, Masha as a little schoolgirl watched children's programs together with a friend every Sunday.[108]

At the end of the 1950s, her parents were lucky to win a lottery prize and bought with the money the TV set *KVN*. Masha's father was a passionate soccer fan who strictly watched over which program would be switched on. What further prohibited her from watching television as she would have liked to was the fact that the *KVN* was a very unreliable TV set. It required a great deal of energy to permanently tinker with the TV set or to repair it. Still, looking back she made clear that watching television was a pleasure:

> Nevertheless, it was intriguing. Every program we watched on TV. How many fine actors could you see in reality? You could count them on the fingers of one hand. But all (were) on the screen. Originally, the programs of Leningrad television—at that time, when Moscow programs could not be received—, when we had only the Leningrad channel, were very fine. The Leningrad artists were of the highest quality. There were plays that were broadcast on television. There

were programs . . . that do not exist any longer. No talk shows, in which people only cry, in order to cause laughter and more arguments so that somebody knocks some other straight in his face or pours fruit juice over his head. Instead, there were people who . . . indeed discussed. Yes, there (were) no phone calls to the television studio. But these talks were intriguing, about very different issues. Such programs have ceased to exist. Of course, not all issues, that we would have needed, were on air, that's self-evident. But that, what existed, was fine. Of course, I don't think of propaganda now.

Masha's appreciation was not least due to the live broadcasts from the Leningrad theaters that created the illusion of immediacy and authenticity. At the samc time, she devalued present-day television with new formats like talk shows compared to the round tables Soviet television had presented. In her view, the televisual performance and even the substance of information are inferior today. During the interview, she came back to television-specific capacity as disseminator of the outstanding Leningrad theaters. In her perception, television did not change the content of the original performance, but rather had the effect of a magnified theater hall of which any living room had become part:

There were very good broadcasts, great programs of Leningrad Television. . . . For example, such a program: Sometime the Leningrad station broadcast Iurskii's "*Fiesta*" on TV. . . . The actors were of such a high quality! By the way, I watched almost all movies on TV. I hardly ever went to the movies. . . . It was not because I would have regret the money, the tickets were cheap. . . . You need to go to buy the tickets, to go in advance in order to buy convenient, good tickets. Then you find yourself in the cinema, where it is either stuffy or hard to listen or you sit next to people who stink or crack pumpkin seeds or chat—that's awful! . . . I realized that these were different people, different worlds, that I never will go to the movie. So that (I watched) everything on television.

We see, at least, two more intriguing points characterizing Masha's construction of cultural boundaries and of the way television changed the practices of cultural consumption. She basically appreciated cinematic arts, but more with regard to the issues these movies covered than with regard to the viewing experience in the dark cinema with its big screen. Elsewhere during the interview, she explained that movies need to tell about a serious problem, a drama. She loved existentialist movies, especially French ones, but also Soviet movies of the thaw period. As cinemas had a democratizing effect and were easily available for all social groups, who obviously watched the same movies, Masha preferred to watch at home. One of television's obvious competitive advantages was that it was allowed to broadcast movies rather soon after their cinema release. The new medium, thus, enabled people like

Masha to socially distance themselves and to avoid getting in contact with people who did not adhere to her concrete ideas about how movies should be watched. Apart from this privatizing and somehow separating effect of television, Masha again highlighted television's propagating effect on the established performing arts, as well as the impact of the regional television stations. In referring to the renowned actor and director Sergei Iu. Iurskii's "Fiesta," she was reminded of a performance that at this juncture could possibly only be realized in a theater scene like Leningrad and in the context of a local station. At least, this was Iurskii's own retrospective interpretation of his work. In 1969, he produced Ernest Hemingway's famous novel at the *Bol'shoi Dramaticheskii Teatr imena M. Gor'kogo*.[109] The play was actually never staged at the theater, because of internal quarrels between Iurskii and the artistic director Georgii A. Tovstonogov. Iurskii then turned to the local Leningrad Television station that even after Firsov's dismissal had a liberal reputation and took a rather experimental approach. According to Petersburg historian and journalist Lev Ia. Lur'e and Sergei Iurskii, Firsov was still involved in staging "Fiesta" as a television play in 1971. True or not, looking back in television documentary about Iurskii, the contemporary protagonists of the "Fiesta" story agree that Leningrad Television had a greater freedom of action than theaters or journals during the 1960s and early 1970s.[110] In particular in Leningrad, it was a prolific coincidence that television directors had more options to choose their topics, even from foreign literature, and could work with the best actors of the renowned local theaters. This period is remembered as the "*zolotoi vek*" (golden age) of the cooperation of highbrow culture and television, as director and actor Aleksandr A. Belinskii distinctively put it.[111]

Undoubtedly, this unanimous narrative of the contemporary protagonists tells much about their self-representation in the lost Soviet and current media and artistic landscape. As retrospective assessments, they suggest a clear decay of high demands toward television and a critical attitude about the prevalence of popular productions on TV after the "golden age." The contemporary enthusiasm about the new medium becomes blatantly obvious, as television opened new opportunities for the performing arts. At this point, the remembrances and assessments of the former "ordinary" TV viewers do not basically differ from the former leading protagonists. They expose post-Soviet television as artistically inferior and intellectually poorer compared to Soviet TV of the 1960s and 1970s. The contemporary appreciation of the cultural and artistic programs was television viewers' stance in the negotiations on the normalization of daily life in the Khrushchev and early Brezhnev years. Thus, these cultural events of live broadcasts and artistic movies became common places in people's memories. This surely was an interplay, as the positive retrospective assessments of the respondents to the quality of Soviet television reveal that the early program policy of the 1950s and 1960s successfully embedded the new medium into Soviet society. Broadcasting the established performing arts attracted educated groups and deconstructed their prejudices toward television. Watching television quickly became a broadly accepted pleasure-giving leisure practice and enforced the normalization of private life and probably the emotional bonding to central values of the sociopolitical order negotiated in cultural productions.[112] At least, television's performances in the field of culture and arts seemed to have even prohibited essentially dismissive emotions among the majority of Soviet viewers. The negotiations between audience and television continued throughout the 1970s. They brought about an officially and societally sanctioned popular culture and improved television's position in the cultural hierarchy. Television, thus, became the most wide-ranging intermediary space in which people negotiated interpretations, meanings, and—according to Roger Chartier—ultimate, power relations. However, the hegemonic interpretation frames in authoritarian regimes were much tighter than in liberal societies.[113] This was true for both sides of television—in front of the screen and behind the cameras. From the point of view of Leningrad Television professionals, Boris Firsov strongly highlighted these conditions of censorship, nevertheless limiting their work.[114]

NOTES

1. The interview with Yuri was conducted by Elena Bogdanova in St. Petersburg in October 2010. Yuri was born in Leningrad in 1938 and graduated from university.

2. I refer to Elihu Katz's meanwhile famous question, "what do people do with media?." See Katz, "Mass Communication Research," 2.

3. I take up here the considerations of David Crowley and Susan E. Reid, "Introduction: Pleasures in Socialism?" in *Pleasures in Socialism. Leisure and Luxury in the Eastern Bloc*, ed. David Crowley and Susan E. Reid (Evanston, IL: Northwestern University Press, 2010), 4–8; Susan E. Reid, "Communist Comfort: Socialist Modernism and the Making of Cosy Homes in the Khrushchev Era, *Gender & History* 21, 3 (2009): 469.

4. Communication scientist Elisabeth Klaus proposes not to contrast the spheres of "entertainment" and "information," but to view them as entangled elements of journalism and media products. See Klaus, "Der Gegensatz von Information ist Desinformation, der Gegensatz von Unterhaltung ist Langeweile," 51–64.

5. See the inspiring piece of Evans, "The Soviet Way of Life as a Way of Feeling," 543–69, and also chapters 1 and 2 of this book.

6. As mentioned earlier, I refer here to French cultural historian Roger Chartier. He objects the research practice that attributes certain characteristics of popular or

highbrow culture to certain cultural products and then deduces social classifications to the people consuming these products. He also rejects the idea that the social background of media consumers predetermines their interpretations or the classification of the respective cultural product. According to Chartier, the creation of sociocultural power relations is a much more fluid process based on consumption practices of texts and media products. See Chartier, *Forms and Meanings*, 83–97; idem., *Lesewelten. Buch und Lektüre in der frühen Neuzeit*, 7–24.

7. Concerning Russia and the Soviet Union, see the ground-breaking study of Stites, *Russian Popular Culture*.

8. Roth-Ey, *Moscow*, 192–280; Evans, *Between Truth*, 21–46.

9. Evans, *Between Truth*, 23.

10. In West Germany, it was the so-called *Frankfurter Schule* (Frankfurt School) that decidedly depreciated the new medium and biased West German debates. See Hickethier, *Geschichte des deutschen Fernsehens*, 178–180.

11. Birgit Menzel and Ulrich Schmid, "Der Osten im Westen. Importe der Populärkultur," *Osteuropa* 5 (2007): 3–22.

12. GARF, f. 6903, op. 48 ch. 1, d. 21, 1971, l. 7.

13. GARF, f. 6903, op. 1, d. 869, 1965, ll. 62–63.

14. Roth-Ey, *Moscow*, 201, 223–280; Simon Huxtable, "The Problem of Personality on the Soviet Screen, 1950s–1960s," *View: Journal of European Television History and Culture* 3, 5 (2014): 119–130; here in this book, chapter 4.

15. GARF, f. 6903, op. 1, d. 542, 1958, ll. 13–14, 50.

16. GARF, f. 6903, op. 1, d. 542, 1958, l. 13.

17. GARF, f. 6903, op. 1, d. 499, 1956, ll. 90–91; ibid., d. 130, 1959: Protokol i stenogrammy konferentsii radioslushatelei i telezritelei g. Permi, Sverdlovska, l. 4; ibid., d. 667, 1960, l. 34.

18. Quoted from Hickethier, *Geschichte*, 291.

19. All quotes are from Igor Il'inskii, "Razmyshleniia u televizora," *Literaturnaia gazeta*, May 12, 1956, 1.

20. GARF, f. 6903, op. 1, d. 542, 1958, l. 14.

21. Il'inskii, "Razmyshleniia."

22. Levine, *Main Street*, 158.

23. Il'inskii, "Razmyshleniia."

24. Despite its complaints, television was broadcasting considerably high amounts of movies as early as in the mid-1950s. In 1956, it broadcast more than 200 movies. In 1958, it aired 174 new artistic movies and 346 movies released in the years before. The number constantly increased, as the number of artistic movies accounted for 186 including 88 newly released in the first half of 1959. See GARF, f. 6903, op. 31, d. 2, 1956, l. 3; ibid., op. 1, d. 578, 1959, ll. 119–120.

25. GARF, f. 6903, op. 1, d. 724, 1961, ll. 22 (quote)–22ob. According to Zdobnov, Czech TV aired movies only after six months, and US TV did so "not at all apart from very old ones."

26. The interview with Yuri Vasil'evich was conducted by Elena Zhidkova in July 2010 in Samara region. Yuri was born in Kuibyshev (Samara) in 1951, lived for years in a small town, and graduated from university.

27. GARF, f. 6903, op. 31, d. 2, 1956, ll. 4, 21 (quote).

28. The interview with Masha was conducted by Elena Bogdanova in St. Petersburg in September 2010. She was born in Leningrad in 1946 and graduated from university.

29. The interview with Liudmila was conducted by Elena Bogdanova in St. Petersburg in November 2010. She was born in Penza region in 1942 and moved to Leningrad in 1962. She finished secondary school.

30. GARF, f. 6903, op. 31, d. 2, 1956, l. 5.

31. GARF, f. 6903, op. 1, d. 499, 1956, l. 78; ibid., d. 578, 1959, l. 118; ibid., d. 724, 1961, ll. 78–83.

32. GARF, f. 6903, op. 31, d. 2, 1956, l. 9; ibid., op. 1, d. 542, 1958, ll. 49–51.

33. GARF, f. 6903, op. 31, d. 2, 1956, l. 3; ibid., op. 1, d. 578, 1959, l. 117.

34. GARF, f. 6903, op. 1, d. 542, 1958, ll. 14–16.

35. Il'inskii, "Razmyshleniia."

36. See for a thorough analysis of West German debates about the relations of traditional arts and TV, as well as of televisual developments with regard to artistic movies theater, music, and arts programs the contributions in Helmut Schanze and Bernhard Zimmermann eds., *Das Fernsehen und die Künste* (München: Wilhelm Fink Verlag, 1994).

37. Cf. Hickethier, *Geschichte*, 142–152.

38. "Sovetskoe televidenie," *Pravda*, March 13, 1960, 1.

39. GARF, f. 6903, op. 1, d. 869, 1965, ll. 26–32, 83–90.

40. GARF, f. 6903, op. 1, d. 542, 1958, l. 50.

41. The interview with Boris M. Firsov was conducted in October 2010 in St. Petersburg by Elena Bogdanova. He was born in 1929 in Saransk, and was professor of sociology and director of Leningrad TV station 1962–1966. Firsov was removed from his job because the roundtable *Literaturnyi vtornik* (Literary Tuesday) in January 1966 ran out of control and he refused to stop the live program. He then turned back the university and became one of the most influential figures in young Soviet sociology. See his autobiographical outline of his years heading Leningrad Television in Boris M. Firsov, *Raznomyslie v SSSR 1940–1960–e gody. Istoriia, teoriia i praktika* (St. Petersburg: Izdatel'stvo Evropeiskogo Universiteta v Sankt-Peterburge, Evropeiskii dom, 2008), 428–440.

42. GARF, f. 6903, op. 1, d. 667, 1960, l. 19.

43. GARF, f. 6903, op. 3, d. 147, 1960: Stenogramma vstrechi rabotnikov Tsentr. studii televideniia s telezriteliami g. Kaluga, ll. 2–5.

44. GARF, f. 6903, op. 3, d. 178, 1962: Stenogramma vstrechi rabotnikov radio i televideniia s trudjashchimisia zavodov imeni Il'icha, gidroagregatov "Kauchuk," "Elektrosvet," l. 9.

45. GARF, f. 6903, op. 3, d. 178, 1962, l. 18.

46. GARF, f. 6903, op. 3, d. 147, 1960, ll. 3, 13, 19; ebd., d. 178, 1962, l. 15.

47. GARF, f. 6903, op. 3, d. 403, 1967: Programma sotsiologicheskogo issledovaniia NMO po teme "Struktura zaprosov i trebovanii radiosl. i telezr. po printsipu intellektual'noi vertikali," ll. 30–34.

48. See also for the ambivalent search of early television producers: Kristin Roth-Ey, "Playing for Cultural Authority: Soviet TV Professionals and the Game Show in

the 1950s and 1960s," in *Pleasures in Socialism. Leisure and Luxury in the Eastern Bloc*, ed. David Crowley and Susan E. Reid (Evanston, IL: Northwestern University Press, 2010), 147–176; for an autobiographical assessment see Firsov, *Raznomyslie*, 428–433.

49. Cf. the interview with Boris Firsov, October 2010.

50. Firsov, *Raznomyslie*, 429.

51. GARF, f. 6903, op. 1, d. 869, 1965, ll. 2–11 (quote from l. 10).

52. GARF, f. 6903, op. 1, d. 869, 1965, ll. 14–25 (quote from l. 20), 102–105.

53. Mikhail Kapustin, "Priglashenie k turniru," *Pravda*, May 11, 1969, 3.

54. Viktor Shepel', "I sokratilis' rasstoianiia," *Pravda*, April 13, 1969, 3. The renowned professor of journalism at MGU, Aleksandr Iurovskii, supported Shepel''s view concerning the inadequate preparation of TV journalists. The professionalization was a "difficult and lengthy" process. The specific courses provided by Gosteleradio and the Union of journalists were by far effective. See Aleksandr Iurovskii, "Vystupaet publitsist," *Pravda*, December 19, 1969, 3.

55. Boris P. Kutyrev, *Budzhet vremeni. Voprosy izucheniia i ispol'zovaniia* (Novosibirsk: Nauka, 1977), 57; Aleksei S. Pashkov, *Obraz zhizni naseleniia krupnogo goroda. Opyt kompleksnogo sotsial'nogo issledovaniia* (Chelovek i obshchestvo, vol. 23) (Leningrad: Izdatel'stvo Leningradskogo Universiteta, 1988), 206–207.

56. David L. Hoffmann, *Stalinist Values: The Cultural Norms of Soviet Modernity, 1917–1941* (Ithaca, NY: Cornell University Press, 2003), 159–164; Stephen Hutchings, *Russian Literary Culture in the Camera Age. The Word as Image* (London and New York: Routledge, 2004), 97–140.

57. Grushin et al., *Die freie Zeit als Problem. Soziologische Untersuchungen in Bulgarien, Polen, Ungarn und der Sowjetunion*, 68–69.

58. See here chapter 2.

59. Gordon and Klopow, *Der Mensch in seiner Freizeit*, 289–308; Weinberg, *The Development of Sociology in the Soviet Union*, 111–112. Especially highlighting the political context of the reintroduction of Soviet sociology: Beliaev and Butorin, "The Institutionalization of Soviet Sociology," 418–435.

60. Gordon, Klopov and Onikov, *Cherty sotsialisticheskogo obraza zhizni*, 131.

61. *Gordon, Klopov, and* Onikov, *Cherty*, 60.

62. Vrääth Öhner, "Die Dystopie Fernsehen," in *Klassenproduktion. Fernsehen als Agentur des Sozialen*, ed. Andrea Seier and Thomas Waitz (Hamburg, Münster: Lit 2014), 101–110.

63. Brigitte Weingart, "Alles (McLuhans Fernsehen im *global village*)," in *Diskursgeschichte der Medien nach 1945: Medienkultur der 60er Jahre*, vol. 2, ed. Irmela Schneider, Torsten Hahn and Christina Bartz (Opladen, Wiesbaden: Westdeutscher Verlag, 2003), 232.

64. Gordon, Klopov and Onikov, *Cherty*, 60.

65. Kolpakov and Patrushev, *Biudzhety vremeni gorodskogo naseleniia*,137.

66. Kolpakov and Patrushev, *Biudzhety*, 148–153.

67. Gordon, Klopov and Onikov, *Cherty*, 60–63.

68. GARF, f. 6903, op. 3, d. 130, 1959, l. 2.

69. GARF, f. 6903, op. 3, d. 208, 1963: Itogovaia spravka o rezul'tatakh anketnogo oprosa radio-slushatelei i telezritelei "O roli radioveshchaniia i televideniia v kul'turnom i esteticheskom vospitanii naseleniia" i dokumenty k nei, ll. 2–3.

70. Gordon, Klopov and Onikov, *Cherty*, 63.

71. Gordon, Klopov and Onikov, *Cherty*, 127–133.

72. For the amount of time people watched television, see chapter 2 and Kolpakov and Patrushev, *Biudzhety*, 209, 212; Gordon, Klopov and Onikov, *Cherty*, 59–66, 149–150.

73. Gordon, Klopov and Onikov, *Cherty*, 65.

74. The rise of the level of education constituted a standard argument of the Soviet time budget research. See, for example, also: Vladimir I. Bolgov, *Biudzhet vremeni pri sotsializme. Teoriia i metody issledovaniia* (Moscow: Nauka, 1973), 238.

75. Grushin et al., *Zeit*, 78.

76. Grushin et al., *Zeit*, 76–77.

77. Grushin et al., *Zeit*, 68–69.

78. Olga S. Bogdanowa, *Freizeitgestaltung* (Ost-Berlin: Volk und Wissen Verlag, 1962), 36–37.

79. Cf. the interview with Boris Firsov, October 2010.

80. See for example for an early positive assessment of television's educational capacities: Adel' S. Stroeva, *Deti, kino i televidenie* (Moskva: Znanie, 1962). The Communist party supported this view and let *Pravda* publish an article of Kiev cinema dramaturge Boris Vinnitskii in the same vein, see "Ekran na puti k shkole," *Pravda*, April 1, 1969, 3.

81. GARF f. 6903, op. 3, d. 277, 1965: Itogovaia spravka o rezul'tatakh sots. issledovaniia populiarnosti radio i telev. program, provedennogo na predpr. g. Moskvy, ll. 1–2.

82. Firsov considered this to be a logical development, as "one has always thought about education in this country."

83. Oleg Pisarzhevskii, "Akademiia millionov," *Pravda*, Iiunia 7, 1964, 3.

84. Cf. the interview with Firsov, October 2010.

85. Evans, *Between Truth*, 186.

86. Michail P. Kashin, "Uchebnomu televideniiu – dorogu v shkolu," in *Televidenie v shkole. Trudy 1-go Vsesoiuznogo seminara po uchebnomu televideniiu*, ed. idem and Lev P. Pressman (Moscow: Akademiia pedagogicheskikh nauk, 1967), 2–7.

87. Lev P. Pressman, "Uchebnaia teleperedacha i trebovaniia k nei," in *Televidenie v shkole. Trudy 1-go Vsesoiuznogo seminara po uchebnomu televideniiu*, ed. Michail P. Kashin and Lev P. Pressman (Moscow: Akademiia pedagogicheskikh nauk, 1967), 47.

88. Anatolii A. Stepanov, "Psikhologicheskie problemy uchebnogo televideniia," in *Televidenie v shkole. Trudy 1-go Vsesoiuznogo seminara po uchebnomu televideniiu*, ed. Michail P. Kashin and Lev P. Pressman (Moscow: Akademiia pedagogicheskikh nauk, 1967), 107–109, 117–118.

89. On *KVN*, see Roth-Ey, *Moscow*, 253–261; Andrew Janco, "*KVN:* Live Television and Improvised Comedy in the Soviet Union, 1957–71," in *Popular Television in Authoritarian Europe*, ed. Peter Goddard (Manchester: Manchester University

Press, 2013), 124–140; Evans, *Between Truth*, chapter 6; for a contemporary presentation, see: Elena V. Gal'perina, ed. *KVN raskryvaet sekrety* (Moscow: Molodaia gvardiia, 1967).

90. Evans, *Between Truth*, 190–201.

91. Zinaida A. Keilina, *Televidenie v semeinom vospitanii (Metodicheskoe posobie k lektsii)* (Voronezh: Izdatel'stvo Kommuna, 1969), 8–14.

92. Boris T. Likhachev, "Rol' televideniia v esteticheskom vospitanii detei," in *Televidenie i deti*, ed. Gosteleradio, Tsentr nauchnogo programmirovaniia and Nauchno-issledovatel'skii institut obshchikh problem vospitaniia APN SSSR (Moscow: Iskusstvo, 1976), 8.

93. Likhachev, "Rol,'" 8.

94. Likhachev, "Rol,'" 11–12.

95. Likhachev, "Rol,'" 12–13.

96. Lapin's arrival and the changes in Central Television's programming policy considerably curtailed NMO's sociological research agenda and reemphasized viewer letters as communication channel with the audience. See chapter 4; Evans, *Between Truth*, 52–54, 76–81.

97. See for example the contributions in *Televidenie i deti*.

98. See in this book chapter 4; Evans, *Between Truth*, chapter 2.

99. See for example GARF, f. 6903, op. 10, d. 70, 1966: Analiz pisem telezritelei za period s 1964 po 1966, l. 31; Firsov, "Srednego zritelia net," 44.

100. Roth-Ey, *Moscow*, 131–145; Lovell, *Russia*, 155–158.

101. Evans, *Between Truth*, 49.

102. With regard to *Igla*, see Maria Zhukova, "Rashid Nugmanov's film *Igla* (*The Needle*): 'Televisionised' Cinema?" in *Television beyond and across the Iron Curtain*, ed. Kirsten Bönker, Julia Obertreis and Sven Grampp (Cambridge: Cambridge Scholars Publishing, 2016): 162–192.

103. See for the lyrics Fedor Razzakov, *Gibel' sovetskogo TV. Tainy televideniia: Ot Stalina do Gorbacheva, 1930–1991* (Moscow: Eksmo, 2009), 77–78.

104. The interview with Antonina was conducted in St. Petersburg by Elena Bogdanova in December 2011. Antonina was born in 1947; he graduated from a technical university and worked as engineer.

105. The theater was renamed as Alexandrinsky Theatre after the fall of the Soviet Union. For Yuri, see footnote 1 of this chapter.

106. The interview with Zina was conducted by Elena Bogdanova in St. Petersburg in March 2012. Zina was born in Leningrad in 1949 and had secondary education.

107. *Vremia* started broadcasting on January 1, 1968, on Central Television's Channel I.

108. Interview with Masha (born in 1946, St. Petersburg).

109. The novel, titled *The Sun Also Rises*, was first published in 1926. The theater today is named after its long-time director Georgii A. Tovstonogov.

110. See Lev Lur'e's TV program on Iurskii and "Fiesta" on the website of St. Petersburg Channel 5 (*Peterburg Piatyi Kanal*): http://www.5-tv.ru/video/501819/ (accessed September 22, 2015). See especially 7.00–11.00 min.

111. See http://www.5-tv.ru/video/501819/, 9.28 – 9.53 min (accessed September 22, 2015).

112. With regard to the emotional influence of entertaining shows, see Evans, "The Soviet Way of Life."

113. I refer to the concept of Hall, "Encoding/Decoding," 163–173.

114. See the interview with Boris M. Firsov, October 2010; Firsov, *Raznomyslie*, 428–440.

Chapter 4

Working with "Emotional Means"

Soviet Television's Relationship with the Audience

"Dear television workers, how grateful we are for your noble work, for your everyday concern to raise the horizon of the Soviet people," declared a male Muscovite viewer in a letter to Central Television in 1971.[1] Many other viewers were similarly enthusiastic about Soviet television's programs. Even *Vremia* (Time), Central Television's nightly news program, reached considerable viewing figures until the 1980s, in contrast to its East German counterpart *Aktuelle Kamera* (Current Camera).[2]

The viewers quickly fostered communication with the TV staff and with the Soviet regime about the medium and its programs. The change of political communication with the rise of the new medium seems obvious but at the same time hard to judge. Assuming that they engaged in an altered field of communication, this chapter examines the communicative relationship between audience and television. This refers to the way television workers imaged the audience, the way they got into contact, as well as the specificity of letters sent to television's editorial desks and the demands of viewers toward the medium. What do these viewers' letters tell us about the role television played in Soviet people's leisure habits and about their viewing practices? What do they reveal about the question of how "ordinary" citizens made use of television programs and how they assessed television content? What does the mail reveal about viewers' emotional bonding toward the medium within an officially and societally accepted popular culture and representation of late Soviet lifestyle? To what extent did television engage its audience in political communication? How might the involvement of the audience have facilitated the communicative stabilization of the Soviet regime up to 1985?

Generally, letters addressed to state and party institutions were a common phenomenon in socialist countries. Letter writing as a form of exchange with elites and political authorities had a long tradition going back to prerevolutionary times. Research has established that letters formed part of the regime's stabilizing communicative mechanisms.[3] They did so at least to some extent because letter writing normally entailed response. This conveyed the impression of a protected space of communication between the public and the private sphere.[4] After Stalin's death, the regime was searching for new ways of engaging Soviet citizens. The aim was to increase citizens' and, in particular, young citizens' commitment to the Soviet sociopolitical order.[5] This made it even more important to engage viewers in communication with the regime. The easiest way to track their opinions on TV programs, to assess their tastes, or to compare ordinary viewers' wishes with intellectual ideas about TV was to encourage them to write letters.[6]

LETTERS AND AUDIENCE RESEARCH

Indeed, Central Television's editorial offices received rising numbers of letters from the 1950s to the demise of the Soviet Union. And perhaps unsurprisingly, viewers often expressed different points of view about the medium than did scientists, artists, or party leaders. Much of this correspondence addressed the departments of information, entertainment, and youth broadcasting that were in charge of game shows and musical programming.[7] The rise of quiz and game shows and of musical request programs changed the nature of communication between viewers and television. On the one hand, television was supposed to be part of the missionary project to inculcate the "New Soviet Person" with "proper" tastes. On the other hand, after the late 1960s, the regime saw the role of television as meeting people's demands for pleasure, entertainment, and rest in their free time. This meant taking viewers' persistent—and ideologically perhaps not ideal—preferences into consideration.[8]

However, it is important not to overload the analysis of audience mail with quantitative expectations. Viewer letters should not be taken as being representative of the TV audience as such. There are no comprehensive statistics on the letter writers. Yet letters offer a variety of values and preferences well worth examining, because Soviet individuals voiced these values and preferences. Individual viewers referred to their "personal and private history,"[9] as American political scientist Ellen Mickiewicz aptly put it. Historical research has demonstrated that in fact a rather small portion of the Soviet population addressed party elites, government representatives, or the mass media.[10]

Although television also received "whistleblowing" letters or petitions expressing private interests, ordinary media consumers addressed TV's editorial departments primarily to negotiate on media issues. At first sight, many letters covered "unpolitical" themes related to entertainment, film, or serial aesthetics and to the hairstyle, clothing, or linguistic styles of female announcers. However, these were exactly the topics TV editors wanted to communicate about with the audience.

As the party, scholars, and TV professionals regarded television as a medium of communist education, the idea of audience participation was ideologically highly charged. Audience participation relied on Soviet logics of getting directly in touch with the "masses" in order to propagate ideological messages and mold the Soviet new man.[11] From the outset, TV producers also tried to use any audience feedback to research the viewers' habits, tastes, and preferences. Letters were an appropriate communication channel to underline the regime's claim that television was a public forum and accessible to everyone all over the Soviet Union. The Communist Party and therewith also *Gosteleradio* attached importance to the fact that Central Television should represent the multiethnic Soviet Union. Hence, the TV workers were demanded to select the letters they presented on the screen carefully from all over the country and not only from European Russia and the capitals.[12]

The recurring criticism of the way TV workers presented letters on the screen documented that the internal use of audience response was diverse and ambivalent. Some of the artistic staff as well as the early "enthusiasts" tended to ignore audience responses out of intellectual arrogance, as they thought their TV work was valuable in itself.[13] Being aware that letters were not representative of the audience, many others, nonetheless, considered audience responses of considerable importance.[14] This was the case when letters could be used to highlight the propagandist influence of television on the audience in order to get more financial sources.[15] Furthermore, program coordinators, editors, and announcers were interested in audience responses to grasp ephemeral "public opinions" and to improve the program. Hence, the processing of incoming mail changed toward the end of the 1950s. In 1957, Moscow Central TV opened a letter desk of its own, separate from the corresponding radio letters office. It was transferred to the corresponding department of the newly founded Central Television's main editorial office in 1960. The letter department was an independent unit among Central Television's editorial departments from 1962. In principle, the bigger regional TV stations like Samara, Irkutsk, and Rostov did the same until the mid-1960s and delegated at least one staff member to deal with the incoming mail. Many of the smaller local TV stations asked Central Television to support professionalization of the work with letters in the mid-1960s.[16]

Looking back, former television employees of regional TV stations described how much time consuming the work with letters was. Liudmila, editor of Irkutsk TV station,[17] Evelina E., television director of Rostov TV station,[18] and Valentina V.[19] and V. V.,[20] both editors of Samara TV station, all reported that apart from writing letters viewers also got in contact with local television by calling the studio. Employees had to register these telephone calls and write them down. The professionalization of local television's letter desks took, however, a while. Despite the gradual establishment of letter desks, employees of any editorial department were occupied with reading letters in some way or another. With a slight trace of nostalgic sentiment about the lost communication channel as well as sighing relief, Evelina E. remembered the high workload of coping with the mail: "Every day. A sack of letters, we poor did not know how . . . [she interrupted] they were distributed among us all. . . . Well, we all read them by ourselves, because then the service even, even the letter desk, such a thing did not exist at that time."[21] As early as in the late 1950s, some of the newly founded editorial desks like those responsible for the music and children's programs received so many letters that the employees had difficulties coping with the amounts.[22]

Looking back, the interviewed former television employees document that letters played an important role in their daily business since the early days of local TV stations. Although we have to consider that especially these respondents sought to attribute significance to their own professional activity, their narratives offer intriguing insights in the handling of viewers' letters. In principle, considering writing and calling as widespread and appropriate communication practices between television and audience, their narratives left only little room for certain ambiguities. Surprisingly, only few differentiated between letters addressing television issues in the broadest sense and complaints referring to personal or societal problems. As a communication channel with the audience and a source of programming, letters were as such of high importance, both from an ideological and a practical point of view. The interviews show that in the late 1950s television workers were not really prepared how to work with the letters and made at first a rather provisional job. Television director Evelina E. explicitly considered the immediately incoming audience response when Rostov television started broadcasting in 1958 to prove that "all were very enthusiastic that television in general started to work. Magic. Nobody asked about any rating [reiting]! In general, we didn't know this word. . . . We simply made the programs and we received letters."[23]

Samara editor Valentina V. remembered that in the beginning they at once thought of basing some programs on letters. They asked viewers to call the TV station and submit requests, interesting issues, and feedback. The response was unexpectedly overwhelming; people "called, called, called." This echo encouraged Samara TV editors to rely broadly on letters.[24]

According to the expert interviewees and the impression we get from the archival materials, the processing of the audience mail professionalized rather quickly. This was not least due to legal requirements, as all letters had to be answered within four weeks. This requirement unfolded a certain impact and kept the communication channel open. The self-representations of the expert interviewees about television's letter practices document, like the narratives of the few respondents who actually wrote letters to mass media, that the logic worked for a certain segment of Soviet society.[25]

Letters referring to concrete programs were internally forwarded to the relevant editorial office which then worked with them. In cases of requests, questions, or complaints that did not directly refer to television programs, they had to be forwarded to the responsible party, administrative or government institutions. Any letter was numbered and registered with name, address, content, date of receipt, and date of response. In cases where the requests were passed on to other institutions, the television staff verified the processing of the letters. Any local television station had to regularly report about the amount of incoming mail and letter contents to the local party committee and to *Gosteleradio*. Sometimes they took the opportunity to shoot a film of the letter author and tell the story of his or her request.[26]

Thus, not only did the Central TV's new letter department start to thoroughly analyze the incoming mail according to letter authors' content and record which editorial department and which programs were addressed; local and central letter departments included its findings in monthly and annual reports. The delivery of letters and surveys is, however, incomplete. Some mailbags are completely missing.[27]

Until the mid-1960s, Central Television tried to complement the information gathered by letters rather unsystematically with the help of viewers' roundtables in factories and *kolkhozy* (collective farms) and via small-scale surveys.[28] This variegated approach documented that the conceptualization of audience research was only randomly framed by scientific demands in television's early days. This changed in the early 1960s, with the reemergence of Soviet sociology—a reemergence closely intertwined with the first studies of mass media and public opinion. Early Soviet sociologists started with surveys on the readership of *Komsomol'skaia Pravda* in 1960, whereas Polish sociologists, for example, had first analyzed the audience of radio and television years before their Soviet colleagues.[29]

Soviet media scholars depicted direct contact and audience response as important for learning how best to adjust artistic and entertainment strategies. The method, however, turned out to be contested. By far, not all TV staff members involved in the audience research considered letters to be "the most perfect channel of communication" in order to analyze viewers' reactions.[30] Although positive audience feedback was definitely needed to legitimize the high costs of entertainment genres, news, and documentary productions to the Communist Party and the State Planning Committee (*Gosplan*), many representatives of the just reemergent Soviet sociology proved to be disdainful of letters as a scientific source. The same was true for Central Television's Scientific-Methodological Department (*Nauchno-metodicheskii otdel*: NMO). As Central Television tried to catch up with international standards, it promoted a more scientifically based approach to audience research. In this context, the NMO was upgraded in 1959 when the regime took initiatives to promote popular engagement in state institutions. From this moment, the NMO gained more money and influence. Its predecessor had already been established in 1944, but had played a rather marginal role in the work of the committee until then. Now, the NMO was to "organize and run, jointly with radio and the Central Television studio . . . meetings with radio listeners and TV viewers to acquaint them with the work and plans of all-Union radio and the Central Television studio."[31] In fact, this way of getting in contact with the audience was not new and had been taking place since 1948. Meetings with radio listeners and television viewers were supposed to gather reliable feedback and to suggest viewers to be directly involved in program changes and plans.

In the context of the restructuring of NMO, Central Television started the Public Council of Television Viewers (*Obshchestvennyi sovet telezritelei*) in January 1959. It consisted of *sovkhoz* workers, employees of factories, collective farms, and scientific institutions from Moscow and Moscow region.[32] Intended to establish an institutional framework to interlink television producers and viewers and to provide a forum for discussion and audience feedback, it was based on the idea of convening representatives of a model audience. The council formed an executive body including the deputy chair of *Gosteleradio*, the program director, and the editor of the letter department.[33] One of its tasks was to become familiar with the audience mail. It should further collect requests and ideas from their colleagues at the workplaces, as well as report there about the work of television. Another purpose was to monitor the programs and give the television producers feedback. All this was supposed to provide ground for representative recommendations to improve the television schedule from the perspective of the viewers.[34]

This concept of public councils was emulated by the regional TV stations to get in contact with the local audience. Central Television often requested local stations to found a public council.[35] Although the meetings seemed to run rather vibrantly, as the present representatives were wholeheartedly committed, a considerable number of members did not participate in the meetings that were taking place twice a month.[36] In its second year, the council approached Central Television to conceive a program to publicize its work to the broad audience.[37] The council, however, remained an intermezzo. It is not explicitly evident why the Public Council of Central Television was in the end rather soon and tacitly discontinued. Basically, its denouement could be interpreted as reducing too-active participation of citizens. Also, many television editors seemed rather reluctant to accept the council's recommendations and often critical remarks.[38] Members of Moscow Public Council criticized that there was no regular collaboration or direct exchange with the editorial departments, as no editors participated in the council meetings and no council members in the desks' briefings.[39] As participating in the council meetings indeed implied a certain workload, the decreasing participation seemed to have been further enhanced by the ignorance of television workers. Debates of the council revealed the disappointment of the continually attending members.[40]

The corresponding local bodies seemed to have functioned somewhat longer. Rostov television even staged council meetings in the mid-1960s. Samara television still aired meetings with viewers at their workplace on the screen in the 1970s.[41] The staff of Central Television audience research considered meetings with viewers then after Lapin's inauguration as head of *Gosteleradio* as especially appropriate means to explore the preferences of the viewers and to get their "solid feedback."[42]

Not surprisingly, NMO contested the significance of meeting the viewers and letters. It explicitly criticized the practice of using them as the main information source about the television audience. Its workers argued that letters and meetings offered no representative data about the viewers' program preferences and watching habits. One of the NMO's basic messages was that programs would gain quality if their producers knew their audiences better.[43] NMO closely observed theory and practice of audience research in foreign state socialist, as well as capitalist countries that had already relied on sociological surveys for a certain time.[44] One report of 1965 underlined the significance of sociological surveys, which, because of the generally reviving sociology, also gained impact on audience research of radio and press after the early 1960s.[45] The report read like a promotion campaign on the NMO's own behalf and aimed to support the changing official attitudes toward sociological media research. Consequently, the tasks of NMO were officially adapted in 1966. The new assignment legitimized an increasing number of surveys but was rather ambiguously formulated. NMO should "actively support the state committee for central and regional broadcasting with the improvement and development of the radio and television program, to advance the role of radio and television in the life of our society . . . , to fulfill all propagandistic tasks the party and the people claim toward radio and television."[46] However, this consolidation enabled the NMO in the following years to strongly insist that Soviet TV could only gather "objective data" by scientific analysis. From the perspective of the NMO, letters, by contrast, only offered vague or even misleading conclusions.[47]

Nevertheless, since only viewers' acceptance of the program prepared the ground for establishing television as the key medium of information and entertainment, the mailbag never lost its significance: After all, letters often represented emotions and assessments and, thus, said more about TV's popularity than any statistics for internal use. Therefore, letters corresponded with Lapin's new program strategy that aimed to normalize the Soviet way of life. He wanted to trigger viewers' affirmative attitude toward it by proclaiming the importance of the ordinary working people for the Soviet society and making the worker the leading figure on screen. The working people as protagonists in movies and shows should represent the dreams and hopes of the ordinary Soviet viewer in front of the TV screen.[48] This strategy was obviously based on the idea of involving the viewers emotionally. It was bolstered by the more qualitative audience research, whereas scientific analysis claiming to provide representative data and solutions for program designs lost ground again. This was partly due to its worrying and partly delicate findings. In 1970, Soviet TV's NMO concluded that television worked with "emotional means." This was surely a very welcome result. However, the NMO also underlined that radio addressed people's intellect, whereas TV attracted less-educated strata which were interested in "entertainment and satisfaction of emotional needs." The NMO even supposed television viewers to be generally "more passive" in life.[49] In contrast to the NMO, Central TV's letter department contented itself with a kind of "thick description." The overviews widely cited from viewer mail and presented the pleasure grateful viewers apparently felt in flowery words.[50]

These flowery words about the alleged pleasure of the audience fit in well with the genre of Soviet reports. If we do not want to take them just as cliché and simply relegate any possible agency to the dustbin of history, they let us wonder if they testify in one way or another to what Soviet viewers did with the medium. At least, it is rather obvious that several viewers consciously wrote to TV departments instead of other possible addressees in order to discuss rules of social order and living together in their letters. Both observations further raise the question as to what extent the audience mail might reveal any emotional bonding of viewers to television. To what extent did the audience mail referring to films, theater features, documentaries on travel and animals, newscasts, or consumer advice programs finally prove any impact on viewers' relations to the Soviet system?

Right from the start, the letter bureau drew up a model audience to tackle who watched which show and which program received which kind of audience response.[51] Thus, it tried to compete with the findings the NMO presented and to defend its position within Central Television. The importance of communicating with the audience rose within internal discussions of TV staff during the 1960s and especially the 1970s. Soviet media scholar and former director of Leningrad Television Boris Firsov painted a very idealistic picture of the communication between television producers and the audience. He argued that television had "the possibility of communicating with all people regardless of profession, education, age, or any other feature."[52]

He adopted the idea of the diversity of opinions that was gaining popularity in the early 1960s. Increasingly, sociological audience research saw the viewer more as an individual and not simply obscured in a collective.[53] A new, ambitious programming policy related to this new theoretical approach by attempting to strike a balance between reaching the largest audience and satisfying specific subgroups.[54] In 1967, the NMO strongly believed that a uniform standard program schedule had become anachronistic because the audience's educational level had risen considerably.[55]

This discussion reflected a contest between the traditional way of exploring viewers' taste by letters and the scientifically based approaches of the new media scientists. The staff dealing with viewer mail considered letters as statements of the "progressive and active part of the audience": letters helped to improve the program, they thought, and were deliberately included in certain broadcasts.[56] These staffers obviously adhered more to the theory of individuals than of representatives of collectives in front of the screen and led them to value responding to audience mail. Responding meant sustaining the communication and aimed to tightly bind viewers to the medium. From this perspective, television became another important communicator of vertical communication between the people and the regime.

Like all other institutions receiving letters, TV had to report on its mailbag management within the internal administration and to party organs. In its own interest, the letter department of Central Television recognized mail as an important indicator of viewers' preferences and always pointed out the high amount of incoming mail in its summaries. However, the reader can only get an impressionistic interpretation of the contents because the department seldom classified the number of letters in relation to representative information about the writers and issues. Describing their findings "thickly," the staff of the letter department rather referred to details they gathered by analyzing the viewer mail.[57] Furthermore, the reports relied on random samples and only sometimes presented information about the social background of viewer-writers. Social scientists deliberately touched upon this obvious weakness of the letter department's analysis. In the late 1960s, Soviet TV journalist and media scholar Rudol'f Boretskii asserted that mainly disabled persons, pensioners, and viewers expressing either very negative or extremely positive judgments addressed the television.[58] This statement obviously contradicted the letter department's repeated claim that letters were sent by people of rather diverse professions.[59] Indeed, in the late 1950s and early 1960s the data the department presented revealed a much broader social background of letter writers than Boretskii suggested.[60] According to an analysis by the letter department in April 1964, 55 percent of the letter writers were blue- and white-collar workers, 20 percent were pupils and students, and 15 percent were pensioners and veterans.[61] In 1973, of the viewers who addressed the editorial office of propaganda, 38 percent were in unskilled work, 19 percent were in rural labor, and 14 percent were white-collar workers.[62] There are similar examples indicating that the social background of ordinary viewer-writers became more diverse during the 1970s.

Gender was not considered a significant analytical category, and thus letter reports mainly used masculine nouns or the seemingly neutral plural. The department staff seldom singled out women writing letters to TV. If they did, they only tentatively suggested that working women constituted a specific interest group of viewers. For example, in 1966 "quite a lot of women" asked for more entertainment features and films after 9:00 p.m. These female viewers wanted to relax in front of the television after their jobs, finishing their housework, and putting the children to sleep.[63] The editorial offices for music, literature, and cinema programming seem to have received slightly more letters from female viewers than those of information or propaganda.[64] We should not suppose the sample of letter writers to be generally representative of all viewers. But judging by these differences, women apparently had different preferences and seemed to have been slightly more interested in entertainment formats than in news. The program schedulers, however, were reluctant to pick out women as a special target group. Furthermore, given that contemporary sociological research found women on average had much less time to watch TV, it does not come as a surprise that women wrote fewer letters than male viewers.

Generational, geographic, and social background had far more influence on the picture the letter department painted of the Soviet viewer than did the category of gender. As early as 1963, Central Television launched *Sel'skii chas* (Village hour), a weekly program addressing viewers living in the countryside and focusing on topics related to the village life. It was complemented by others like *Selo: dela i problemy* (The village: issues and problems). With the spread of TV sets, writing letters to the television did not remain an urban practice. In 1966, the internal report still mentioned that village viewers rarely wrote letters, and when they did, they criticized the airtimes of entertainment broadcasts and demanded more movies or music programs.[65] Although the letter department staff gave no concrete figures, they emphasized that village viewers increasingly participated on a low level in the communication with television.[66]

As letter writers normally did not indicate their age, the question of how TV consumption varied by generation is difficult to answer. The relatively high numbers of letters going to the editorial offices of children and youth programming may tell us something about the rising interest of young viewers in the medium especially addressing their concerns. Some shows like *KVN* (*Club of the Merry and Quick-Witted*), *Ot vsei dushi* (With All My Heart), and *Chto? Gde? Kogda?* (What? Where? When?) were popular with viewers of all ages, but especially triggered very enthusiastic responses among young viewers.[67]

The percentage of viewers who chose to write to TV editors is hard to assess. There are no representative statistics regarding the number of TV sets, the audience size, or viewing figures for the different Soviet republics and regions. The reports create the impression that letters were registered foremost for the sake of counting. This does not come as a surprise, as any increases of incoming post underlined the department's importance. A certain use of language seemed to imitate what Simon Huxtable has called a "sociological aesthetic."[68] It was characterized by descriptions of constant progress in regard to the increase in letters received, of TV's steadily growing popularity, or program improvements resulting from viewer participation. The editors of the reports often employed adjectives like "many" and "numerous," adverbs like "more often," or phrases like "in respect of the further development of TV, letters came in from all over the country."[69] This style concealed questions of representativeness of the letters, of the relation between quantity and quality, or TV's specificity compared with the press or radio.

According to my own cautious estimation, not more than 3 percent of all viewers wrote letters to TV's editorial departments. Moreover, as historians, we are reliant on the selections made by the letter department for their reports. These aspects constitute severe heuristic constraints: viewer mail cannot be regarded as representative of viewers. Nevertheless, I would argue that even this small sample offers some informative value in two respects. The first relates to our understanding of public communication: generally, letters to mass media interconnected private and public communication in a new way. Contemporary Soviet sociologists found that people included television content, in particular, into their family talks.[70] Letters, thus, almost naturally drew on private life experiences, on relations, or on contradictions between private and public life.

My second point as to why it is worth studying audience mail refers to the abovementioned observation of Boretskii. The media scholar underlined the extreme diversity of opinions presented in letters. A brief glance at the letter reports shows that they indeed comprised a broad range of views. This should not be seen as fault from our point of view but rather as a challenge for contemporary Soviet ideas on collective consensus. Furthermore, it does not come as a surprise that people holding rather moderate points of view often did not feel the need to express their opinion.

THE AUDIENCE MAIL'S CHANGING TOPICS AND VIEWERS' EMOTIONAL COMMITMENT

Central Television had more than ten editorial offices that received audience mail including, just to name a few, the editorial offices of propaganda, information, youth programming, literature, music, and sport. Each of them had to forward the figures of incoming letters to the letter department that registered them for every office. It usually compiled and often even published reports about every month and year. For the purpose of analysis, I summarized the numbers from the different reports according to the given classifications. In Table 4.1, I have selected some of these editorial offices to show the quantitative development of letter distribution.

The quantitative development of the audience post addressed to Soviet mass media reflected the rising importance of television toward the beginning of the 1970s. As TV was not yet the key medium during the 1960s, some of the important national newspapers as well as Central Radio received considerably more letters than television. National newspapers like *Pravda*, *Komsomol'skaia Pravda*, and *Trud* got between 300,000 and 700,000 letters a year, although they reached a much smaller audience than television. The letter office of Central Radio counted ten times as many letters as Central Television in 1963 (675,000 letters compared to 67,990).[71] After this, Central Television's mailbag began to swell quickly. With the growing audience, the amount of viewers-writers also grew in absolute terms. The number of letters multiplied tenfold up to 400,000 from 1960 to 1970. At the same time, the amount of letters going to the radio slightly decreased, so that in 1970 Central TV registered almost as many letters as Central Radio: 405,984 compared to 421,000.[72] TV's incoming mail continued to rise during the following years until it stagnated at a level of 1.3 to 1.8 million letters after the late 1970s, a time when nearly all Soviet households were equipped with TV sets.

The reports show that the topics and the amounts of letters going to certain editorial departments varied. In TV's early days, viewers often wrote about the new audiovisual experiences TV offered. In the 1950s, there were obviously few characteristic protagonists, so that viewers rather addressed TV as an anonymous medium like an acting person. The interest in reflecting upon its capacities and deficiencies as a medium remained an important motive of writing at least until the early 1970s. In 1959, with party officials in mind a viewer hoped that "TV technology is in trustworthy hands" because it reached millions of viewers. Others deemed television as having an important impact on building communism and stressed its greater propagandist possibilities in comparison to the press and radio.[73] Comments such as these which bore on matters of content and ideology easily tended to criticism of the technical equipment, organization, and design of the TV program. In the 1960s and still in the early 1970s, viewers complained about technical faults, the small scale of transmitting power, inaccurate newspaper reporting about the daily TV program, the shortcomings of announcers and moderators, or simply boring programs. These letters give us an idea of how troublesome it may have been in TV's early days for the staff to satisfy both viewers' demands and the party-state's claims to ideological accuracy. In 1963, workers at a *sovkhoz*

Table 4.1 Letters to selected editorial offices of CT in absolute numbers and percentage

Years	*Total*	*Information*	*Propaganda*	*Cinema TV Film*	*Literature Drama*	*Music*	*Children Youth*	*Letters not referring to TV/complaints*
1960	41,632	n/s	ca. 7,000	n/s	n/s	n/s	n/s	n/s
			ca. 17.0					
1964	75,016	2,936	9,156	14,419	5,922	10,214	20,519	6,449
		3.9	12.2	19.2	7.9	13.6	27.4	8.6
1965	148,966	5,429	6,558	26,171	44,972	16,416	30,008	13,475
		3.6	4.4	17.6	30.2	11.0	20.1	9.0
1966	210,070	8,028	47,759	n/s	38,446	21,962	59,212	28,259
		3.8	22.7		18.3	10.5	28.2	13.5
1968	195,973[1]	8,325	5,377	n/s	30,303	49,126	60,420	19,777
		4.2	2.7		15.5	25.1	30.8	10.1
1971	1,781,090	36,912	26,867	2,379	51,371	1,283,355	230,821	7,470
		2.1	1.5	0.1	2.9	72.1	13.0	0.4
				(TV film)				
1973	1,270,757	109,424	124,465	61,333	8,158	308,070	401,096	5,370
		8.6	9.8	4.8	0.6	24.2	31.6	0.4
1977	2,056,696	29,834	197,517	865,586	6,615	230,440	208,730	10,249
		1,5	9,6	42,1	0,3	11,2	10,1	0,5
1978	1,558,295	19,677	197,763	465,109	7,957	325,181	236,869	9,245
		1.3	12.7	29.8	0.5	20.9	15.2	0.6
1980	1,498,835	30,669	235,152	337,850	38,742	189,311	215,619	14,108
		2.1	15.7	22.5	2.6	12.6	14.4	0.9
1984	1,197,957	27,220	75,667	261,687	29,496	61,619	448,811	25,467
		2.3	5.3	21.8	2.5	4.3	37.5	2.1

[1]*Obzor pisem za 1971*, 3.

(state farm) in Lipetsk region mourned that they worked a lot and wanted to relax at home watching television. The program, however, they argued, had a narcotic effect. An engineer from Kaluga observed that the programs evoked discontent among many viewers because they lacked content and meaning. In his view, television was on its way to becoming a "tele-newspaper" (*Televidenie prevrashchaetsia v telegazetu*).[74] Another typical complaint also referred to the impression that television too often ignored its specific capabilities of visualization. A viewer from Kalinin complained that "the majority of TV programs are more to be listened to than to be watched." Programs that offered lectures should be broadcast on radio: "You need to consider the specificity of television," he claimed.[75]

The attempts of producing and editing staff to improve TV programs seem to have had considerable success during the 1960s. A viewer from Lugansk mirrored these comments considering TV's special capacities. In 1969 he wrote:

> We obviously watch your programs, like millions of other viewers, with great attention and interest. In comparison with other sorts of information, they are easily understood and impressive because they do not only tell a story, but also explain events well. We certainly understand that this is not only a question of technical possibilities, but also depends on the art of selecting, representing, and commenting on interesting and important events. Even the youth, who normally do not like to listen to radio news, watch your program with interest.[76]

By and by, the visible protagonists on screen as well as the unknown producers in the background found more interest among viewers. Often, coverage of political events quickened interest on the part of the audience. Such events included the Congress of the Communist Party in 1971, international talks, trips of party and state representatives to foreign states or of foreign government representatives to the Soviet Union, the enactment of the new constitution in 1977, and the disarmament summit in Vienna in 1979. Not surprisingly, the overviews of viewers' letters happily picked out those replies that were close to official propaganda. By the 1970s, many letter writers—and obviously not only party members—explicitly perceived television as the best facilitator of party politics.[77]

The reports also observed an increase of letters to the editorial departments that oversaw entertainment programs. This was the case, for example, when Central Television introduced new features, released new films, aired music programs, quiz shows, or international sport events. For example, in 1964, the office for the propaganda editorial department received more than 9,000 letters and, in 1966, more than 47,000. In 1964, almost half of the mail referred to a sports quiz rising to 85 percent in 1966, something one would not perhaps expect to be supervised by the propaganda department.[78] But this allocation did very well fit in the logic of television's societal function and its program policy of the late 1950s and 1960s, as quizzes proved to be a perfect way to stimulate people's commitment to public life. In the calmed-down times after intensified de-Stalinization and the "learning experience" (Kristin Roth-Ey) of the first famous game show, *Vecher veselykh voprosov* (Evening of Merry Questions), quizzes remained an appropriate means to trigger viewers' response.[79] They enabled the television producers to stage in a rather defined framework drama and performance at once, as well as to screen rather ordinary people on television. Generally, television professionals often used *viktoriny* (quiz games) to trigger audience response as engaging elements of programs that otherwise would have received less letters.[80] Integrating viewer letters into the broadcasting turned out to be another new strategy related to audience participation. Thus, the editors of game show, musical programming, and advice programs brought viewer input directly on air.[81]

In the 1960s, topics like World War II, in particular, triggered viewers' feelings and prompted them to write to the editors. In 1965, the number of letters almost doubled in comparison to the previous year. This was due to the new feature *Podvig* (Heroic Deed), a TV almanac on the defense of the Brest fortress in 1941. It was led by the very popular writer Sergei S. Smirnov and received a quarter of that year's mailbag. In the case of *Podvig*, the party made deliberate use of the capacities of television as an unprecedented instrument of memory politics. Twenty years after the end of the war, television vividly visualized the radical turnaround in Soviet regime's memory politics in films and documentary features. The new strategy anticipated Lapin's idea that TV should represent normal biographies because *Podvig* not only focused on the victory over fascist Germany, but also deliberately glorified ordinary war veterans. Veterans sent their memoirs that were included in the script. Many viewers highly appreciated *Podvig* for praising the unknown war heroes and presenting authentic stories. Viewers also applauded other programs on World War II, illustrating "patriotism, bravery, and the humanism of the Soviet people."[82] However, the interest in *Podvig* somewhat weakened from the late 1960s on after Smirnov had been criticized for his work and resigned.[83] After that the feature used less material sent by veterans.[84]

As of the early 1970s, the letter reports stressed the increasing volume of letters, in the characteristic Soviet style of "higher–faster–further." The rise was said to indicate that Central TV's programs "more and more 'conquer' the viewer, they enter the life of any working person in the country." Further, TV's growing acceptance was said to have altered the content of viewer mail. Critical letters grew more factual and offered more advice. The letter department also insisted that viewers generally expressed more respect for the television staff than before.[85]

An invariable claim of most viewer-writers, however, remained the appreciation of television as the most important source of entertainment. Until the end of the 1960s, viewers repeatedly called for more entertainment shows, more motion pictures, quiz shows, programs of art, and music but for less opera. Most of these critics complained that television aired too many societal-political programs and newscasts. Viewers typically complained that Central Television aired too few "happy" broadcasts and comedies.[86] Blue-collar workers from Sverdlovsk region described watching TV as "a way of passing time when there is nothing [else] to do" that brought "enjoyment." Many eagerly strove to schedule their working days in order not to miss interesting programs.[87] The viewers' idea of television as a foremost pleasure-giving medium was also reflected by findings of Central TV's audience research about viewers' reception of the political program *Leninskii universitet millionov* (Leninist university of millions). It was—at least from Central TV's perspective—somewhat disappointing that almost no "ordinary" viewer, that is, someone who did not participate in the educational system of the Communist Party, regularly watched these lectures on Marxism-Leninism. But only a minor part even of Communist propagandists and listeners otherwise participating in lectures on political education regularly switched on this program.[88]

TV, however, was not only an entertaining leisure practice. Corresponding viewers increasingly perceived it as source of information and facilitator of political communication. In 1962, the letter department categorized almost 80 percent of all letters as requests concerning motion pictures, stage plays, and music, whereas the share of this kind of material fell to 11 percent in 1965. In comparison, the volume of letters that the department characterized as a "continuation of the conversation" with the audience on news, social, and political issues went up to 33 percent.[89] The statistics which the editors compiled for their reports do not immediately suggest this ideological interpretation. They relied on content analysis from all the editorial departments. Looking at individual topics, however, this interpretation becomes comprehensible: viewers responded to themes that touched upon socialist values and societal cohesion or propagated the achievements of the Soviet Union. They seem to have associated media consumption with the right to comment on the social order and the problems of living together and to point out flaws in societal issues.[90] Therefore, the letter department saw the reason for the rising amounts of viewers' letters as the steadily growing "socio-political activity of the working people." To underline this interpretation, editors paid special attention to comments, questions, and letters of thanks that referred to Soviet domestic and foreign policy.

Apart from politics, editors were most interested in viewer-writers' interpretations of modern Soviet life and socialist morals and education. Letter editors gathered viewers' comments on motion pictures, newscasts, advice programs for consumers, and documentary features.[91] More or less, all of the editorial offices received letters that touched upon social life in one way or another. The letter department did not, however, classify its reports systematically by topics, but rather according to Central TV's organizational structure. If we scan the reports according to these classifications, those letters commenting on the social order in its broadest sense accounted for, in my estimation, approximately 20 to 25 percent of the post after the mid-1960s.

One of the most striking characteristics of the mail, in my view, is its insights into how television caught many viewers' fancy on an emotional and intellectual level. The letter department staff tried to illustrate audience commitment by quoting relevant statements from viewer mail. The staff must have been very excited about the contents of a letter sent by a paralyzed and housebound war veteran. In 1959, he wrote about his wife's "great idea to buy a television." It had become his friend, assistant, and best medicine, he declared. The television had pushed apart the apartment's walls, so that his apathy, pessimism, and despair faded away. The television enabled him to believe in life again.[92]

The letter department assessed letters that referred to politics in its broadest sense as "stimulation of societal-political activity of the working people in our country."[93] The reports highlighted viewers' positive response to TV's broadcasting on official state and party affairs. They emphasized the close connection between the television as facilitator of politics, the audience, and the Soviet party-state.[94] In a ritual fashion, all published letter reports started with quotes illustrating how viewers praised the role of the party and the Soviet government. They never failed to mention that viewers were thankful to television "for the splendid elucidation of all important events this year in the country and abroad, of visits of party and governments leaders."[95] The message of the letter department was clear: the images of Soviet politics that TV spread all over the country involved viewers in the sociopolitical order. Winning their support seemed hardly possible without TV broadcasting. We can trace the obvious self-representation of the letter department in these representations.

To survive the internal rivalry over the question of who provided the most meaningful information on the audience, the letter department could best rely on its capacity not only to illustrate but also to build an emotional relationship with the audience. More than a few writers stressed that television strengthened a feeling of social cohesion and trust through letter communication. In 1973, a woman wrote: "I only needed to write a letter with the request to answer my question and you immediately responded. That touched me deeply. Your answer has given me back the belief in people, humanity and sympathy."[96]

Indeed, this direct communication with viewers by responding to letters constituted the trump card of the letter department within the internal discourse of TV staff on the audience. Although the letter department relinquished "objective" categories, Soviet TV benefited from viewers' emotional connections to this communication channel. Because letters revealed the subjective insights of their authors, some editors considered them a means to research the composition and interests of the audience. Unsurprisingly, the letter department's assessment of audience mail changed little until perestroika. Still, the letter report of 1980 insisted that the post was "an inexhaustible source of topics for the program" because the writers commented on all sorts of everyday issues and proposed program changes.[97] TV producers, therefore, deliberately encouraged letter writing in quiz and music shows.[98] Moreover, Central Television and local TV stations developed new programs based on viewers' letters. They were called *Nasha pochta* (Our post), *Po vashim pis'mam* (Following up on your letters), *Na voprosy telezritelei otvechaet ministr . . .* (Minister . . . responds to viewers' questions), *Chelovek i zakon* (Person and Law), *Subbotnaia pochta* (Saturday mail), and *Chas pis'ma* (Hour of letters).

These programs based on complaints and requests played an important role in the retrospective self-representation of former TV editors because of its immediacy, the active involvement of the audience, and, last but not least, the way television seemingly impacted on sociopolitical life by drawing on viewer letters. Samara TV editor Valentina V. drafted a narrative of television's commitment to defend Soviet citizens' interests toward administration and state institutions. Valentina V. contrasted television with newspapers: "The television word," she asserted, "comes somehow faster, probably, more emotional. . . . That's why I joined television." She traced television's higher impact for the benefit of ordinary people to the fact that many state and party officials were afraid of publicity and would not have dared to refuse television's involvement:

> How should I say towards what we oriented. Where we felt that support was needed. If my support was needed, precisely, if they did not get by without it. You sit and cry and went there. And it happened, as I mentioned earlier, it happened that I call [*sic*], let it be supposed, the ZhEU (*Zhilishchno-ekspluatatsionnaia kontora*: Housing and utilities administration) and say: You know that, comrades, if you don't do this within ten days, I will invite you to television. If you don't come, I will say that you didn't come. Either you do it or you explain, why you don't do it. You watch—they did it. They didn't do it for years—they did it.

Valentina V. highlighted the way the chair of the town executive party committee Rosovskii exerted his influence in order to back her journalistic strategies toward local party and state officials. Consequently, local officials seldom refused her invitation: "Well, declining was a bit difficult, because I say this on TV, I present the topic and say that, unfortunately, they had declined. There was Rosovskii, I had Rosovskii [the chair of the town executive party committee] behind me. And all knew about that."

According to her perception Rosovskii's support allowed her to televise critical and sensitive letters, as well as in general to cover "publicly significant topics, in particular, consumer issues. . . . This means central heating in the city, the readiness of the beaches, public spaces in the city. Well, everything that concerned the life of the city. . . . Certainly, current [consumer issues] were present, but only presented them based on letters."

Looking back, she underlined her critical journalistic attitude exploiting television's capacities in favor of ordinary people's interests. This was why societally significant topics were in her view always the most important ones on television. This seemed to characterize the common strategy of Moscow and local television stations.

As I will elaborate in more detail in chapter 5, consumer problems, in particular, became a favorite topic on Central and local Soviet television. Viewers welcomed this, as some like party member N. A. Makarov from Leninsk-Kuznetskii considered it to be television's task to report on authorities' commitment to economic problems in response to viewers' complaints.[99] Programs such as *Our Post, You Asked–We Respond to your Table, Hostess*, and *Let's Talk About Your Letters* reported on flaws in local services, the food industry, and retail trade.[100] Viewer-consumers appreciated these programs and thanked television for reports on new products, advises on taste and style, features on production processes, on development of new products, on price trends, and on product flaws.[101] Television responded to viewers' communication and called representatives of industrial enterprises to TV studios to account for faulty consumer items that had been signaled in viewer mail. Another strategy of involving viewers was to organize public discussions like in the case of the consumer program *Bol'she khoroshikh tovarov* in the *Klinskii zavod termometrov*.[102] It was common for these programs to display Central TV's postal address and invite the viewers to send complaints and to write about their troubles and problems.[103] Thus, television directly responded to viewers and sustained the letter communication by using letters in documentary features or presented them as occasion for journalistic work. By steadily representing its seemingly close contact to the audience, television positioned itself in the field of political communication not only as a medium of entertainment but also as agent of societal interests and improvements. Former Samara editor Valentina V. summarized this development: "We criticized sort of those store units, store units' shop clerks. We criticized. . . . It means that the government established the trade, but who could stand up against the government in our time? This is now possible."[104] Apart from the equally interesting comparison with the alleged possibilities of actions at the time of the interview, she perfectly analyzed the basic mechanism of possible critic during Soviet times: Anybody was allowed to criticize local shortcomings and persons in charge but never ever the top level. In this context, consumption became an "object" in Niklas Luhmann's sense: a topic that became an anchor point of further communication without forcing people to accept any detail of it.

Central Television's letter reports reveal that many viewers provided feedback on these kinds of programs, and similar to the interviewed former TV employees, the letter department took this as a reason to judge them to be very popular.[105] Not surprisingly, the editors picked out the comment of one Moscow viewer who praised the program led by political commentator and former correspondent based in Paris Anatolii G. Potapov as "the best and perfect."[106]

It should be emphasized that central and local TV stations regularly used letters to visualize the direct involvement of the viewers. It occasionally even presented letters to the camera, thus suggesting close contact between viewers and medium.[107] All this did not serve to present "objective" findings about the audience. Rather it suggested that letter writers, the television staff, and the viewers watching these features formed a community in the first instance bound together by common interests in the topic. Extrapolating from the case of the female viewer quoted above who felt touched by letters as a form of communication, we can suppose that presenting physical letters on screen even reinforced the impression of emotional relations between the audience and the TV. The emotional connection probably had two overlapping dimensions. As we have seen in chapter 1, the TV set gained its place in the living room, where it became a reference point of communication with family and friends. Moreover, because viewer-writers normally got a "private" response to their letters, they might have perceived letters initially as a purely "private" communication channel. When TV editors chose to screen physical letters, they interconnected the private and the public by promoting private interests and embedding them in societal references.

With the help of audience mail, Soviet television appeared as an authority speaking for viewers' interests. The screening of letters kept the communication with the writing part of the audience going and, into the bargain, involved nonwriting viewers who watched these features. The letter department itself put forward the opinion that the "broad use of letters" contributed to the general increase in the volume of mail in the early 1970s.[108]

The postal communication between TV and its viewers demonstrated another interesting aspect of the development of media use in the late Soviet Union. We should see this communication in the broader context of writing to party-state institutions and local and national deputies, all of which were accepted ways to get in contact with the regime on personal terms. If we take this alongside evidence of committed readers writing to newspapers and journals, we have a portrait of TV's letter writers as enthusiastic media consumers.[109] A good number represented themselves as active, demanding, and self-confident media consumers, who deliberately used information presented by Soviet media. In particular, letter writers addressing state and party representatives often referred to information they gathered from newspapers, radio, and television.[110] In these cases they perceived the mass media as an objective observer. The letter writers obviously held the view that media consumption implied a right to employ the presented information in order to complain or to make proposals. In return, they reported on personal experiences and instances which they considered to be of public interest. Thus, they wanted to provide television with information for reporting, for example, about bribery to get a better place for recreation or to study at university, about deficient construction, or about parasitism.[111]

Again, the fact that they inscribed themselves into official discourses interconnected mass media coverage with communication based in the private sphere. It also documented that letter writers considered television to impact on public affairs in pursuing people's interests. A group of Rostov-on-Don citizens hopefully turned to the local TV station after having addressed the local city council without success.[112] The files document that a rising number of viewers resorted to this communicative strategy of appealing to television after having written to party or state authorities. Many requests referred to abuses of authority, bribery, or consumption flaws.[113] Others took a pragmatic stance hoping to benefit from television's supposed influence. In order to bypass waiting lists, people appealed for television's help by presenting themselves as eager and enthusiastic TV consumers.[114]

However, despite the contemporary and retrospective self-representation of members of the letter desks, this kind of letter writing and complaints to television was surely no mass phenomenon as such. As the sample of interviewees is not representative of former Soviet citizens' practices of letter writing and complaining, it is hard to judge on the basis of their remembrances what role television played as the addressee of complaints from the citizens' point of view. Nevertheless, television expanded the space of communication by its immediacy and range. As a "third space" it filtered representations, interlaced them, and initiated new ones. In several respects, television conveyed the impression that it responded to viewers' private concerns. Thanks to the possibility of putting forward personal wishes regarding programs or broadcasting times, television provided a platform for seemingly unpolitical topics of communication and gave a new emphasis to private life. Due to most viewers' perception of TV as an entertainment medium, it became a communicator on values, morals, and taste. That TV's rapidly rising impact on popular culture contributed to a predominantly positive emotional assessment of Soviet life might also be concluded from rather small quantities of letters criticizing motion pictures and serials. This capacity to involve viewers and to familiarize and normalize topics was something that concerned only a small number of viewers, however. A letter from a female viewer from Penza is representative of how viewers criticized Soviet TV. In 1969, she complained that many programs gave the impression that "life was easy. There are few programs on problems or serious broadcasts that inspire you to think."[115] Some contemporaries seemed to think that television involved the audience in a special way. This Penza viewer obviously saw a relation between watching television and people's attitude toward the social order. She suspected that light entertainment distracted the audience from serious deliberations about societal improvements. Her point of view also revealed strong criticism of contemporary popular culture, a view that might not have been very common among the average TV audience. Most of the viewers did not accuse television broadcasts or films of being far from reality.

LETTERS AND THE STAGING OF AUTHORITATIVE DISCOURSE

We can learn from the audience mail about several aspects of communication that television influenced. First, until perestroika, television absorbed communication and kept negotiations on airtimes, programs, or the outfit of a newsreader within the unpolitical realm. Television expanded the space and topics of communication by picking up everyday questions, observing and reflecting the Soviet lifestyle. Features like *Bol'she khoroshikh tovarov* (More good products), on which I will focus in the next chapter, or cinematic representations of taste and lifestyles, offered viewers topics of conversation. It becomes obvious that even if many letter authors pointed to grievances, they should not be interpreted as disagreement with the existing order and official values. Rather, letters confirmed the emotional entanglement and interplay of the community of viewers with television and the regime. The entry threshold was lower and more informal than for people's participation in other more ritual acclamations. For the regime, however, it was much more challenging, as the official party and state representatives had to take the challenge of being televised.[116]

Second, in the context of authoritative discourse letters to television have functioned as a channel of criticism and participation. The language boundaries of communication practices were in some respects strictly limited and highly ritualized, as criticism must never contest the regime's basic legitimacy. Nevertheless, letters to television documented how people's agency energized the frozen authoritative discourse. Letters to television revealed that people's private practices and "the regime" were mutually dependent and influenced each other. Television provided a "third space" in which the new materialistic icons of late socialism became visible and negotiable.

Third, the audience mail points to the audiovisualization of politics and the performing of party officials on television. Media producers, media scholars, media commentators, and party representatives agreed that television gave "the possibility to speak to the broadest audience. The heads of a city now get into hundreds of homes, into thousands of families at once."[117] This statement of a provincial *Gorkom* secretary was published in *Pravda* in 1972 in the section *Party life: the committee's working style*. It documents well that the party had discovered the communicative potential of the television after the late 1960s and tried to a certain extent to adapt its own communication practices to it. Politics should be visualized, and politicians should perform interestingly on TV and communicate with the audience via the medium and letters. In this case, the *Gorkom* secretary quoted from a citizen's letter who sought for more information on a certain construction project. Instead of responding to him in writing or inviting him to a talk at the *Gorkom*, like it had been frequent practice, the secretary then favored to respond to such letters via television. Once a month, local party functionaries discussed and answered viewer letters in the local Krasnodar TV program *We inform, comment, explain*. This programmatically titled broadcast was based on the ideas that television generated immediacy and that it maintained viewer participation through integrating the traditional letter communication. Television was even supposed to replace face-to-face conversations and written personal responses. The individual request should gain collective importance by presenting it to the television audience of millions. Thus, television should not only propagate political messages, but genuinely establish a dialogue with the audience about the topics viewers brought up in letters: "There are peripheral issues, but more often they are of public interest. Therefore, we try to respond addressing a large audience. . . . We are convinced: The meetings on TV are a very efficient form to inform the populace. This permits us to keep the people informed about all events going on in the city, the region, the country, to react to them in a timely way."[118]

Television energized the authoritative discourse by promoting an adapted concept of participation and communication. In the particular context of television, letters could evoke the impression of immediate relations among the medium, the audience, and the regime, as complaints about consumption grievances, for example, provided the occasion in consumer programs to track them on the spot by looking for the responsible persons. Central and local TV staffs gathered incriminating materials on the basis of consumers' complaints. Supported by the "Regional Board of State Inspection on Quality of Goods and Trade" Rostov broadcast staff, for example, confronted the retail sales and companies in question with the grievances. The program then proudly presented the impressive results of their efforts to defend consumers' interests. In one case they accused more than 200 companies of having produced goods of minor quality. TV viewers were informed that 10 careless workers were fired, 10 lost their position, more than 250 lost their bonus, almost 500 were imposed an administrative punishment, and more than 100 had to compensate for costs of the defect goods.[119] An equally important strategy consisted in documenting the successful cooperation with the responsible government agency. For this, *Nashi telereklamatsii* interviewed the inspector who was charged to report on the investigation brought about by consumers' complaints. Thus, the viewer-consumers were put in the position of being able to trace concrete improvements and to judge persisting problems.[120]

Visiting factories in order to report on defective goods or to invite politicians for interviews could, moreover, be interpreted as a trigger for change or new meaning. At least, this kind of criticism, although basically reproducing authoritative discourse by adhering to the official boundaries of speaking, generated a much broader publicity than letters to newspapers, party, or state authorities.

In principle, the televisual adaption broadened the range of performative, discursive, and aesthetic forms and relativized the loss of authoritative discourse's literal meaning that Yurchak has described as a general phenomenon: The viewer was located at home in the private sphere and enabled easily to ignore communication or to refrain from participation. Party and state officials were encouraged to perform on television; the political message was submitted by voice and face simultaneously. With this, television was fraught with high risk, as not all state and party officials were telegenic and performed appealingly on the screen. In any case, this specificity of the medium represented a particular challenge to officials who were not accustomed to perform on the screen. Being on television could mean speaking live or without notes, taking part in a round table, or responding to questions in a way comprehensible to the audience. Especially in the 1960s, officials seldom met the high expectations of the audience for this mode of visualized politics. It was a typical viewer request of a member of Central Television's Public Council who criticized the performances of some ministers, because their talks were "little expressive and consisted of general phrases." According to the Public Council, the actual issues, the ministers covered, were, however, highly interesting. Central Television's editor N. I. Sokontikov agreed and promised to rehearse with the ministers their performance in front of the camera in order to stage their talks in a more lively fashion. To avoid "boring performances that are difficult to listen to" in the future, he proposed to include illustrations and examples.[121] Whether this verbal undertaking was actually realized is not clear, but that television workers conceded the room for improvement in performance demonstrated the will to involve the audience at this moment of development.

Fourth, apart from party and state officials, television professionals also staged letters and represented participation on the screen instead of viewers participating themselves in meetings, votes, and so on. The reproduction of the "correct" phrasing and audiovisual form surely still mattered a lot for all presenting agents. It seemed to be rather successfully reenacted by a much more flexible performance on television, as the way both party officials and television professionals responded to audience letters was new. The televised interplay of the different actors contributed to reinforce the new significance of Soviet way of life. Thus, from the early 1960s onward, television disseminated and considerably renewed the ideological message that private life mattered. The audience could feel entitled to mold their private lives and negotiate with the regime on the range of agency.

Fifth, the focus on the private sphere seemed to have, however, also the effect that people did not demand for systematic and binding solutions of problems. According to the earlier-mentioned argument of Niklas Luhmann, the audience contributes to the stability of a social order as long as it does not openly oppose interpretation frames and could still negotiate on certain topics. Following this argumentation, Soviet television's impact on political communication probably built upon negotiations of the social order, socialist values, or domestic politics. The oral history interviews, which are explored in greater detail in chapters 3 and 6 with regard to television's pleasure-giving effect, reveal today that television in the first instance contributed to people's commitment to their Soviet lifestyle. Its narratives and interpretation frames were ambivalent enough to obey ideological limits and to allow the majority of viewers to feel entertained.

NOTES

1. GARF, f. 6903, op. 36, d. 1, 1971: Godovoi obzor pisem i obzory pisem telezritelei za ianvar'-dekabr' 1971, l. 168.

2. Ellen Mickiewicz, *Split Signals. Television and Politics in the Soviet Union* (New York: Oxford University Press, 1988), 8; Jost-Arend Bösenberg, *Die Aktuelle Kamera. Medienpolitische Untersuchungen zu den Lenkungsmechanismen im Fernsehen der DDR, in der Zeit von 1952 bis 1990* (PhD manuscript FU Berlin, 2002), 273–285; Franka Wolff, *Glasnost erst kurz vor Sendeschluss. Die letzten Jahre des DDR-Fernsehens (1985–1989/90)* (Köln, Weimar, Wien: Böhlau, 2002), 147.

3. With regard to the political culture, most scholars emphasized "whistle blowing" letters because the very act of writing a denunciation helped to legitimize the

Soviet regime. Yet, denunciations were only one, albeit politically very important, practice of writing. See for a longer perspective Margareta Mommsen, *Hilf mir, mein Recht zu finden. Russische Bittschriften von Iwan dem Schrecklichen bis Gorbatschow* (Frankfurt a.M.: Propyläen Verlag, 1987); with special regard to petitions and denunciations since the seventeenth century, see Valerie A. Kivelson, *Autocracy in the Provinces: The Muscovite Gentry and Political Culture in the Seventeenth Century* (Stanford, CA: Stanford University Press, 1996); Svetlana Boym, *Common Places: Mythologies of Everyday Life in Russia* (Cambridge, MA: Harvard University Press, 1994), 168–173; Jeffrey Burds, "A Culture of Denunciation: Peasant Labor Migration and Religious Anathematization in Rural Russia, 1860–1905," in *Accusatory Practices: Denunciation in Modern European History, 1789–1989*, ed. Sheila Fitzpatrick and Robert Gellately (Chicago, IL: University of Chicago Press, 1997), 40–72; recently with regard to post-Stalinist times, see Gleb Tsipursky, "'As a Citizen, I Cannot Ignore These Facts.' Whistleblowing in the Khrushchev Era," *Jahrbücher für Geschichte Osteuropas* 58 (2010): 52–69.

4. Merl, *Kommunikation*, 82–110. Denis Kozlov recently analyzed the debates that were argued out between the literary journal *Novyi Mir* (New World), its readers, and its writers via letters. Cf. Denis Kozlov, *The Readers of Novyi Mir: Coming to Terms with the Stalinist Past* (Cambridge, MA, London: Harvard University Press, 2013).

5. Christine E. Evans, *From Truth to Time: Soviet Central Television, 1957–1985* (PhD manuscript University of Berkeley, 2010) (http://escholarship.org/uc/item/2702x6wr), 115–183; idem, "Song of the Year and Soviet Culture in the 1970s," 617–645.

6. The Soviet press faced similar problems of information about the specific reader. Cf. Simon Huxtable, "In Search of the Soviet Reader. The Kosygin Reforms, Sociology, and Changing Concepts of Soviet Society, 1964–1970," *Cahiers du monde russe* 54, 3 (2013): 623–642.

7. At the beginning, the editors compiled an overview of the incoming mail for party organizations and internal use. From 1970 onward, Central Television published them, however, with a relatively low number of copies. Cf. with regard to the published overviews, for example *Obzor pisem telezritelei za 1970 god* (Moscow: Gosudarstvennyi komitet po televideniiu i radio, 1971).

8. Evans, From Truth to Time, 10.

9. Mickiewicz, *Split Signals*, 197.

10. On the late Soviet period see Merl, *Kommunikation*, 82–110.

11. GARF, f. 6903, op. 3, d. 306, 1965, l. 20. Kogan, "Buch und audio-visuelle Medien," 151; Alex Inkeles, *Public Opinion in the Soviet Union. A Study in Mass Persuasion* (Cambridge, MA: Harvard University Press, 1951), 122.

12. GARF, f. 6903, op. 48, ch. 1, d. 10, 1971: Osveshchenie podgotovki k XXIV s''ezdu KPSS v informatsionno-publitsisticheskikh programmakh TsT za dekabr' 1970, l. 19.

13. Roth-Ey, *Moscow*, 270–271.

14. Mickiewicz, *Split Signals*, 81.

15. GARF, f. 6903, op. 1, d. 623, 1960: Pis'ma v TsK KPSS o rabote komiteta, ll. 4, 53, 67.

16. The capacities of smaller TV stations to delegate a staff member for working with letters was usually rather constrained: GARF, f. 6903, op. 1, d. 894, 1966: Protokoly NoNo 17, 18 zasedanii Komiteta i materialy k nim, l. 25.

17. The interview with higher-educated television journalist Liudmila was conducted in October 2010 in Irkutsk by Michail Rozhanskii. Liudmila was born in Irkutsk in 1955 and started to work at the local TV station on the basis of an internship in 1974. She still worked as editor in chief of local Irkutsk television when the interview was conducted.

18. The interview with Evelina E. was conducted in October 2010 in Rostov-on-Don by Galina Orlova and Sof'ia Kontorovich. Evelina E. was born in Rostov in 1935, graduated from Moscow Institute of Culture, and served as television director of Rostov television station from 1958 until the early 1990s.

19. The interview with Valentina V. was conducted in November 2010 in Samara by Elena Zhidkova. Valentina V. was born in Orenburg region in 1937 and graduated from the Alma-Ata University Department of Journalism. She first worked for several newspapers before she was hired as editor at Samara television in late 1965.

20. The interview with V. V. was conducted in November 2010 in Samara by Oksana Zaporozhets. V. V. is a man born in 1948. He graduated from Kuibyshev Pedagogical Institute and worked at Samara television from 1973 to 1999.

21. Interview with Ėvelina E. (Rostov, born in 1935).

22. GARF, f. 6903, op. 1, d. 666, 1960: Protokoly zasedanii obshchestvennogo soveta telezritelei pri tsentral'noi studii televideniia, l. 4.

23. Interview with Evelina E. (Rostov, born in 1935).

24. Interview with Valentina V. (Samara, born in 1937).

25. The interviewees are not representatives of people who wrote letters to mass media. But party members like Svetlana Vladimirovna (Samara, born in 1947) internalized very well the idea "to enter in a discussion, in a dialog with them. And I wrote then a letter to Pravda." She here referred to the events of the Prague Spring and to the government which she addressed. She was, however, not the only respondent who formulated that from the audience's point of view, letters were the main way to get in contact with television and to submit any concerns.

26. Interviews with Liudmila (Irkutsk) and Svetlana V. (Samara, born 1937).

27. Letters addressed to Central television are completely missing for the years of 1957, 1959, 1961 and 1962, and partly for 1963 and 1965 to 1967. See the survey in: GARF, f. 6903, op. 10, 1952–1970 gg., l. 1.

28. GARF, f. 6903, op. 3, d. 185, 1962: Spravka NMO o neobkhodimosti izucheniia auditorii radioslushatelei i telezritelei Vsesoiuznogo radio i televideniia, ll. 1–12.

29. Vladimir Shlapentokh, *The Politics of Sociology in the Soviet Union* (Boulder, CO: Westview Press, 1987), 167–170; Weinberg, *The Development of Sociology in the Soviet Union*, 82–107.

30. Shonin, "Radio and Television as Art Forms in the Soviet Union," 87–88.

31. The quote is from the preface of the inventory: GARF, f. 6903, op. 3.

32. GARF, f. 6903, op. 1, d. 667, 1960, l. 12.

33. GARF, f. 6903, op. 1, d. 724, 1961: Protokoly zasedanii obshchestvennogo soveta telezritelei pri tsentral'noi studii televideniia, l. 4.

34. GARF, f. 6903, op. 1, d. 612, 1959: Stenogramma zasedaniia obshchestvennogo soveta telezritelei pri tsentral'noi studii televideniia, ll. 2–3, 22; ibid., d. 613, 1959: Protokoly zasedanii obshchestvennogo soveta telezritelei pri tsentral'noj studii televideniia, ll. 1–18; ibid., d. 666, 1960: Protokoly zasedanii obshchestvennogo soveta telezritelei pri tsentral'noi studii televideniia, ll. 2–31; ibid., d. 667, 1960, ll. 2–10; ibid., d. 724, 1961, ll. 1–2.

35. GARF, f. 6903, op. 31, d. 14, 1959, ll. 13–15.

36. GARF, f. 6903, op. 1, d. 666, 1960, ll. 14–16; ibid., d. 667, 1960, ll. 11–12; ibid., d. 724, 1961, l. 17.

37. GARF, f. 6903, op. 1, d. 666, 1960, l. 15.

38. Cf. e.g. GARF, f. 6903, op. 1, d. 724, 1961, ll. 11–14.

39. GARF, f. 6903, op. 31, d. 14, 1959, ll. 21–25; ibid., op. 1, d. 667, 1960, ll. 7–8.

40. See the rather open discussion of the council when faced with a very bad attendance: GARF, f. 6903, op. 1, d. 724, 1961, ll. 17ob–22, 31–34.

41. GARO, f. 4237, op. 1 (prod.), d. 877, 1963, ll. 166-174; Interview with V. V. (Samara, born in 1948).

42. GARF, f. 6903, op. 48 ch. 1, d. 22, 1971, l. 24.

43. GARF, f. 6903, op. 3, d. 185, 1962, l. 1; GARF, f. 6903, op. 3, d. 403, 1967: Programma sotsiologicheskogo issledovaniia NMO po teme "Struktura zaprosov i trebovanii radiosl. i telezr. po printsipu intellektual'noi vertikali," l. 3.

44. GARF, f. 6903, op. 3, d. 185, 1962: Spravka NMO o neobkhodimosti izucheniia auditorii radioslushatelei i telezritelei Vsesoiuznogo radio i televideniia: ll. 1–12. Soviet sociologists generally studied the theory and research of bourgeois sociology to establish their own distinct approach on the basis of Marxist ideology. Cf. Weinberg, *The Development*, 12–20. The famous Pskov study of 1965 was linked to the UNESCO international time budget research. See the decision of the TsK of the Communist Party published in Lev N. Moskvichev et al., eds., *Sotsiologiia i vlast'. Sbornik 1. Dokumenty 1953–1968* (Sociology and power. Collection 1. Documents 1953–1968) (Moscow: Academia, 1997), 71.

45. GARF, f. 6903, op. 3, d. 277, 1965, Itogovaia spravka o rezul'tatakh sots. issledovaniia populiarnosti radio i telev. program, provedennogo na predpr. g. Moskvy, ll. 16–19.

46. GARF, f. 6903, op. 3, d. 319, 1966: Polozhenie o NMO. Kopiia, l. 1.

47. GARF, f. 6903, op. 3, d. 320, 1966: Stenogramma sovmestnogo soveshchaniia NMO i sotsiologicheskoi gruppy Akademii obshchestvennykh nauk pri TsK KPSS na temu "Izuchenie radio i televizionnoi auditorii i effektivnosti veshchaniia," ll. 5, 7, 8.

48. Vilen V. Egorov, *Televidenie: Stranitsy istorii* (Moscow: Aspekt Press, 2004), 36–39.

49. GARF, f. 6903, op. 3, d. 563, 1970: Otchet NMO ob issledovanii temu: "Effektivnost' tsentral'nykh i mestnykh programm radioveshchaniia i televideniia po chetyrem oblastiam. Zaprosy i trebovaniia radioslushatelei i telezritelei v zavisimosti ot urovnia intellektual'nogo razvitiia," l. 72.

50. For example: GARF, f. 6903, op. 10, d. 64, 1965: Obzor pisem telezriteli za ianvar' 1965, l. 1; GARF, f. 6903, op. 10, d. 101, 1970: Godovoi obzor pisem telezritelei za 1970 g., l. 4.

51. How the program policy relied to letters, see for more details Evans, *Between Truth and Time,* chapter 2.

52. Firsov, Televidenie, 108.

53. Weinberg, *The Development*, 82–83. A similar development could be observed for Soviet press journalism: Huxtable, "In Search of the Soviet Reader," 624. With regard to TV, this development is analyzed for the 1970s by Mickiewicz, *Split Signals*, 196–199.

54. Evans, From Truth to Time, 83.

55. GARF, f. 6903, op. 3, d. 403, 1967, ll. 18, 19.

56. GARF, f. 6903, op. 3, d. 277, 1965, l. 16.

57. As a typical example: GARF, f. 6903, op. 10, d. 101, 1970, ll. 4-34ob.

58. Rudol'f A. Boretskii, *Televizionnaia programma. Ocherk teorii propagandy* (Moscow: Gosudarstvennyi komitet SSSR po televideniiu i radio, 1967), 82–83.

59. GARF, f. 6903, op. 10, d. 44, 1963: Obzor pisem radioslushatelei i telezritelei, postupivshikh v dni podgotovki k plenumu TsK KPSS 18 iiunia 1963 g., l. 1; ibid., d. 101, 1970, l. 4.

60. E.g., GARF, f. R-6903, op. 10, d. 18, 1959: Obzory pisem telezritelei i spravki o rabote s pis'mami za 1958–1959 gg., ll. 1–13.

61. GARF, f. 6903, op. 10, d. 70, 1966: Analiz pisem telezritelei za period s 1964 po 1966, l. 3.

62. *Obzor pisem telezritelei za 1973 g.* (Moscow: Gosudarstvennyi komitet po televideniiu i radio 1974), 17.

63. GARF, f. 6903, op. 10, d. 70, 1966, l. 32.

64. See for example *Obzor pisem za 1979*, 11–42, 88–116, 134–145.

65. GARF, f. 6903, op. 10, d. 70, 1966, l. 31.

66. See GARF, f. 6903, op. 10, d. 101, 1970, l. 4.

67. Christine Evans explains how the show "From the Bottom of My Heart" created emotional bonds with the younger audience. Cf. Evans, "The Soviet Way of Life."

68. Huxtable, "In Search of the Soviet Reader," 624.

69. GARF, f. 6903, op. 10, d. 101, 1970, ll. 4, 7.

70. Kogan, "Buch und audio-visuelle Medien," 162. A study on rural viewers found out that already in 1971 more than 40 percent of the people who also regularly read literature discussed the programs after watching. The survey, however, was only interested in the relation between watching television and reading. Its aim was to prove that TV did not have any negative influence on reading practices and that reading TV users were more critical than nonreading ones. One could probably suppose that actually more people talked about the programs and that talking about TV became even more involving with the increasing amount of TV sets and the growing popularity of TV movies and serials during the 1970s.

71. GARF, f. 6903, op. 10, d. 34, 1963: Godovoi obzor pisem i obzory pisem za ianvar'-mai radioslushatelei Tsentral'nogo Vnutrisoiuznogo Radioveshchaniia, Tsentral'nogo Radioveshchaniia na zarubezhnye strany i Tsentral'nogo televideniia za 1963, l. 1.

72. *Obzor pisem radioslushatelei za 1970 god* (Moscow: Gosudarstvennyi komitet po televideniiu i radio, 1971), 5; *Obzor pisem telezritelei za 1977 g.* (Moscow: Gosudarstvennyi komitet po televideniiu i radio, 1978), 3.

73. GARF, f. 6903, op. 10, d. 18, 1959, l. 3.

74. GARF, f. 6903, op. 10, d. 46, 1963: Obzor pisem telezritelei ob uluchshenii programm Tsentral'nogo televideniia, podgotovlennyi nauchno-metodicheskim otdelom, l. 3.

75. GARF, f. 6903, op. 10, d. 46, 1963, l. 6 (quotation); ibid., op. 48 ch. 1, d. 22, 1971, l. 17.

76. GARF, f. 6903, op. 10, d. 89, 1969: Obzory pisem telezritelei za ianvar'-dekabr' 1969 g., l. 112 ob.

77. *Obzor pisem telezritelei za 1973 g.*, 3–5.

78. GARF, f. 6903, op. 10, d. 70, 1966, l. 7.

79. On the story of the game show *Vecher veselykh voprosov* see Roth-Ey, *Moscow*, 246–253 (quote on 253); Evans, *Between Truth*, 27–31.

80. Evans, *Between Truth*, 53.

81. See chapter V; Evans, *From Truth to Time,* 166–183.

82. GARF, f. 6903, op. 10, d. 66, 1965: Obzor pisem telezritelei za aprel'-avgust, oktiabr' 1965, ll. 132ob, 148 (quote); ibid., d. 70, 1966, l. 7; ibid., op. 1, d. 883, 1965: Spravka otdela pisem Komiteta o rabote s pis'mami zritelei, poluchennykh Tsentral'nym televideniem v 1965 g., l. 1.

83. See Roth-Ey, *Moscow*, 271–272.

84. GARF, f. 6903, op. 10, d. 101, 1970, ll. 9, 10.

85. GARF, f. 6903, op. 36, d. 1, 1971, ll. 461 (quote), 462.

86. GARF, f. 6903, op. 1, d. 542, 1958, l. 50; ibid., d. 612, 1959, l. 23; ibid., d. 666, 1960, l. 22; ibid., d. 724, 1961, ll. 2, 40; ibid., op. 10, d. 46, 1963, l. 3; ibid., op. 31, d. 14, 1959, l. 13.

87. GARF, f. 6903, op. 10, d. 46, 1963, l. 7.

88. GARF, f. 6903, op. 48 ch. 1, d. 22, 1971, ll. 7, 8, 11, 12.

89. GARF, f. 6903, op. 1, d. 883, 1965, l. 2.

90. For example: GARF, f. 6903, op. 36, d. 18, 1973: Obzor pisem telezritelei za aprel' 1973 g., ll. 1, 4–17; *Obzor pisem telezritelei za 1979 god* (Moscow: Gosudarstvennyi komitet SSSR po televideniiu i radio 1980), 6.

91. GARF, f. 6903, op. 36, d. 1, 1971, l. 461; *Obzor pisem telezritelei za 1981 god* (Moscow: Gosudarstvennyi komitet po televideniiu i radio, 1982), 167.

92. GARF, f. 6903, op. 10, d. 18, 1959, l. 26.

93. GARF, f. 6903, op. 36, d. 1, 1971, l. 461.

94. GARF, f. 6903, op. 10, d. 89, 1969, ll. 57, 57 ob.

95. *Obzor pisem telezritelei 1973 g.*, 5.

96. GARF, f. 6903, op. 36, d. 18, 1973, l. 15.

97. *Obzor pisem za 1980*, 9.

98. Evans, "Song of the Year and Soviet Culture in the 1970s," 635–636.

99. *Obzor pisem 1979*, 6.

100. GARF, f. 6903, op. 36, d. 1, 1971, 462; GARO, f. R-4237, op. 1 (prod.), d. 1051: Materialy televizionnykh peredach redaktsii otdela pisem, tom III, 1964, ll. 142–146.

101. For example: *Obzor pisem telezritelei, ianvar' 1972 god*, Moscow: Gosudarstvennyi komitet po televideniiu i radio 1972, 17; *Obzor pisem 1979*, 9.

102. GARF, f. 6903, op. 36, d. 40, 1975: Obzor pisem telezritelei. Ianvar' 1975 g., ll. 10–11.

103. GARO, f. R-4237, op. 1 (prod.), d. 876: Materialy redaktsii pisem, tom II, 1963, ll. 114, 116, 120; GARO, f. R-4237, op. 1 (prod.), d. 877: Materialy redaktsii pisem, tom III, 1963, l. 99; GARO, f. R-4237, op. 1 (prod.), d. 1051, 1964, l. 147.

104. Interview with Valentina V. (Samara, born in 1937).

105. GARF, f. 6903, op. 36, d. 1, 1971, l. 462.

106. GARF, f. 6903, op. 10, d. 89, 1969, l. 130ob.

107. See a typical example of the popular music program "Utrenniaia pochta" (Morning post): http://www.youtube.com/watch?v=e5s_-MdQrhI [15:20] (accessed September 1, 2014).

108. *Obzor pisem telezritelei 1973 g.*, 15.

109. Cf. Kozlov, *The Readers*.

110. RGANI, f. 5, op. 90, d. 90, 1984/85, ll. 40–41, 48, 65, 69, 99–100.

111. GARF, f. 6903, op. 1, d. 724, 1961, ll. 40, 86–87; RGANI, f. 5, op. 67, d. 133, 1974/75, ll. 19-23.

112. GARO, f. R-4237, op. 1 (prod.), d. 1051, 1964, l. 146.

113. TsGA SPb, f. R-7384, op. 42, d. 332, 1963: Zhaloby i zaiavleniia trudiashchikhsia i otvety na nikh Ispolkoma Lengorsoveta o nepravil'nosti deistvii uchrezhdenii, organizatsii, predpriiatii i otdel'nykh rukovoditelei, t. 1, ll. 190, 193, 193ob; RGANI, f. 5, op. 90, d. 90 (1984/85), ll. 32–37.

114. TsGA SPb, f. R-7384, op. 42. d. 332, 1963, ll. 84, 85, 89, 89ob. However, the files in most cases do not report on potential successes.

115. GARF, f. 6903, op. 10, d. 89, 1969, l. 169.

116. Yurchak has described participation in Komsomol meetings, elections, parades, or voting for resolutions in order to confirm being "a good Soviet citizen" as a complimentary phenomenon that, however, represented the frozen side of authoritative discourse and its performative shift. Alexei Yurchak, "Soviet Hegemony of Form: Everything was Forever, Until it Was No More," *Comparative Studies in Society and History* 45, 3 (2003): 499–504; Bren, *The Greengrocer*, 206.

117. Pravda, No. 223, 10.8.1972, 2.

118. Pravda, No. 223, 10.8.1972, 2.

119. GARO, f. R-4237, op. 2 (prod.), d. 677, 1969: Dokumenty televizionnykh peredach promyshlennoi redaktsiia, l. 67.

120. GARO, f. R-4237, op. 2 (prod.), d. 677, 1969, ll. 68-76.

121. GARF, f. 6903, op. 1, d. 724, 1961, ll. 39, 40ob.

Chapter 5

Addressing the Viewer-Consumer

Consumer Issues on Soviet TV

"*Obrashchaius' k vam za pomoshch'iu*" (I ask for your help) was a very familiar phrase in the letters Soviet citizens addressed to state authorities, government and party officials, and mass media.[1] In the following case from October 1964, TV viewer Valentina G. Saenko from Novoshakhtinsk chose to contact the letter department of the regional Rostov TV station. Saenko was inspired to write after watching a feature on the Rostov domestic services center (*Biuro dobrykh uslug*). She reported about problems she experienced at her local services center, where she had brought two items of clothing for drying and cleaning. Saenko complained that the center had not kept its "ready by" dates. She received back only one piece of clothing, without its hood and belt. She turned to the regional TV program because she thought that it represented the most effective way of pursuing her request: "It would not be bad, if the employees of the service agencies of our city adopted the practice of the Rostov service people. In fact, we have many shortcomings in this business . . . I ask you for help . . . remind the domestic services center in the city of Shakhty, which has an intake location in Novoshakhtinsk, that one cannot work this way."[2]

Saenko's complaint was broadcast on the TV program *Pogovorim o vashikh pis'makh* (Let's talk about your letters). The Rostov television's letter department created this twenty-minute program to communicate with the audience about local matters in an involving and engaging way. The head of the department Georgii S. Morozov presented the program, giving a face to the mostly anonymous TV staff behind the camera.

According to the letter department, many letters explicitly thanked the staff of local service facilities, industries, and the retail sector. The letter department aimed to further reinforce this seemingly strong relationship between employees of local companies and local consumers. Therefore, Morozov typically applied the control strategy of criticism and self-criticism. He first praised the importance of the staff and called them a "huge army" that was becoming more and more visible, because the life and standard of living of the Soviet people were steadily improving, making headway toward communism. After presenting two stories that expressed the appreciation and gratitude of some Rostov consumers for the engagement of local employees, Morozov turned to certain aspects that needed improvement. He quoted from letters that "rightly criticized flaws in the area of consumer services." TV viewer Saenko's demands seem to have been successful in that Morozov sternly asked the Shakhty domestic services center to put things right: "We expect the workers of the Shakhty domestic services center to take the most efficient measures."[3] However, the files do not reveal whether the letters really brought about improvements or just remained appeals. At the very least, they generated publicity that was hitherto unheard-of and opened a site of communication between viewers, letter writers, TV staff, factories and retail services, and party and state institutions responsible for consumer issues. The letters promoted and sometimes even launched televised discussions of consumer issues.

The fact that media consumers addressed a mass medium on consumer matters was as such not a new phenomenon. It is remarkable, however, that the Rostov television station evoked viewers' trust in advocating consumers' interests as early as the beginning of the 1960s in a way that seemed to distinguish TV from other media. This raises questions of how Soviet television stepped into the long-established discourse on consumer issues on the one hand and how it related to and engaged in "authoritative discourse" (Alexei Yurchak) on the other. Against a background of overwhelming public rituals and stifling routine procedures, TV was supposed to inform and educate the viewer, but also to entertain the audience. Therefore, the authoritative discourse that might work well at meetings, elections, in speeches, or in press releases needed to be staged differently on television. It perhaps needed dramatization and narration, so the audience could be prevented from switching off their sets. These aspects have barely been examined up until now, so that we are able to present an initial hypothesis.

As we have seen in the previous chapters, TV was a rather ambivalent force of privatization, promoting the new home-centered Soviet lifestyle and the nuclear family. As starting point for negotiations between "ordinary citizens" and the regime about new meanings of everyday life, television made home, consumption, and leisure part of a continuum between the public and the private spheres whose boundaries were permeable and shifting. But in what way did TV transform these negotiations on the new Soviet way of life within the shifting continuum of private and public life? To what extent did

TV allow for more ambiguities and ambivalences in the authoritative discourse that framed and interlaced this continuum?

This chapter is an inquiry into the communication between Soviet television and its audience about consumer issues. It refers to the local TV stations of Rostov and Leningrad, as well as Moscow Central Television. I highlight the communicative strategies TV stations employed to address viewer-consumers. How did Soviet TV aim to involve the audience? How did TV offer viewer-consumers information on consumption issues? How did viewer-consumers take up this discourse, and to what extent did they claim their rights on consumption via letters to television stations?

The next section gives a rough idea of the relations among material consumption, politics, and media coverage in the Soviet Union after Stalin's death. This conjuncture constituted the space of communication into which television stepped, taking up the discourse on consumption with its own program features. The second and third sections explore how TV programs like Rostov Television's *Pogovorim o vashikh pis'makh* and Central Television's *Bol'she khoroshikh tovarov* (More good products) visualized consumer issues, how viewers reacted to them, and how television interlaced concrete cases with broader debates on consumption.

CONSUMPTION, POLITICS, AND MEDIA COVERAGE

Consumer policies during the Khrushchev and Brezhnev eras were part of a variety of political strategies that aimed to provide the Soviet system with new sources of legitimacy. The Communist Party tied the legitimacy of the socialist system even more strongly than before to the idea of material abundance for all.[4] Recent research has demonstrated that consumer rights were a central topic of political communication in socialist societies.[5] Promising to catch up with the United States, Nikita Khrushchev introduced into the public discourse the claim of a rising living standard for all Soviet people. The downside of this was that the regime not only promised more goods, better services, comfortable apartments, and improved household technology, but also expected "ordinary" people to become "proper" Soviet consumers equipped with the "right" socialist tastes, wishes, and consumer practices.[6] Consumption and choice, with all the related aspects of housekeeping, furnishing, dressing, styling, or watching TV, became matters of ordinary people's interest. In the 1970s, Soviet consumers learned to be selective, paying more attention to design, technical aspects, and brands.[7]

At the same time, consumer interests were mostly ignored by Soviet trade and industry. Among other things, the quality of many consumer durables hardly improved, which remained a problem until the fall of the Soviet Union. Thus, staging consumer interests on TV also had the pragmatic dimension of instructing citizens how to claim warranty in cases of defective commodities or bad consumer services. The Soviet government developed a voluminous legal framework to regulate consumer protection.[8] This, however, functioned as a fig leaf, but reflecting the growing societal claims for consumer protection and for more and better goods. Consistently, newspapers and journals began to cover consumer problems more emphatically and the fight for consumer rights.[9]

Print media coverage of consumption dated back to the campaigns of the 1920s and the early 1930s.[10] As soon as the Communist Party realized that radio offered new opportunities not just to transmit political information but also to convey the cultural mission of bringing *kul'turnost'* and the new Soviet lifestyle to Soviet homes, the Radio Committee launched the program *Peredachi dlia domashnikh khoziaek* (Broadcasts for Housewives).[11]

These campaigns gained considerable further quantitative and qualitative strength throughout the 1960s and became more and more diversified through new media genres. Both the formal enforcement of consumer rights and increasing media coverage made this period fundamentally different from Stalinist times.

Media representations also corresponded with the invigorated genre of advice literature, popular science periodicals, housekeeping books, and commodity dictionaries during the Khrushchev period. These publications aimed to educate "new" socialist consumers.[12] They covered many fields of daily life such as hygiene, health, cooking and nutrition, childrearing, public etiquette, and furnishing. The advice publications instructed the consumer in how to use commodities and organize their life around them.[13] While trying to engage the reader, they consolidated the illusion of an attainable perfect consumer world and visualized material abundance.[14]

After Stalin's death, newspapers and journals like *Rabotnitsa, Krest'ianka, Zdorov'e*, and *Ogonek* presented visual representations of consumer issues. In the mid-1950s, *Ogonek* had several illustrated editorials with pictorial motifs that concentrated on the theme of the paternalistic welfare state providing good vacation, cultural, and educational facilities. From the late 1950s on, the journal increasingly depicted happy people pursuing leisure activities and accentuated the rising mass production of consumer goods. *Ogonek* also highlighted the "cultured" way of shopping with respectful saleswomen and no queues.[15] In contrast to their East German colleagues, *Ogonek's* journalists refrained from visually criticizing conditions for the consumer. Although they described certain grievances in written form, they only rarely used photographs or illustrations to criticize shortages. And if they did, they contrasted any negative examples with cases of excellent sales and service practices.[16]

These observations should suffice to indicate the political implications of the long-standing media discourse on consumption into which television entered with its own programs.

Exploring Moscow Central Television's schedules, one discovers several programs that sought to advise viewers on problems of everyday life and raise consumer issues.[17] The first of these programs were already launched in the late 1950s. American radio and television correspondent Irving R. Levine depicted several advice programs on Soviet TV that ranged from regular dance lessons to hints on housekeeping and gardening.[18]

Television took up the established traditions of print media. It adopted the advice genre and stepped into the discourse on consumption and everyday life by presenting viewer-consumers' letters on screen. These programs covered a variety of topics; consumer letters, however, became an integral part of their schedules.[19] *Nasha pochta* (Our post), *Po vashim pis'mam* (Following up on your letters), *Na voprosy telezritelei otvechaet ministr . . .* (Minister . . . responds to viewers' questions), *Stranitsy vashikh pisem* (The pages of your letters), *Otvety na pis'ma zritelei* (Answers to viewers' letters), and *Govorim po vashim pis'mam* (Let's talk about your letters) were just a few programs broadcast by Central Television. It would, however, be inaccurate to suppose that Central Television was unique in developing the letter genre and presentation of consumer issues. The abovementioned example of the Rostov feature *Govorim po vashim pis'mam* demonstrates that regional TV stations also televised consumer issues by presenting letters focusing on these topics.

Besides these local and central letter programs, program planners developed advice programs that also addressed the consumer. As early as in the mid-1950s, Central Television and local Leningrad Television launched the program *Dlia vas, zhenshchiny* (For you, ladies), which covered questions TV editors assumed to be of interest to women: fashion, cosmetics, furnishing, raising children, and housekeeping.[20] In 1960, the Public Council of Television Viewers welcomed this program, but reprimanded, its producers for specifically including advice on housekeeping and sewing for women.[21] The council members debated in which way a television magazine could address women with issues that were interesting and political at the same time. A solution turned out to be to show the new Soviet lifestyle: "In our women's magazine we need to tell and show how to live in a new fashion."[22]

Further, *Dlia doma, dlia sem'i* (For the home, for the family) was one of Central Television's first advice programs.[23] A serious problem facing all these programs was that until the early 1970s they were broadcast only sporadically and infrequently.[24] *Dlia doma, dlia sem'i* was apparently aired only twice on Saturday evenings in early 1963. Subsequently, the program planners gradually consolidated the program to reach a more stable audience.[25] *Dlia doma, dlia sem'i* offered advice on how to arrange a new flat, how to choose the right materials, colors, furniture, and home appliances.[26] Leningrad Television's advice program *Sovety molodym materiam* (Advice for young mothers), broadcast as early as 1962, also demonstrated that women constituted an important target audience from TV's earliest days. The program was presented by a professor of medicine who responded to questions about childrearing, hygiene, and the body that viewers addressed to him.[27]

Early audience research found that these topics sparked the interest of certain segments of the TV audience. Although audience research could not claim representativeness at that stage of its development, it constitutes an important source for exploring viewers' tastes and demands. As early as 1962, an overview of more than 2,500 questionnaires completed by employees of factories located in Moscow, Moscow Oblast, and some nearby cities like Tula, Kaluga, or Ryazan' disclosed that at least some respondents requested more programs on housekeeping, fashion, and domestic topics. Karpova, a saleswoman from Novomoskovsk was, however, perhaps one of only a few viewers who wanted more reports on "new trade organizations."[28] Central Television's 1966 internal report on letters confirmed viewers' growing interest in these topics. The report recorded increasing requests to "resume special programs for women with advice on household and family questions, cosmetics, fashion, cookery, and so on; realize a series on cultured behavior."[29]

This report unfortunately did not specify the actual number of these requests. But the interesting point is that the editors mentioned them, because they obviously considered such requests to have strategic importance for future program planning. Viewer preferences were of special interest, as the TV station's incoming mail was still rather sparse compared with radio stations or some national newspapers.[30] On the one hand, the idea to establish new advice programs gives us an indication that Central Television was yet to find its place in the media landscape. It was still struggling to adapt existing radio and print formats to its own possibilities and to stand out with its own profile. One strategy was to find out the best way to satisfy the audience's existing demand for features on consumer issues. Moreover, growing popularity among media consumers would have provided further arguments for television to claim its political impact and demand more financial support from the state. On the other hand, viewer requests show that consumer issues were a suitable topic with which to involve the audience, because viewers obviously believed that television promoted the "right" consumer practices and represented a kind of emancipatory claim in advocating for consumer interests.

The modern consumer advice genre was a very interesting media form that developed throughout postwar Europe on both sides of the Iron Curtain. In the 1950s and 1960s, it existed in the context of universal visions of social engineering and aimed to shape a modern society of mass consumers.[31] Thus, it was not only a socialist sociopolitical strategy or media genre and would benefit from closer examination in a comparative perspective, though such a comparison is beyond the scope of this chapter.[32]

Communicating with the audience via letters became a central strategy of Soviet TV because letters could be used not only in advice programs but also in other genres. As early as the beginning of the 1960s, local Leningrad Television broadcast eight programs per month based on viewers' letters.[33]

The practice of using letters was based on theoretical assumptions that were promoted by Soviet media experts and program planners, who attached importance to audience reaction because they perceived it as a way to tailor TV's communication strategies. From their perspective, letters seemed to be "the most perfect channel of communication" through which to analyze viewers' demands.[34] Viewer mail emphasized direct involvement with the audience and conveyed the idea of immediate communication with the viewers. Contemporary media producers, media scholars, media commentators, and even party officials were fascinated by the capacity of this new medium to bridge time and distance. They shared the view that television gave "the possibility to speak to the broadest audience. The heads of a city now get into hundreds of homes, into thousands of families at once."[35] *Pravda* published this statement by a provincial Gorkom (City Party committee) secretary in 1972 in a section called "Party Life: The Committee's Working Style." It documents that by the early 1970s party officials had discovered the communicative potential of television. Politics should now be visualized on the screen, and politicians were supposed to perform on TV programs in order to communicate with the audience. In the case presented by *Pravda*, regional politicians regularly discussed and answered viewers' letters on local Krasnodar television program *We inform, comment, explain.*

This new strategy was based on the idea that television generated immediacy, that it established viewer participation via letters and thus promoted a dialogue with the audience about the topics it brought up in the letters. Television was also an important tool for political propaganda: "We are convinced: The meetings on TV are a very efficient form of informing the populace. It allows the people to be kept informed about all events going on in the city, the region, the country, to react to them in a timely way."[36]

The television concept of talking to the audience via letters should represent close relations to the party and the government. To consider the often-cited "voice of the working people," the "signals from below," to hear people's recommendations, reports, problems, and requests still constituted the guiding principle of late socialist politics, albeit one that was given little more than lip service. Nevertheless, television, like any other public institution, was explicitly expected to meet this credo.[37] Television was, for example, supposed to teach "socialist competition" in its broadest sense. It should cover efforts to raise labor productivity; the conservation of raw materials, electric energy, fuel, and so on; the quick implementation of achievements in science and production technologies; and the improvements in the quality of consumer goods. The editorial departments—including those of youth and news and documentary programs—were supposed to report on these topics.[38]

Central Television realized the content-related demands in connection with the theoretical approach of viewer involvement, participation, and authenticity by presenting letters in a wide variety of features. Most of them were assigned to the editorial department of propaganda that was required to set consumer issues as one of the central sections of the letters programs and focused on the efforts to improve the quality of consumer goods.[39] Central Television's letter department characterized viewer mail as "serv[ing] as a source of content for new programs."[40]

Thus, the creation of new TV genres was closely related to the swelling mailbag, which in turn challenged TV staff to design new programs to engage with the audience response. During the 1970s, advice programs regarding lifestyle, body care, and material consumption gained further ground. In the early 1970s, television engaged in the expanding discourse on do-it-yourself practices launching the TV show *Eto vy mozhete* (You can make it). It became an evergreen show and was broadcast until 1991.[41] Also *Zdorov'e* (Health), *Chelovek i zakon* (Person and law), *Spravochnoe biuro* (Information office), *Dlia vas, roditeli* (For you, parents), and *Nash sad* (Our garden) responded to viewers' request for "practical advices."[42] Some of them were extremely popular. The letter department perceived thc "interesting, important topics" of these "necessary programs" to "provoke the activity of the audience." From the point of view of the TV staff, the advice programs contributed to viewers' "organization of instructive leisure."[43]

As letters became the all-purpose weapon of Soviet television to convey viewer participation in the 1970s, Central Television's department of propaganda designed new programs that used letters to publicize consumer requests. Let us first turn to the more classic letter programs that, among other topics, established a dialogue about consumer issues with the audience.

"LET'S TALK ABOUT YOUR LETTERS": TELEVISING CONSUMER INTERESTS

Presenting viewers' letters on screen was not just an exercise in reading them out in the manner of a radio presenter. Judging by Central Television's internal reports on viewers' letters from the 1960s, the letter genres faced certain problems and had to be carefully developed. At this time, the internal reports did not refrain from criticism. The editors of Central TV's letter department repeatedly criticized their colleagues in the editorial propaganda department who supervised the editorial subdivision of culture and way of life. The subdivision received a considerable number of consumer complaints, but the colleagues did "not show any concern about strengthening the connection to the viewers. They respond to the letters with great delay. . . . Such a way of working with the letters has become traditional in the department."[44]

Improvement in the processing and presentation of letters took a while to satisfy Central TV's letter department. The department was established in 1957 to register and answer incoming mail that did not directly refer to the program. It also became the main channel of communication between Central TV and its audience, because it monitored the work with the letters and drafted summaries of incoming mail. The delay the letter department criticized referred to the law that letters normally had to be answered by TV staff or should be forwarded to the responsible party or government institution within four weeks.[45] Still, in 1968 some sections of the propaganda department were criticized for their careless handling of viewers' post.[46] The criticism referred not only to the editors' work with letters behind the scenes, since more than a few viewers repeatedly demanded a better presentation of letters on screen. A locksmith from Volgograd Oblast, for example, pointed out that

> you read out only three—four comments. You did not talk about the category of your listeners precisely enough. Who responds more often to your broadcasts—younger or older people? Which topics and problems do viewers worry about? . . . You know, we television viewers would not only like to see the pile of letters on your copy desk but to hear more about their concrete contents.[47]

It was only by 1971 that the letter report admitted that editorial work with letters on television had generally improved during the previous six months, because it had been performed "more extensively and seriously."[48] Along with new programs created on the basis of viewers' letters, this signaled the increased importance of audience feedback.[49] Obviously, the TV editors learnt to cope with the challenges and demands of the audience, as well as with the guidelines of Central Television's management during the early 1970s. Part of this new strategy of more actively employing letters was that television—like other mass media—tried to capture people's opinions on social, cultural, and political topics. Generally, TV presenters and anchors encouraged viewers to report their grievances and address the corresponding program departments.[50]

As early as the beginning of the 1960s, many viewers welcomed television's efforts to incorporate consumer issues into programming. Although people were still occupied with rather pressing problems concerning housing, pensions, health care, or the search for relatives and friends lost during the war, TV viewers started to write about consumer demands, report product flaws, and raise questions about material consumption that exceeded basic needs.[51] This led to a noticeable increase in incoming mail during the 1960s. Central Television's propaganda department that supervised the consumer complaints programs recorded a swelling mailbag until 1981. This was not least due to new formats like *Po vashim pis'mam*, *Na voprosy telezritelei otvechaet ministr . . .* , or *Chelovek i zakon*, all launched in 1971. These programs obviously made social, political, and economic issues the subject of discussions that were central to the official ideology and aimed to legitimize the Soviet regime. Nevertheless, they quickly became popular because they related the issues to the television viewers' daily life, repeatedly broached consumer issues, and thus involved the audience.[52] The letter department asserted that *Chelovek i zakon* was initiated by viewers' demands to propagate knowledge of Soviet law. The episode that got the most responses (367 letters) during its first year of airing was the broadcast of December 1971. This led to a press conference by the attorney general who discussed the legal struggle for the better quality of contents.[53] Central Television systematically covered the topic of raising the quality of consumer goods over this initial period. The editors of *Na voprosy telezritelei otvechaet ministr . . .* invited the deputy attorney general to answer viewers' questions about the measures his agency had taken to control product quality. Like *Chelovek i zakon*, this series received by far the most letters (571) during the first year it was aired. In one of the first broadcasts, the editors of *Na voprosy telezritelei otvechaet ministr . . .* also invited Nikolai V. Timofeev, minister for the timber and wood-processing industry. Directly responding to questions viewers had raised in their letters, he explained how his ministry tried to meet increasing consumer demand for furniture.[54] This appearance inspired fifty-one viewers to address the editorial offices, whereas the following series that presented the minister for consumer services received only four letters.[55] The internal report on television viewers' letters did not explain these fluctuations. The higher number of incoming letters to advice programs like *Zdorov'e* or *Chelovek i zakon* could be due to the fact that these programs more easily engaged in communication with viewers. This is because they offered more reliability with regard to the presenters, the televised topics, and airtime. Health and law were topics broad enough to attract a variety of questions. Viewers could address editorial offices whenever they wished and were not bound to a particular episode to raise their question. Thus, these programs established a continuous flow that, in the case of *Chelovek i zakon,* regularly included consumer issues. TV staff had further learned during the 1960s that viewers much more readily accepted programs when they were regularly presented by the same hosts at a regular time.[56] Recognition value thus became the key to higher audience response and was perhaps the missing element in a format like *Na voprosy telezritelei otvechaet ministr . . .* that constantly changed its topics and panelists. The lower number of incoming letters did not, however, necessarily mean that viewers did not watch this program, but the lacking response might have pointed to the program's insufficient ability to create credibility.

In 1973, the editors of the propaganda department were proud of the highest gain in viewer mail of all editorial offices.[57] This was reflected by a rising number of letters presented on screen: The editorial department of propaganda presented 11.8 percent of its mail in its programs in 1968, only 9.0 percent in 1972, but 20.8 percent in 1973.[58] The new letter policy corresponded with the general tendency that most of Central Television's editorial departments brought more letters to the screen.

In this way, editors tried to materialize the idea of immediacy and viewer participation. Letters were not only thought to denounce all kind of problems but were supposed to establish a real-time dialogue between viewers and the medium in front of a regional or—depending on the program—even national audience.[59] This assessment still prevails among media scholars. They argue that television as the agent of bridging time and space conveys the "immediacy" of connection and communication.[60] British cultural sociologist John Tomlinson reminds us that television actually hides its practice of mediation and artifice of its presenting modes.[61] The practice of mediation would provoke a dramaturgical effect of a pristine and immediate communication between television and audience via viewers' letters. And indeed, exploring the way Soviet TV programs televised letters, one could agree with Tomlinson's idea that television promoted the illusion of an "untouched" communication.[62] The impression of close bonding between viewers and the medium derived in no small part from Soviet TV's practice of presenting letters to the camera. That these letters at first sight often covered individual consumers' interests allowed the communication to appear even more authentic. Televised authenticity reached a much broader audience than any newspaper and was the main asset on which television based its strategy to relate individual and societal interests.

As in the case of TV viewer Saenko from Novoshakhtinsk mentioned at the beginning of this chapter, editors normally aimed to embed individual cases into a societal context and identify them with broader political and economic problems. The presenter of Rostov Television's *Pogovorim o vashikh pis'makh* placed the complaint in the politically delicate context of the general improvement of the Soviet standard of living on the way toward communism: "The Soviet government invested more and more money during the last years to expand and develop the cultural and consumer services in the city, as well as in the countryside. And the people have the right to demand that this enormous amount of money causes maximal effect and is for their benefit."[63]

Similarly, this local TV program used other grievances about local consumer services, food industries, or retail trade to make official discourse visible and tangible to ordinary consumers.[64] Letters addressed to television thus communicated with politics, because the immediacy of television also provoked an impression that the letters would lead to action.

Rostov Television was not the only local station that aired its own feedback program connecting local consumer matters to the official discourse on consumer issues. Letter programs like *Otvechaem na voprosy Moskvichei* (Responding to the questions of Muscovites), aired by Central Television's editorial department for the Moscow region starting in 1973, and Leningrad Television's *Televizionnyi ezh* (Television hedgehog) from the early 1960s also framed dialogues with local color. *Televizionnyi ezh* was based on letters and often used satire, parody, and caricatures in covering consumer issues. The program's host during the 1970s and 1980s was popular writer and columnist Aleksandr Matiushkin-Gerke. In 1971, Leningrad Television's *Televizionnyi ezh* received the second highest number of letters among all local broadcasts and 5.1 percent of all viewers' mail. Leningrad Television's letter programs enjoyed the privilege of being occasionally broadcast on Central Television's first and second channels.[65]

This widespread airing of consumers' letters inspired dynamics between different segments of the public—the complainants, the indirectly involved television audience, the accused actors in trade, industry, or consumer services, and the various state and party authorities responsible. TV was the only medium that gave authoritative discourse voices, faces, and perspectives simultaneously. TV confronted the rather static, immutable, and predictable language of written and spoken authoritative discourse with its specific immediacy and authenticity. The effect of TV was to visualize the communication between viewer-consumers and the regime. Television allowed it to appear immediate and authentic, as well as to give it an interactive quality. This was the case when a group of citizens of Rostov-on-Don first addressed the local city council and then turned to the local TV station after waiting for a response to their request for half a year. The anchormen of *Pogovorim o vashikh pis'makh* presented their letter and concluded that "we [he and the audience] hope, that the city council will nevertheless answer their request."[66]

The host demonstrated television's typical strategy of creating an interpretive community of television viewers, letter writers, and television staff who stand up for consumers' interests. During programs *Pogovorim o vashikh pis'makh* and *Sprashivali—otvechaem* (You ask, we respond), the presenter confidentially addressed the audience as *Rostovchane*, *tovarishchi* (comrades), and *druz'ia* (friends). He also mentioned the names of the letter writers and occasionally even read out their addresses. The TV announcers also tried to consolidate the bonds between the audience and the local TV station by ending the program with the invitation to send complaints, to write about their troubles and problems: "*Pishite nam, tovarishchi, zhdem vashikh novykh pisem*" (Write to us, comrades, we wait for your new letters).[67]

Whereas *Pogovorim o vashikh pis'makh* emphasized dialogue with the audience, Rostov Television's letter department also broadcast *Sprashivali—otvechaem* that, as early as 1963, followed a more active and interactive script. This fifteen-minute program aimed to suggest that the audience was the judging authority that could voice its requests on TV. However, we can observe a certain development of the concept over the 1960s. The episode *Dlia vashego stola, khoziaiki* (For your table, hostesses) in September 1963 covered the problem of vegetable supply in the Don region. The broadcast was introduced by an invisible announcer whose colorful narration about the upcoming harvest framed the picture of autumnal Rostov. The camera moved on to farmlands in the Don region, showing workers harvesting their crops. According to the script, the images should convey that "joyful, enthusiastic work is to be felt everywhere." Then the TV camera showed a line of cars, tracing the route of fruits and vegetables to the consumers. Back in Rostov at the "Proletarian market," the audience watched how busy the vegetable trade was. The consumers appeared satisfied with the large selection of fruits and vegetables. In the background, the announcer quoted Nikita Khrushchev, who asserted that vegetables were a must for every table. The anonymous announcer went on praising the party and government for further improvements in the food supply. He also applauded the hardworking staff of retail outlets who were busy processing the harvests. Meanwhile, the viewers saw a large Rostov warehouse where all the fruits and vegetables were stored and then delivered to the shops. Then the script inserted a cut and the program's host, the abovementioned head of the letter department Georgii S. Morozov, took over. He asserted that retail employees did a lot; the television viewers' letters, however, demonstrated that there was still much to improve. As the viewers had requested, he invited the heads and managers of local retail outlets to directly respond to questions certain viewers raised in their letters and to generally explain which measures would be taken to improve the situation for Rostov consumers. *Sprashivali—otvechaem* further emphasized the involvement of the audience, critical investigation, and the claim to link separate segments of the public by presenting a roundtable with the director of a local vegetables distribution center, the deputy head of the regional trade department, a local journalist, and a member of the Rostov Television viewers' council. The camera focused alternately on individual participants when they rose to speak but also took a long shot of all speakers to stage a dynamic discussion. Criticism was brought up by letters, the journalist who presented his observations, and the council member who had visited several local shops in order to talk to the salespeople and consumers about their experiences.[68] The viewers heard a lot of information on production data and the measures party and government authorities were about to take. As always, the ideological practices of critique and self-criticism were prominent, but ultimately praise of achievements prevailed, painting a hopeful, optimistic picture of the future of food consumption.

We will now take a brief look at a later episode of *Sprashivali—otvechaem.* In 1969, the editors produced the episode "Vegetables in the shops," claiming that they would not "show the piles of envelopes and to say that we can't read out all the letters." Instead, they gave the screen over to the viewers themselves. To stage a more engaging discussion, the editors interviewed consumers in front of grocery stores, asking "whether residents of Rostov satisfied with the supply of vegetables in the city this summer."[69] This was no live event, although the report offered much more immediacy and authenticity than print media version could: the audience heard selected audio dialogues with consumers, while still photographs depicted the interviewees who were identified by names and workplaces. They were asked what they had just bought, if they regularly shopped in this store, and if they were satisfied with the supply. Afterward the editors submitted the comments of interviewees and letters received to the head of the local trade administration, Vsevolod Kirillovich. The program's host let him answer eleven questions that had been prepared in advance. The head of the trade administration used many typical linguistic markers of progress like "increase of capacities and choice," "fulfillment of schedules," "extension of the trade network," and "the speed of delivery increases from day-to-day."[70]

This conversation was a perfect staging of authoritative discourse. Its literal content was rather predictable, boring, and formulaic. The production data could well have been printed in the local party newspaper and would probably have been ignored by a larger audience. That they were presented on screen by a local official, however, created the possibility for new meanings. The presentation of a local official was itself new and interacted with the innovative way of presenting local consumer voices. Thus, TV personalized, localized, and more strongly disseminated the authoritative discourse. Television also moved the discourse much closer to the everyday life of local viewers. The statements of consumers could be critical as well as enthusiastic, but they only seldom reproduced such explicitly ideological markers. Thus, the utterances in the interviews and letters mirrored the increased spaces that, in this case, the local TV opened for voluntary and subjective input into the discourse.

"MORE GOOD PRODUCTS": ADVOCATING CONSUMER INTERESTS

From the early days of Soviet television, viewers steadily complained about the insufficient or boring visual content of news, informational, and documentary programs. An overview of viewers' letters from 1968 assured that "the viewers, who understand well the specificities of television, ask as much as possible for illustrations of the information."[71] By the end of the 1960s, the rise in these complaints referring to the character of the medium convinced TV editors that they could further increase viewers' enthusiasm by better exploiting the technical possibilities of television. Viewers were especially critical of the fact that programs often simply repeated newspaper materials. In turn, several viewers commended those editors, journalists, and commentators who "prepared their own original performances, interviews, and reports."[72]

This criticism had an impact on the development of new genres. Television of the 1970s was generally characterized by the professionalization of documentary and news programs, as well as by diversification of the schedule. Although consumer issues were not central to program planning as such, they became an integral part of the programming sector that covered advice, information, and education. The diversified screening of consumer topics also reflected the new self-representation of television as a journalistic authority: henceforth, television deliberately presented more critical materials and disclosed all kinds of flaws, shortcomings, mismanagement, or bad organization of the production on screen. This new strategy also included advocating consumer interests and addressing product flaws.

Diversification of programming was not only determined by technical developments, party policies, and initiatives of TV staff, but also resulted from the interplay of these factors with audience responses and expectations. Thus, new programs that included consumer issues reflected attempts at reshaping television to meet audience demands. The rising interest in consumer issues was illustrated by the fact that in 1968 viewers sent more than 3,000 packages containing poor-quality household goods to the "satire" section of the propaganda department. A viewer from Lugansk Oblast highly praised this section in his correspondence: "Television does a great thing; it struggles with bunglers and other problems."[73]

To better meet these kinds of expectations the editors of *Po vashim pis'mam* created the series *Tovary—narodu* (Consumer goods for the people) that first aired in January 1972. It was one of the first broadcasts that focused completely on consumer issues and paved the way for the even-more sophisticated program *Bol'she khoroshikh tovarov*, launched not long after in February 1972. Both were intended as political propaganda and were produced by the propaganda department. The first episode of *Tovary—narodu* presented how the resolutions of the XXIV Party Congress were being realized to meet people's demands for mass consumer goods. The self-promotion of the series' editors was not surprising: they asserted that the program would cover "important, current problems"[74] and urged viewers to engage in discussions about the quality and selection of goods and services.

Although the program was based on letters, the audience response of only fifteen letters to the first two airings of *Tovary—narodu* in 1972 was disappointing. However, the numbers consistently grew over the first year. The editors received 422 letters over the course of 1972. Viewers who addressed the editorial offices welcomed the program: "It is right that television airs those sorts of important questions such as the rising production of consumer goods and quality improvements," wrote a viewer from Naro-Fominsk near Moscow. Other viewers who lived in villages in the Urals and in Crimea had more practical concerns, asking where they could buy all the commodities that were presented on screen.[75]

Tovary—narodu experimented with new formats. In the beginning it combined answers to viewers' letters with elements of advice, as well as investigations into viewers' complaints about bad product quality. The editors took their camera and went on location to interview the heads of the accused factory or service office. The program well represented the new practice of critically addressing sociopolitical and economic problems. The Soviet journalist and television scholar Aleksandr Iurovskii proposed that the ideological aim of this practice was to "affect the audience" and engage viewers in the struggle against flaws and mismanagement.[76]

Viewers especially welcomed this new active strategy of confronting the responsible persons with criticism and thus advocating for consumer interests. A woman from Omsk commended that "your programs inspire confidence that, although rejected goods are still produced, one could strive for positive results in this question and that only high-quality products will get on the shop counters."[77]

Interestingly enough, some viewers admonished the editors for forgetting to report on the quality goods that were out there. Others suggested that the program should further broaden its scope. They proposed that the editors should report on new products and attempt to educate viewers' tastes.[78] A black-and-white clip on YouTube called *Tovary—narodu* gives us a notion of how television took viewers' requests and presented new commodities to the audience. We see a gray-haired host wearing a dark suit in the midst of a variety of consumer goods produced by Khar'kov factories in 1980. He hardly moves and only occasionally takes an umbrella, feather duster, or thermos flask in his hands, presenting them to the camera and praising their qualities. It is only the camera that moves by panning between the shelves on which TV sets, radios, cameras, a slide projector, a music system, a fan, a lamp, remote-controlled space vehicles, or chessboards with historical figures for pieces are placed.

The TV viewer might have received the impression of choice but could barely see any details of the presented goods in the long shot. The host names certain goods and the camera zooms in. The presentation lacks dynamism, as the host does not explain how to handle the electric devices. Instead, he assiduously praises the latest achievements of Khar'kov industry. From today's perspective one imagines a more colorful and vivid presentation demonstrating how to use the commodities, with commentary on how they would improve consumers' everyday lives. At the end of the clip, the host turns to the explicit political significance of his presentation. He embeds the promotion of the local products in the current party campaign that promises to raise Soviet living standards in the run-up to the Eleventh Five-Year Plan and the XXVI Party Congress in February 1981.[79]

This kind of performance was rather typical for the Soviet discourse on consumption that aimed to combine quantity and quality, although consumers steadily complained about limited access to and poor quality of products. In all likelihood, this presentation did not exploit the technical possibilities of the medium and thus did not cause real excitement among the audience, as it appeared like a televised product nomenclature. Yet, a TV program was easier to access and at least offered a quick overview.

Bol'she khoroshikh tovarov was also supervised by the propaganda department and was from the very beginning connected to political campaigns to raise Soviet living standards and product quality. The following anecdote illustrates this relationship to authoritative discourse even better: renowned Soviet TV critic and film scholar Vasilii G. Kisun'ko is quoted as saying that Leonid Brezhnev himself proposed the title of the program to Sergei Lapin, the head of *Gosteleradio* (the State Committee for Radio and Television).[80]

The rising importance of consumer issues for the editorial work of Soviet television was further mirrored by a reorganization of this office in 1973. From this time, programs like *Bol'she khoroshikh tovarov, Po vashim pis'mam*, and *Na voprosy telezritelei otvechaet ministr . . .* were assigned to the editorial department covering issues of the Soviet industry.[81]

Bol'she khoroshikh tovarov was a thirty-minute program broadcast twice a month on Saturdays at 11:00 a.m. in the morning on Channel 1. It adopted the sections that *Tovary—narodu* had already introduced but benefited more from the technical possibilities of dynamic visualizations. Involving the audience and referring to topics raised in the viewers' letters, the editors aimed to create immediacy and authenticity. Therefore, they regularly took TV cameras to production plants and factories, filmed the production process, and interviewed employees, directors, and managers. The audience response obviously welcomed this strategy. *Bol'she khoroshikh tovarov* seemed to prove that the swelling mailbags documented an improved, diverse program. The editors of Central Television's letter department affirmed that the editorial offices paid greater attention to viewers' letters and tested "different forms to work with them on the screen."[82] Thus, in 1973 Central Television's propaganda department traced its threefold letter increase back to the longer programs, as well as new programs like *Bol'she khoroshikh tovarov* that emphasized letters in order to create an "embedded viewer."

The viewer correspondence sent to *Bol'she khoroshikh tovarov* in 1973 grew five times in comparison to 1972 but accounted only for 1.6 percent of the whole propaganda department's mail, placing it in the fourth place. The two other extremely popular programs—Sergei S. Smirnov's *Poisk* (The search) and the earlier-mentioned advice program *Chelovek i zakon*—were unrivaled and together received 85.3 percent of the entire department's mail.[83] Central Television's internal reports on letters from 1973 explained the rising success of *Bol'she khoroshikh tovarov* with reference to its "very broad use of viewer letters" that "impart massiveness, significance, fighting spirit. It was exactly with the help of the public (*obshchestvennost'*), using audience letters, that the department succeeded in contributing to the solution of economic tasks concerning the increase in production and the improvement in quality of mass consumer goods."[84] One could conclude from this kind of statement that the internal reports on letters were attempts by editors of letter departments to promote their own interests within the authoritative discourse. They aimed to solidify their own position within television with its different interest groups—artistic, technical, financial staff working behind or in front of the camera, and so on—as preservers of the dialogue between television and its audience. However, many viewers engaged in the dialogue via their letters and took the opportunity to participate in the field of communication that TV opened. Internal reports certainly represent only a small proportion of any program's mail. We cannot prove that they constituted representative samples or demonstrate how audience response related to the "silent" viewers who did not send letters to the editorial offices. Nevertheless, I suggest that the amount of incoming mail also hints at the number of "silent" viewers.

The upward curve of incoming mail reflects the audience's strong and rising interest up to the mid-1970s. Compared to *Tovary—narodu, Bol'she khoroshikh tovarov* quickly multiplied the audience response. It received 825 letters in January 1975 alone and more than 10,000 in 1979.[85] Judging from its mailbag, *Bol'she khoroshikh tovarov* was one of the most popular programs on Central Television. There were certainly programs that received much more mail, but there were also many broadcasts that received minimal viewer response.[86] The audience response to *Bol'she khoroshikh tovarov* remained constant until 1980 and then began to decline until the beginning of perestroika.

That viewers felt welcome to report their ideas and complaints seemed related to the efforts of TV editors to improve the program according to their perceptions of the audience requests. For this purpose, they asked viewers, for example, to respond to a questionnaire and organized a collective viewing with factory workers in Klin, not far from Moscow.[87] In 1976, viewers proposed that a regular segment should be included about repair services. This, at least in the editors' own portrayal, was the reason that motivated the editors to launch the recurring segment *Aktual'nye problemy kachestva* (Current quality problems).[88]

During the first two or three years after launching the program, the editors gradually refined the script. Up to 1975 they created segments such as *Otvechaem na vashi pis'ma* (We respond to your letters), *Telezriteli rasskazyvaiut* (Viewers tell), *Iarmarka brakodelov* (Trade fair of bunglers), *Televisionnyi fel'eton* (Televised feuilleton), and *Gosudarstvennyi znak kachestva prisuzhden* (State quality mark awarded) that were solely based on viewers' letters. In 1974, the editors added a segment to *Iarmarka brakodelov* that was called *Braku—zaslon* (Backstop against defective goods). It was presented by a member of the State Committee on Standards (Gosstandart SSSR) and was another example of how TV brought authoritative discourse to the screen. It was not simply a lecture on future improvements, but the presenter referred to specific complaints—a strategy that viewers welcomed and responded to by talking about their own experiences.[89]

Later the editors also included the segment *Komandirovka po vashim pis'mam* (Assignment to investigate your letters). It aimed to suggest that viewers were not only participants but also initiators of activity, as the editors took letters as an opportunity to visit accused factories and retail shops. The rubric *Posle nashei kritiki* (After our criticism) presented feedback and reactions of the blamed organizations on the screen. As many viewers asked to learn about concrete improvements and follow-up measures taken after criticism, the program introduced *Mery priniaty* (Measures taken) in 1975. *Mery priniaty* staged viewers' complaints and suggestions alongside the ensuing confrontation with the accused factories and the results. This was done by screening the reply of a state committee member in response to a critical letter sent in to Central Television. The staging, however, was not very attractive,

Table 5.1 Viewers' letters to *Bol'she khoroshikh tovarov*, 1972–1977

1972	*1973*	*1974*	*1975*	*1976*	*1977*
412	2,024	5,875	10,808	11,072	9,502

Source: Obzor pisem za 1973, 15; *Obzor pisem za 1975*, 151; *Obzor pisem za 1980*, 18; *Obzor pisem za 1982*, 26.

Table 5.2 Viewers' letters to *Bol'she khoroshikh tovarov*, 1980–1984

1980	*1981*	*1982*	*1983*	*1984*
10,571	8,645	7,762	5,661	6,609

Source: Obzor pisem za 1980, 18; *Obzor pisem za 1982*, 26; *Obzor pisem telezritelei za 1984 g.* (Moscow: Gosudarstvennyi komitet po televideniiu i radio, 1985), 15.

because the camera just focused on the letter and the voice-over read it out monotonously. Nevertheless, the message was obvious: the program claimed to protect consumer interests on the basis of audience participation.[90]

Internal reports on the letters reveal that the topics the viewers were concerned about and the expectations they had hardly changed until perestroika. The audience of *Bol'she khoroshikh tovarov* demanded critical journalism that made clear the gap between promises and reality and reflected the viewers' own experiences. The presentation of critical materials became the benchmark with which the audience judged the program. A viewer from Kuibyshev wrote in 1975: "The program *Bol'she khoroshikh tovarov* is topical, necessary, and very important. However, its criticism has lessened recently."[91]

The program complemented the broadcasts that covered consumer issues not only because of its critical attitude, but also because of its own specific claim to find solutions to the indicated consumer problems, to help viewers, and to reveal the persons responsible. The incoming letters demonstrate that the viewers who addressed the program readily accepted TV's self-representation as a critical authority throughout the 1970s and early 1980s. In this sense, a viewer from Baku reminded the editors in 1975: "I think that your program is not conceived as advertising for good commodities (for this, there is the magazine *New goods*), but to reveal the reasons for substandard products."[92] A viewer from Krasnodar region made the claim even clearer: "I wish, I strongly wish that public opinion from the side of the television turns out to be stronger than any administrative barriers in the struggle for product quality."[93]

The program's audience maintained these high expectations until perestroika. Even in 1985, a viewer from Tomsk encouraged the TV editors by saying that he considered the program to be "extremely necessary as it covers severe problems of our daily life."[94] Statements like this suggest that viewers still reproduced authoritative discourse. These statements, however, were also slightly ambiguous, as they reminded the regime of its own claim to improve the supply of consumer items and product quality. Nothing would have been more destructive to the Soviet system than breaking off this communication channel between audience and TV.

Likewise, the viewers were consistent—also until perestroika—with respect to the topics about which they wrote to the program. They focused on complaints about unsatisfactory consumer services, the badly functioning retail trade, inferior product quality, or insufficient supply of goods.[95] The editors of the internal reports probably had little scope to alter authoritative discourse, as they were themselves constitutive parts of it. However, slight ambiguities appeared even in the course of the formulaic reproduction of the authoritative discourse. The editors did not try to find any reasons when mail coming to the program started to decline in 1981. Also, they did not propose any new strategies to win back the interest of the audience. Nevertheless, with declining audience response the internal reports on letters addressed to *Bol'she khoroshikh tovarov* also lost their dynamism and empathy. They became processed more briefly, simply, and automatically. Whereas the editors portrayed all their serious efforts as causing positive audience response, after 1981 they typically started with a loveless routine sentence like this: "As usual, complaints dominate the mail to the program *Bol'she khoroshikh tovarov*." Thus, the year 1981, and the generally deteriorating economic situation in the Soviet Union, might have been a watershed with regard to the immutability of authoritative discourse on consumer issues.

Until then, the way viewers were quoted on the program suggests that these kinds of broadcasts were relatively successful in conveying the impression that television defended consumers' interests or at least contributed to the improvement of consumer issues. The editors blew their own trumpet and referred to letters that expressed gratitude in cases when television had offered practical help. These letters of thanks often sounded like this one sent by a couple from Karatau in Kazakhstan who successfully complained and then enjoyed their new TV set very much: "Dear employees of the program *Bol'she khoroshikh tovarov*! After we asked you for help, our rejected TV set *Gorizont* was replaced by a new television *Izumrud* in the stationary shop within half a month. This time the TV set is good, with excellent sound and picture. Many thanks for your help."[96]

These words of gratitude suggest not only that a certain number of viewers regarded TV as an advocate of consumer interests. The overviews of letters also painted a normalized picture of Soviet material culture that embraced its flaws and deficiencies. The televisual coverage further emphasized the significance of consumer items for Soviet everyday life. Programs like *Tovary—narodu* and *Bol'she khoroshikh tovarov* propagated a variety of images of consumption, consumers, and commodities, and constituted at the same time a space in which consumers communicated positive as well as negative emotional bonds toward consumer goods. The viewers did this in different ways. Whereas the couple from Karatau endowed their TV set with emotions, others drafted an almost-affectionate relationship toward a long-serving household item and converted its reliability to praise for the factory staff. Viewers repeatedly charged the individual significance of a consumer item in their private life with societal relevance. Letters like that of a woman from the Georgian city of Tkvarcheli show that viewers supposed their emotional bonding to Soviet lifestyle to be so common that TV seemed to be the obvious medium to convey their affirmative attitude to the audience:

> Twenty years ago, the refrigerator of the brand Saratov-II appeared in our family. It worked without any need for repair for all these years, on weekends,

> without any leave. We value this refrigerator very much, as we received it as a gift on our son's birthday. Our son turns 21, and the refrigerator works faultlessly. One needs to tell about such workforce collectives on TV, to give the floor to the management of the factory, to leading workers, that is to those who need to be thanked for honest work, for profound respect for the people. I ask the program's editorial office to report on the work of the Saratov Refrigerator Plant, about its employees, about their life and honest work, about how they succeed in putting out such a durable product that brings joy to every house.[97]

The choice of these quotes in the internal reports on letters followed the traditional Soviet rationale of criticism and praising of achievements. The editors gave viewers a chance to not only demand that successful factory staff be reported on but also request a ceiling on food prices. The explicit mention that viewers reproached the program for "publicizing consumer items that are not available for retail sale (e.g., sewing machines)"[98] and seemed to consciously reveal television's balancing act between two contradictory constraints: television had to satisfy viewers by advocating consumer issues and illuminating consumption flaws, but it also had to avoid systematic criticism of the sociopolitical system. This was a delicate task, as viewers themselves embedded the consumer issues of the program in the political context of the general improvement of the Soviet standard of living on the way toward communism: "Our life improves from year to year; the consumer demand for nice furniture, good coats, durable shoes increases. And it is very pleasant that our factories try to satisfy the growing needs of the population." In a similar way to this viewer from Chelyabinsk, a woman from Krasnoyarsk asserted that one could be "very happy to see how our welfare grows, how the product quality improves. Here in Krasnoyarsk we have not had any particular difficulties buying a television set, a washing machine, or a refrigerator for a long time now."[99]

If we take these statements as viewers' voluntary input to the discussion of consumer issues, we see how some viewers reproduced the linguistic markers of progress, better consumer goods, and an improving standard of living that characterized authoritative discourse. They also interlaced them with personal experiences, with local events and observations. The audience response to *Bol'she khoroshikh tovarov* demonstrated that the program made consumer issues an anchor point of communication. It enabled viewers to interact with authoritative discourse, to embed parts of it into their own perception, to reproduce it, but also to benefit from it. In this sense a viewer from Vladivostok applauded the segment *Aktual'nye problemy kachestva* that was said to have been inspired by viewers' ideas: "I did not know until your broadcast that factories plan warranty repair. You were right to say that the factories should be economically accountable to the consumer. . . . We have to find ways that make factories take a real economic responsibility towards the consumer."[100]

We could conclude from this kind of statement that the relationship between audience and television was quite interactive, although it began to decline in this program genre during the early 1980s. Until then, television had shaped the demands of consumers by visualizing consumer issues and by engaging in consumer politics.[101] TV programs like *Bol'she khoroshikh tovarov* suggested that television became an advocate of consumers' interests and created interpretative communities in front of the screen.

The curve progression of the incoming mail reflects the strongly rising interest until the mid-1970s. The audience response remained constant until 1980 and then started to decline slightly until the beginning of perestroika. Looking back, former Soviet viewers identified the clear change of communicative rules perestroika brought about for television. A user of a Russian internet forum commented that "against the background of the information boom of the perestroika" programs like *Televisionnyi ezh* "were perceived as anachronism."[102] The removal of communication taboos, the new possibilities to meet and discuss more openly than before, let letter programs quickly appear outdated.

The perception of this significant turning point further supported the observation that television played an important part in normalizing consumer issues by interlacing home, consumption, and leisure in a public-private continuum presented on screen. The televising of letters and the involvement of the audience quickly became an important strategy of Soviet television. In the 1960s, television adopted the strategy that newspapers and radio had already established by printing and reading out letters. Television thus did not revolutionize communication with the audience as such, but it gradually refined the advice genre and complemented the discourse on consumer issues throughout the 1960s and 1970s.

Reproducing the genre of consumer advice and information, it interlaced authoritative discourse with tangible questions of lifestyle and consumer taste, with personal experiences and local events in a more interactive, perhaps even intrusive way compared with print media and radio. Television did not simply replicate authoritative discourse. Mass media could push the commitment of its users but could not ensure or hardly control it. Boring television programs could easily be switched off. This meant that television as an audio-visual medium needed to stage authoritative discourse much more at the viewers' requests and to embed it into viewers' everyday life. It therefore aimed to involve the audience by inviting viewer-consumers to communicate with the editors about their experiences, ideas, and suggestions.

Television's offer to communicate was welcomed by many viewers and, to a certain extent, energized the seemingly fixed form of authoritative discourse. Television complemented the ritual practice of letter writing that allowed for criticizing local flaws with its potential dynamic of sound and images. It pushed the normalization of consumer issues by augmenting the practice of letter writing with elements of interaction and participation. The audience got used to watching how a popular TV program's presenter took a viewer's complaint about defective products and interviewed, for example, the deputy minister for light and food industries about measures to improve the situation of Soviet consumers. Viewers took part when journalists asked a local head of the administration of consumer services about problems of public housing, when they visited shoemakers and garment workers as representatives of local consumer services or asked the collective of a shoe factory about product improvements.

Soviet television's specific benefit was that its energized forms of audience participation further disseminated authoritative discourse about consumer issues because it was embedded in viewers' everyday life. This strategy was not without risk regarding its credibility. As an unintended consequence, it could have reinforced irritation because of the perception of difference between the personal experiences of everyday consumer life and official claims. The answer to this problem will, however, hardly be clear but remain ambiguous and manifold, as the perception and use of TV contents depended on many factors such as profession, education, gender, and age.

As we have seen in chapter 4, the viewers' letters rather suggest that the television's capacities of staging topics in a new way and of building a "third space" to negotiate with the audience constituted its most important communication effect. To put this observation simply: It had long been more decisive that certain topics were covered at all than how they were exactly staged. Letters and television, thus, diversified the space of communication and interconnected the public and private anew. Television posed a certain risk for authoritative discourse, but also offered considerable opportunities to involve the audience without contesting the basic mechanisms of communication.

It is obvious that writing letters was no mass phenomenon as such, but it was a commonly accepted practice to communicate with the regime. No Soviet viewer would have wondered about television programs based on letters or about the way these programs staged solutions to the issues raised by viewers. The retrospective views of the interviewees complicate this image in several ways highly depending on their former social position. The interviews also show that former TV experts as well as higher and averagely educated "ordinary" viewers rather supposed television to have broadened communication in favor of the audience. Although many of the respondents assure to have already contemporarily perceived taboos and strict rules of criticism, most of them even retrospectively do not openly accuse the medium of having created a "false reality" or intentionally deceived the audience. The common explanation comes, however, close to the accusation of a parallel reality as it bases on the assumption that the television had intentionally left out information. Typically, the sixty-six-year-old Russian German Waldemar underlined: "they never talked about problems, but only about achievements" until perestroika broke the taboos.[103] Whereas he seemed to have accepted this rather indifferently, the even critical and reserved viewer Aleksandra Apollonovna from Tyumen was much more upset about her perception "there was abundance everywhere on television."[104]

Regarding television's covering of consumer issues and consumption as topics of everyday significance, the narratives reveal huge differences of albeit retrospective perceptions and assessments. The archival materials clearly suggest that televising consumer issues was a substantial strategy to address the audience. Looking back, the experts paint a much more ambivalent picture. Rather reflective and partly very critical Rostov television director Ėvelina E. painted a different picture compared to early-quoted Kuibyshev TV editor Valentina V.[105] Whereas Valentina V. stressed television's agency in favor of consumer interests and the critical stance television took, Ėvelina E. was at first even astonished about the question if Rostov TV had staged consumer issues: "No, my dear, there were no goods and no products . . . well, we couldn't show empty shop counters? Who should authorize that? We did not touch this topic, on principle did not touch it."

Obviously, Ėvelina E. on the one hand thought of consumer issues as critical moments that were tabooed due to Soviet communication rules. On the other hand, she revealed a rather nostalgic notion about Soviet equality. It does not come as a surprise that television did not present empty shelves and people queuing. The strategy was more sophisticated and somewhat clumsy, as well as simple at the same time. Ėvelina E. summed up the inflexible strategy like this: "Our television certainly was 'Hurray!,' with the sign 'plus,' with the exclamation 'Hurray!,' what was basically on air were all good topics."

Her retrospective characterization meets many of the clichés about Soviet media contents. It was associated with the deep resentments she felt toward Soviet elites and because of the lack of transparency. Probably she was a bit disappointed because she herself was to a certain extent convinced of the "Hurray!" slogan:

> Well, you understand, something even went because of sincerity . . . because of sincerity, first, and certainly because . . . people lived all the same. . . . All had the same opportunities. Therefore, we did not have this sense that "oh here you have . . . and I have not." Well, we knew that the party clique steals and milks and all, but we couldn't talk about? I was a director who said: "I detest this, your Soviet authority"—another director grasped me: "How could you say this on the corridor?!—Are you crazy?" And I say: "I detest these Bolsheviks."[106]

Her husband Roman R., a director and documentary filmmaker, joined the interview and diverted the remembrances of consumer issues on Soviet TV in a somewhat different direction.[107] He stressed that consumer issues were not treated because of any informative value, but to communicate that there were no problems: "Everything was there and everything was fine," as Ėvelina E. intervened. She also then amended her former statement somewhat cynically and recollected how Rostov television screened factories fabricating fashion, food, "everything was characterized by 'plus'—hurray! But we did not screen the shop where people were queuing and where was nothing [to buy]."[108]

In contrast to Ėvelina E. disavowing narrative, Valentina V. underlined that Kuibyshev television aired many programs on consumption which were "problem-oriented" and aimed to show how certain companies solve their problems. Although she seemed lastly disappointed by the outcomes of unannounced investigations, she remembered well how the program editors of Rostov *Chelovek i zakon* (Person and law) collaborated with the *State Department Against Misappropriation of Socialist Property* (OBKhSS). In her view, the critical potential of Soviet television was uncontested but not fully exploited.[109]

Thus, even if Ėvelina E. and Valentina V. raised different perspectives on television programs, they would have probably agreed about the communicative limitations and taboos that prohibited systemic improvements. From the point of view of former viewers, most respondents would have consented to Ėvelina E.'s account of missing critical reports about real consumer deficiencies. Samara engineer Rudol'f N., however, reported that he was interested in "where we, nevertheless, go, will it become a bit easier in the country, will there be a deficit of foodstuff, will the deficit of things, that are most needed, vanish? Will they when they start to build apartments rent them out? They talked much. They generally promised us least . . . [he paused] . . . they promised us communism!" He categorically denied the idea that television ever staged consumer deficits. They pretended "as if there was nothing. They only presented the foremost workers. It proved to be that in our country, the productivity of labor increased, the yield always increased. They carefully protected us against such urgent problems." Thus, according to him, there were no programs raising any burning issues, no broadcasts about new consumer items or the situation of consumer services.[110]

What we could conclude from the fact that many respondents did not even remember consumer issues on television is that looking back Soviet practices of criticism appear outdated, boring, and particularly ineffective. Further, the experience of post-Soviet consumerism blurs the memories of Soviet everyday consumer practices. Moreover, the narration of the chronological sequence of watching TV and of shifts and continuities against the background of profound political changes proves to be challenging. The interviews reveal that communicative changes, especially the removal of taboos perestroika brought about, were severe and exciting for both the viewers and the television producers. These changes, however, seem to conceal how viewers actively perceived the everyday impact of TV's offers to communicate and to inform. As only a small minority of respondents experienced television's early days as adult persons, we cannot fully rely on their narrations to highlight the longer development.

Nevertheless, I argue that already the fact that television included certain nonfictional and documentary topics of highly political relevance into its programming supported their normalization of the new Soviet lifestyle. People either just embedded such topics into everyday life or were able to ignore them. But such issues like consumption were addressable and present in the communicative space. Furthermore, watching TV became as such a Soviet consumer practice. It provided ground for television to impact in some ways rather subliminally, as the Rudol'f N.'s spouse commented: "Well, but what? Principally, what they aired, we watched. What they aired, we watched."[111]

NOTES

1. GARO, f. R-4237, op. 1 (prod.), d. 1051, 1964, l. 143.
2. GARO, f. R-4237, op. 1 (prod.), d. 1051, 1964, ll. 142–143.
3. GARO, f. R-4237, op. 1 (prod.), d. 1051, 1964, l. 144.
4. Reid, "Cold War in the Kitchen," 211–252, 217, 221.
5. Reid, Cold War in the Kitchen; Chernyshova, *Soviet Consumer Culture in the Brezhnev Era*; Bönker, "Depoliticalisation of the Private Life?" 207–234.
6. Research has broadly covered the role of consumption in the late Soviet society. Susan Reid paved the way for research in this area: Reid, Cold war in the kitchen; see also the recent works of Chernyshova, *Soviet Consumer Culture*; Anna Tikhomirova, "Soviet Women and Fur Consumption in the Brezhnev Era," in *Pleasures in Socialism: Leisure and Luxury in the Eastern Bloc*, ed. David Crowley and Susan E. Reid (Evanston, IL: Northwestern University Press, 2010), 283–308; Larissa Zakharova, "Dior in Moscow: A Taste for Luxury in Soviet Fashion under Khrushchev," in *Pleasures in Socialism: Leisure and Luxury in the Eastern Bloc*, ed. David Crowley and Susan E. Reid (Evanston, IL: Northwestern University Press, 2010), 94–119.
7. Chernyshova, *Soviet Consumer Culture*, 203.
8. Western observers regarded these laws as insufficient to fulfill consumers' needs. See, for example, Joseph J. Darby, "Product Liability in the Soviet Union," *International Lawyer* 11 (1977): 179–192, 182–184.
9. Cf. Elena A. Bogdanova, "Gazetnye zhaloby kak strategii zashchity potrebitel'skikh interesov: pozdnesovetskii period," *Teleskop: nabludeniia za povsednevnoi zhizn'iu peterburzhtsev* 6 (2002): 44–48; idem, "Konstruirovanie

problem zashchity prav potrebitelei. Retrospektivnyi analiz," *Rubezh* 18 (2003): 167–177, 170–171.

10. Catriona Kelly and Vadim Volkov, Directed Desires: *Kul'turnost'* and Consumption, in *Constructing Russian Culture in the Age of Revolution:1881–1940*, ed. Catriona Kelly, Vadim Volkov and David Shepherd (Oxford: Oxford University Press, 1998), 291–313.

11. Tatiana M. Goriaeva, ed., *"Velikaia kniga dnia . . ." Radio v SSSR. Dokumenty i materialy* (Moscow: Rosspen, 2007), 97–98; Lovell, *Russia*, 100.

12. Catriona Kelly, *Refining Russia: Advice Literature, Polite Culture, and Gender from Catherine to Yeltsin* (Oxford: Oxford University Press, 2001), 312–367; Lynne Attwood, *Creating the New Soviet Woman: Women's Magazines as Engineers of Female Identity, 1922–53* (Houndmills, UK, and New York: Macmillan/St. Martin's Press, 1999); Ol'ga Smoliak, "'Abazhur,' 'abort,' 'abrikos': reprezentatsiia sovetskoi povsednevnosti v knigakh po domovodstvu," in *Razryvy i konventsii v otechestvennoi kul'ture*, ed. Oleg Leibovich (Perm: PGIIK, 2011), 138–184.

13. Cf. Baiar and Blashkevich, *Kvartira i ee ubranstvo*; Efremov, *Domovodstvo*; Vorob'eva et al. (eds.), *Domovodstvo: VII i VIII klassy*.

14. This strategy would have been familiar to consumers from publications of Stalinist times. Cf. Helena Goscilo, "Luxuriating in Lack: Plenitude and Consuming Happiness in Soviet Paintings and Posters, 1930–1953," in *Petrified Utopia: Happiness Soviet Style*, ed. Marina Balina and Evgeny Dobrenko (London, New York, Delhi: Anthem Press, 2009), 53–78; Serguei Alex. Oushakine, "'Against the Cult of Things': On Soviet Productivism, Storage Economy, and Commodities with No Destination," *The Russian Review* 73 (2014), 198–236, 212–21.

15. Isabelle de Keghel, "Konsum im Blick. Visualisierungsstrategien in sowjetischen und ostdeutschen Printmedien (1953–1964)," *Comparativ* 21, 3 (2011): 79–96, 82–85.

16. Keghel, Konsum, 86–87.

17. *Pravda* started to publish the daily television schedule on its back page in April 1959; *Izvestiia* in November 1959. From January 1960, all national newspapers printed the television schedule. From 1965 on, *Pravda* printed weekly schedules every Saturday.

18. Levine, *Main Street*, 155.

19. Viewers' letters were also prominent in entertainment features, especially in quiz and music shows, which are not further considered here. In these cases, viewers could, respectively, vote for contestants and request music tracks.

20. Aleksandr Ia. Iurovskii, *Televidenie – Poiski i resheniia. Ocherki istorii i teorii sovetskoi telezhurnalistiki* (Moscow: Iskusstvo 1983), 85.

21. GARF, f. 6903, op. 1, d. 666, 1960: Protokoly zasedanii obshchestvennogo soveta telezritelei pri tsentral'noi studii televideniia, l. 11; ibid., d. 667, 1960, l. 5.

22. GARF, f. 6903, op. 1, d. 724, 1961, ll. 90–104 (quote on l. 102). In the late 1950s, the government newspaper *Izvestiia* introduced an extra page for home and family issues that was also apparently related to the new phenomena of moving into one's own flat and the prospering of a new Soviet lifestyle.

23. Reid, "The Khrushchev Kitchen," 289–316, 297.

24. Iurovskii, *Televidenie - Poiski i resheniia*, 85–87.

25. Pravda, No. 5, 5.1.1963, 4; Pravda, No. 33, 2.2.1963, 4; Pravda, No. 249, 5.9.1964, 4; Pravda, No. 331, 27.11.1971, 6.

26. Reid, "Communist Comfort," 465–498, 475–477.

27. TsGALI SPb, f. 293, op. 3, d. 132, 1963: Obzory o rabote s pis'mami radioslushatelei i telezritelei, l. 14.

28. GARF, f. R-6903, op. 3, d. 184, 1962: Itogovaia spravka o rezul'tatakh anketnogo oprosa telezritelei o dnevnykh programmakh tsentral'nogo televideiia, ll. 10–11.

29. GARF, f. R-6903, op. 10, d. 70 (1966): Analiz pisem telezritelei za period s 1964 po 1966, l. 8.

30. See here chapter 4.

31. Thomas Etzemüller, "*Social engineering* als Verhaltenslehre des kühlen Kopfes: Eine einleitende Skizze," in *Die Ordnung der Moderne: Social Engineering im 20. Jahrhundert*, ed. Thomas Etzemüller (Bielefeld: transcript, 2009), 36.

32. The television advice genre was developed in the United States during the 1950s. The West German ARD introduced the first health advice feature in 1953 and added several programs addressing women and housewives. The programs covered house and garden, raising children, fashion, dance courses, and cooking tips. They were based on the claim that TV should be educational, which was supported by contemporary television producers, politicians, and social scientists. The programs envisioned modern lifestyles and tried to craft social behavior. West German television complemented the genre with a series of consumer advice programs covering a wide variety of different topics after the early 1960s. These programs covered consumer issues, money, and everyday life. Cf. Hickethier, *Geschichte des deutschen Fernsehens*, 163–164, 227–229; Stephan Habscheid, "Ratgebersendungen: Gesundheitsmagazine," in *Der sprechende Zuschauer: Wie wir uns Fernsehen kommunikativ aneignen*, ed. Werner Holly, Ulrich Püschel and Jörg Bergmann (Wiesbaden: Springer, 2001), 173–186, 173; with regard to the GDR: Wilke, 230.

33. TsGALI SPb, f. 293, op. 3, d. 132, 1963, l. 41.

34. Shonin, "Radio and Television as Art Forms in the Soviet Union," 83–92, 87–8.

35. Pravda, No. 223, 10.8.1972, 2.

36. Pravda, No. 223, 10.8.1972, 2.

37. Pravda, No. 128, 7.5.1972, 1.

38. GARF, f. 6903, op. 48 ch. 1, d. 175, 1973: Mironovich G. A. "Programmy TsT, osveshchaiushchie khod i razvitie sotsialisticheskogo sorevnovaniia v strane," ll. 1–3.

39. Pravda, No. 71, 12.3.1966, 4; GARF, f. 6903, op. 36, d. 1, 1971: Godovoi obzor pisem i obzory pisem telezritelei za ianvar'-dekabr' 1971, l. 462. *Otvety na pis'ma zritelei* was broadcast on Central TV's first programs on Saturday early evening: Pravda, No. 284, 10.10.1964, 4.

40. GARF, f. 6903, op. 48 ch. 1, d. 175, 1973, l. 4.

41. Alexey Golubev and Olga Smolyak, "Making Selves through Making Things: Soviet Do-It-Yourself Culture and Practices of Late Soviet Subjectivation," *Cahiers du monde russe* 53, 3–4 (2014): 517–541.

42. *Obzor pisem telezritelei za 1975 god* (Moscow: Gosudarstvennyi Komitet SSSR po televideniiu i radioveshchaniiu, 1976), 58, 60–61. The department for popular scientific and educational programming, for instance, received 1,760 letters in 1970, 4,470 letters in 1971, and more than 15,000 in 1974. Almost 20,000 letters were sent in 1975 to *Zdorov'e* alone. *Nash sad* received around 3,500 letters in 1974 and 3,100 in 1975.

43. *Obzor pisem za 1971*, 10.

44. GARF, f. 6903, op. 10, d. 58, 1964: Obzory pisem radioslushatelei (sovetskikh i zarubezhnykh) i telezritelei za janvar'-mart, iiul'-dekabr' 1964, l. 111.

45. The letter department was transferred to the newly founded Central Television's main editorial office in 1960. In addition, it was an independent unit within Central Television from 1962 to 1970.

46. GARF, f. 6903, op. 10, d. 79, 1968: Godovoi obzor pisem telezritelei za 1968 g., l. 9.

47. *Obzor pisem za 1971*, 33.

48. GARF, f. 6903, op. 36, d. 1, 1971, l. 462.

49. *Obzor pisem za 1971*, 33.

50. Thus, viewers' letters often referred to domestic and foreign political events or local politics. They also contained general reflections on Soviet society, societal values, cultural practices, or consumer problems.

51. For a typical proportional distribution of letter topics sent to the much more addressed radio in the early 1960s, see GARF, f. R-6903, op. 10, d. 34, 1963: Godovoi obzor pisem i obzory pisem za ianvar'-mai radioslushatelei Tsentral'nogo Vnutrisoiuznogo Radioveshchaniia, Tsentral'nogo Radioveshchaniia na zarubezhnye strany i Tsentral'nogo televideniia za 1963, l. 8.

52. *Obzor pisem za 1971*, 20–21, 31–33. *Chelovek i zakon* was launched in June, *Po vashim pis'mam* in September, and *Na voprosy otvechaet ministr . . .* in October 1971. Each program was broadcast once a month. Measured on the basis of the mailbag, *Chelovek i zakon* became one of the most successful programs of Soviet television: it received 4,841 letters in 1972, its first full year of broadcast. The number of letters received quickly grew to 160,161 in 1979 and 172,790 letters in 1980.

53. *Obzor pisem za 1971*, 31.

54. *Obzor pisem telezritelei, ianvar' 1972 god* (Moscow: Gosudarstvennyi komitet po televideniiu i radio, 1972), 17.

55. *Obzor pisem telezritelei, fevral' 1972 god*, (Moscow: Gosudarstvennyi komitet po televideniiu i radio, 1972), 22.

56. Iurovskii, *Televidenie—Poiski i resheniia*, 85–87.

57. GARF, f. 6903, op. 10, d. 79, l. 3; *Obzor pisem telezritelei za 1973 god* (Moscow: Gosudarstvennyi komitet po televideniiu i radio, 1974), 6–7.

58. GARF, f. 6903, op. 10, d. 79, l. 3; *Obzor pisem za 1973*, 9.

59. In 1963, GDR television launched the very popular program *Prisma*. Based on letters, *Prisma* covered grievances in the fields of local politics, the economy, and not least of consumption and consumer services. *Prisma* aimed to compete with West German political features. The editors always tried to seek out those who were to blame. Cf. Ina Merkel and Felix Mühlberg, "Eingaben und Öffentlichkeit," in "*Wir*

sind doch nicht die Meckerecke der Nation": Briefe an das Fernsehen der DDR, ed. Ina Merkel (Berlin, 2000), 11–46, 30–32; for the CSSR cf. Stefan Lehr,"'Pište nám!': Dopisy diváků a posluchačů Československé televizi a rozhlasu," *Marginalia Historica. Časopis pro dějiny vzdělanosti a kultury* 3, č. 2 (2012), 71–82; idem, "Eingaben von Bürgern an das Tschechoslowakische Fernsehen und Radio: Ein Beitrag zur Kommunikation zwischen Staat, Bürgern und Medien im Staatssozialismus," *Bohemia* 59 (2019) (forthcoming).

60. Raymond Williams, *Culture* (London: Fontana, 1981), 111.

61. John Tomlinson, *The Culture of Speed: The Coming of Immediacy* (Los Angeles et al.: Sage Publications, 2007), 98–100.

62. Tomlinson, *The Culture of Speed*, 99.

63. GARO, f. R-4237, op. 1 (prod.), d. 1051, 1964, 142.

64. GARO, f. R-4237, op. 1 (prod.), d. 1051, 1964, 144–146; GARO, f. R-4237, op. 2 (prod.), d. 754, 1969, l. 4.

65. TsGAIPD SPb, f. 24, op. 145, d. 20, 1972: O rabote s pis'mami i izuchenii ėffektivnosti peredach Leningradskogo televideniia v 1971 godu, ll. 1–3.

66. GARO, f. R-4237, op. 1 (prod.), d. 1051, 1964, l. 146.

67. GARO, f. R-4237, op. 1 (prod.), d. 1051, 1964, l. 147. The programs also publicized the mailing address of the letter department. See GARO, f. R-4237, op. 1 (prod.), d. 876, 1963, l. 120; GARO, f. R-4237, op. 1 (prod.), d. 877, 1963, l. 99.

68. GARO, f. R-4237, op. 1 (prod.), d. 876, 1963, ll. 113–120. This became an established strategy, as other programs also invited managers supposed to be accountable for local grievances: See GARO, f. R-4237, op. 2 (prod.), d. 754, 1969, ll. 4–5.

69. GARO, f. R-4237, op. 2 (pr.), d. 759, 1969, l. 8.

70. GARO, f. R-4237, op. 2 (pr.), d. 759, 1969, ll. 8–19.

71. GARF, f. R-6903, op. 10, d. 79, l. 7ob.

72. GARF, f. R-6903, op. 10, d. 79, l. 7ob.

73. GARF, f. R-6903, op. 10, d. 79, l. 10ob.

74. *Obzor pisem telezritelei za 1972 god* (Moscow: Gosudarstvennyi komitet po televideniiu i radio, 1973), 13.

75. *Obzor pisem telezritelei ianvar' 1972 god* (Moscow: Gosudarstvennyi komitet po televideniiu i radio, 1972), 16; *Obzor pisem telezritelei fevral' 1972 god* (Moscow: Gosudarstvennyi komitet po televideniiu i radio, 1972), 23. The two series broadcast in January received fifteen letters, the two series which aired in February received fourteen.

76. Iurovskii, *Televidenie—Poiski i resheniia*, 209.

77. *Obzor pisem za 1972*, 25.

78. *Obzor pisem 1972*, 25.

79. The clip is available at: https://www.youtube.com/watch?v=4tZ_dD5SEWw (accessed June 20, 2015). Unfortunately, Central Television's internal reports on letters give no information about the fate of *Tovary—narodu* after 1972. Thus, we cannot be sure that this clip actually represents the original program.

80. Fedor Razzakov, *Gibel' sovetskogo TV. Tainy televideniia: ot Stalina do Gorbacheva, 1930–1991* (Moscow: Eksmo, 2009), 85; http://www.tvmuseum.ru/catalog.asp?ob_no=4734 (accessed June 20, 2015).

81. *Obzor pisem telezritelei za 1973*, 20.

82. *Obzor pisem za 1973*, 14.

83. *Obzor pisem za 1973*, 14–15, 23, 29–30. Writer Sergei S. Smirnov was one of the most popular Soviet TV hosts and invited interesting people of all professional fields to talking about their "creative search" in life. His program *Poisk* alone received almost 62,000 letters in 1973.

84. *Obzor pisem za 1973*, 23–24.

85. *Obzor pisem za 1972*, 25; *Obzor pisem za 1979*, 27; GARF, f. 6903, op. 1, d. 40, 1975, ll. 6–7.

86. The quantity of letters sent to Central Television multiplied tenfold, up to 400,000 from 1960 to 1970. Television's incoming mail continued to rise until it stagnated at a level of 1.3–1.8 million letters in the late 1970s, when nearly all Soviet households were equipped with TV sets (*Obzor pisem za 1977*, 3).

87. *Obzor pisem telezritelei za 1975 g.* (Moscow: Gosudarstvennyi komitet po televideniiu i radio, 1976), 20.

88. *Obzor pisem telezritelei za 1976 g.* (Moscow: Gosudarstvennyi komitet po televideniiu i radio, 1977), 28.

89. *Obzor pisem telezritelei za iiul' 1974 goda (i 1 polugodie)* (Moscow: Gosudarstvennyi komitet po televideniiu i radio, 1974), 6–7.

90. http://www.youtube.com/watch?v=yspv1m44l5A [26:55] (accessed September 1, 2014).

91. GARF, f. 6903, op. 1, d. 40, 1975, ll. 10–11.

92. GARF, f. 6903, op. 1, d. 40, 1975, ll. 10–11.

93. *Obzor pisem telezritelei za fevral' 1977 g.* (Moscow: Gosudarstvennyi komitet po televideniiu i radio, 1977), 11.

94. *Obzor pisem telezritelei za fevral' 1985 g.* (Moscow: Gosudarstvennyi komitet po televideniiu i radio, 1985), 18.

95. Cf. e.g. *Obzor pisem telezritelei za oktiabr' 1975 g.* (Moscow: Gosudarstvennyi komitet po televideniiu i radio, 1975), 9; *Obzor pisem za 1980*, 22.

96. *Obzor pisem za fevral' 1977*, 11 (quote); *Obzor pisem za 1977*, 20.

97. *Obzor pisem za fevral' 1977 g.*, 11.

98. *Obzor pisem za 1980*, 23.

99. *Obzor pisem za oktiabr' 1975*, 9.

100. *Obzor pisem za 1976* g., 28.

101. Susan Reid established that consumers' criticism of consumption flaws, negotiations on aesthetics, form, or durability gained strength in and through the official media discourse during the 1960s. She argues that these practices document the communicative interaction between mass media shaping demanding consumers who then took up and reflected on this discourse. Cf. Reid, "Khrushchev Modern. Agency and Modernization in the Soviet Home," 252; for newspapers see Bogdanova, "Gazetnye zhaloby"; idem, "Konstruirovanie," 171.

102. http://www.ljpoisk.ru/archive/1619069.html (accessed June 20, 2015).

103. Interview with Waldemar (Germany, February 2012).

104. Interview with Aleksandra Apollonovna (Tyumen by phone, November 2011).

105. See chapter 4.
106. All quotes from interview with Interview with Ėvelina E. (Rostov-on-Don, October 2010, born in 1935).
107. Interview with Roman R. and Ėvelina E. (Rostov-on-Don, October 2010).
108. Interview with Roman R. and Ėvelina E. (Rostov-on-Don, October 2010).
109. Interview with Valentina V. (Samara, born in 1937).
110. Interview with Rudol'f N. (Samara, born in 1940).
111. Interview with Rudol'f N. (Samara).

Conclusion and Epilogue

> In different corners of the world, authorities started to think, they started to be afraid of television. To be afraid! That means that they realized that they should primarily be engaged with it and not give it away to anyone. All countries of the Communist bloc, plus all developing countries adopted such a totalitarian standard of organizing television from above. (. . .) In the USA, well, also Khrushchev sees [*sic!*] that television as it should be. Powerful, round-the-clock, several channels, competing, absolutely free. There, he delivers his speech without any limitations. But Khrushchev is [*sic!*] Khrushchev. And when he returned from America, instead of "aha, I could rule the country with the help of television, if I engaged in dialog with the people," it is clear, "one could see me every day in every corner of the Soviet Union, I could transmit my ideas," he says—I tell the truth, he came back to Moscow, he gathered the politburo, he would speak a long time about his impressions. . . . But everything had turned into fear. Well, you will say, where did the fear come from? He asked the politburo: "But it is good to keep our television like an institution?" (. . .) In this sense any bloc of countries—the West, the developing countries, the socialist bloc—will find its version of using television for strengthening authority.[1]

Renowned sociologist Boris M. Firsov retrospectively analyzed how the Khrushchev regime handled television in its early days and what a reasonable policy might have been in order to stabilize Soviet rule—namely to hold a dialog with the audience. Firsov obviously supposed Soviet television to have missed a great opportunity. One could agree with him in many respects but perhaps not in all. The Soviet Union existed for a long time after Khrushchev's return from the United States in 1959 and his alleged adjustment of the official television policy. This points to the system's stability even without adhering to the more liberal or interactive television policy that Firsov would have favored.

Indeed, Soviet television did a good deal to stabilize the sociopolitical system, to establish a Soviet pop-culture, and, in relative terms, it rather succeeded in getting into a dialog with its audience. During the 1960s, television became the most important technical and cultural device that rendered home a serious issue for negotiations between ordinary citizens and the regime about new meanings of everyday life. The medium made home, consumption, and leisure part of a continuum, in which the boundaries of the public and private spheres overlapped, in which they became permeable and shifting. The broad acceptance of television and the way most people focused their consumer desires on buying a TV set, the way they made watching TV a typical habit of everyday life, the way they even perceived their environment through watching TV—all these aspects give us a notion of television's potentially stabilizing impact and its capacity to establish emotional bonds between the audience, the medium, and, to a certain extent, also the system it represented. Television brought about an essentially new kind of private communication that was rather complex for the regime to control. Nevertheless, the new private communication television it entailed never really threatened the system until perestroika. Television did not turn out to become a threat, as Boris Firsov suspected Khrushchev and the politburo of fearing, because the medium bound the viewers by entertaining them and informed them by opening a "window to the world," as many respondents characterized their idea of television. The literal meaning of being a "window to the world" pointed to the viewers' perception that television offered new topics to talk about and gave them a notion of remote places they would have never been able to travel to. The medium, thus, constituted what Mikhail Bakhtin calls a "third space," that is, an intermediate space of negotiation, in which authoritative discourse on the one hand and people's ideas about societal norms and values on the other hand encountered each other as two different "voices." Following Bakhtin, their dialog could have set a process of everyday hybridization of official meanings in motion.[2] Indeed, the respondents' narratives about cheering Soviet athletes taking part in international sport events show that many viewers felt "Soviet" in the context of the Cold War confrontation but did not transfer this support into identification with the Soviet regime as such. This at least ambivalent or sometimes even missing identification with the Soviet regime did not, however, necessarily imply an opposing attitude. Homi K. Bhabha's more dynamic and interacting understanding of hybridization highlights this complex and multifaceted process of making sense of authoritative discourse in relation to nonofficial interpretations probably more precisely than Bakhtin's concept. In contrast to Bakhtin, Bhabha does not suppose that any hybridization necessarily contests authority but attributes a hybrid potential to authoritative discourse itself.[3] This correlation can be traced in many narratives. The interviews demonstrate that the "voices" bridged simple binaries and closely interacted. Even if critical and reserved viewers rejected and somewhat undermined authoritative discourse, they never erased it. For example, the narratives of the Soviet sports fans and the results of the recent studies on Soviet identities point to situational contradictions in a transethnic Soviet identity: Watching international sports events on television enabled the viewers to connect a non-Russian ethnic affiliation with the emotional belonging to the Soviet Union and the feeling of being proud of the Soviet athletes. However, watching television in the Baltic periphery also gave the viewers the opportunity to reject any Soviet identity and undermine Moscow's authoritative discourse.

Referring to Bakhtin, the idea that the television might hybridize authoritative messages becomes tangible in the respondents' narratives about the way the medium impacted on their new private life. In a certain respect, the hybrid potential of the "third space" even allowed for reflection on different representations of the new Soviet lifestyle. Television quickly became a rather ambivalent force of privatization. On the one hand, Soviet consumers accepted it in its promoting of the new home-centered Soviet lifestyle and the nuclear family. The Soviet citizens gained formerly unknown opportunities to shape their private lives, to discover personal emotions and relationships. Television merged time and space in a new way and helped in constituting the Soviet audience as "emotional communities" across social, cultural, economic, and even language boundaries. Media scholars relate this process to the capability of television to mediate between the individual and the collective. Switching on the TV gives the viewer a sense of belonging to an imaginary community. Television could not only stimulate a process of individual reflection but also give the viewers the feeling of sharing emotions and experiences with others. Thus, inclusion and a sense of community become the social gratification of watching the "right" programs.[4] All of this let many people develop a feeling of nostalgia for the Soviet life.

On the other hand, people's investment in the private sphere bolstered the regime. Oral history interviews with people who grew up in Soviet times show that positive memories of former Soviet life are no rarity today. They prompt us to consider the positive ascriptions that bear witness to former cohesive factors within Soviet society. The stability of the social and political setting in the Brezhnev era may have been promoted to a considerable extent from these stable "emotional communities" shaped in front of the TV screen. Concluding from the interviews, many viewers really enjoyed watching TV; at least, it only seldom triggered open rejection and condemnation. Viewers' affirmative commitment was considerably due to the television's pleasure-giving perception. Soviet television served as a forum that offered a steadily increasing amount of people access to highbrow cultural products. Soviet television also merged traditional ideas of highbrow culture with popular elements bringing about a new televised pop-culture with movies, games, and music shows.

The current retrospective assessments of many people older than fifty years suggest a clear devaluation of post-Soviet entertainment television. The contemporary enthusiasm about the opportunities it offered to the performing arts was great. The remembrances and assessments of former "ordinary" TV viewers do not basically differ from those of former leading protagonists of Soviet television. In the view of both groups, post-Soviet television appears artistically quite inferior and intellectually poor compared to the Soviet TV of the 1960s and 1970s. The contemporary appreciation of the cultural and artistic programs characterized the way many TV viewers engaged in the negotiations on the normalization of daily life in the Khrushchev and early Brezhnev years. Many live broadcasts and artistic movies became cultural events; they figure as common places in people's memories today. The positive retrospective assessments of many respondents about the entertainment quality of Soviet television reveal that the early program policy of the 1950s and 1960s successfully introduced the new medium into the Soviet society. Broadcasting the established performing arts attracted educated groups, mitigated, or even deconstructed their prejudices toward television.

Watching television quickly became a broadly accepted pleasure-giving leisure habit. This created an intermediary space of communication in which the regime and the people negotiated the emotional basis of the Soviet society. The television offered emotional events that attracted the viewers to gather in front of the screen and, in various artistic and documentary TV genres, presented them with politically and officially sanctioned principles, norms, and values. In this context, the normalization of private life became a central issue of what American philosopher Martha Nussbaum calls "public emotions."[5] Talking about this normalization had a strong political import, as it impacted on the stability of the society. More precisely, people's intensifying compassion for this normalization of the private life challenged but also enforced the emotional bonding to central values of the sociopolitical order that was now negotiated in sports broadcasts, documentary programs, and (pop-)cultural TV productions. Considering the narratives of the respondents, this interplay of challenge and enforcement can apparently not be perceived as a simply routinized approval of official values ensured by authoritative discourse. This may be concluded from still-persistent cultural attitudes and habits among former Russian-Soviet viewers. The way authoritative discourse was often hybridized by the TV viewers who felt entertained in front of the screen vanished neither with the changes of perestroika nor with the demise of the Soviet Union. This becomes obvious if one compares post-Soviet political and cultural developments and reinterpretations of the past. Shortly before the current political crisis evolved, Ukrainian historian Katerina Khinkulova argued that today's Russian television culture is still much more imprinted by traces of the Soviet popular culture than the Ukrainian. As the main reason she stated different attitudes toward the past. Whereas the Russian TV is much more driven by nostalgia and clings to an idea of a Soviet high-quality TV culture including movies and television formats of the 1970s, the latter followed Western trends of reality TV in order to become part of a "Western European" popular culture. These trends tell us much about TV producers' perception of their audiences and the political settings. They also witness the persistent thinking about a superior Soviet popular culture even among the younger Russian cultural elites who devaluate the "Western" mass culture.[6]

However, today's highly complex interplay of producers' perceptions of the audience, the expectations of the audience toward the program, the financing of the TV stations, interests of advertisers, and so on could not be directly linked to the "ordinary" citizen's historical experiences and consciousness without further ado. But, nevertheless, the interviews reveal Soviet TV as still-persistent source of emotional commitment to former Soviet life. This is particularly true for people older than fifty years. Therefore, TV's mediation and use of the Soviet past has become a lucrative business strategy.[7]

The Russian television, especially the *Nostal'giia* channel, has refined its program strategy to constantly update Soviet viewing habits and experiences that again involve feelings related to the Soviet way of life. Deliberately combining the strategies of a mediated nostalgia for the Soviet past and a cultural and aesthetic nostalgia for Soviet TV formats, Russian television is steadily repeating Soviet movies and serials. The *Nostal'giia* channel even reproduces Soviet Central Television's daily schedule to give viewers an authentic feeling of the Soviet past.[8] This program strategy obviously nourishes a nostalgic yearning for the alleged Soviet cultural and probably even political superiority, as the statements of many respondents show. Many of my interviewees would have agreed with the idea of the Soviet cultural superiority compared to Western-style programs and the post-Soviet television. The interviews with former Soviet TV viewers perfectly document that the presumptions of current Russian TV producers about audience preferences do not totally mislead program policies. The narratives of the respondents reveal that the television's performances enriched the field of culture and arts. Most Soviet viewers assessed these contributions positively and did not develop any essentially dismissive emotions. The negotiations between viewers and Soviet TV on genre preferences, airtimes, the share of entertainment programs, and so on continued throughout the 1970s. This kind of communication brought about an officially and societally sanctioned popular culture which improved the television's position in the cultural hierarchy of the media. Television, thus, became the most wide-ranging intermediary space in which people negotiated interpretations, values, and meanings of Soviet life that referred not only to moral and social questions of living together but partly also to the representation of the Soviet Union as a superpower in Cold War times. Although the hegemonic interpretation frames in authoritarian regimes were much tighter than in liberal societies and core parts remained nonnegotiable, people were to a certain extent able to accept political and cultural meanings or to ignore them by switching off the TV set or at least the sound. Preserved in the form of media nostalgia, the former attitude toward Soviet TV still provides the basis for a skeptical attitude, if not open rejection, toward the current Russian television culture. A main issue turns out to be that especially many viewers of the older generation consider it to be "Americanized," and bemoan a qualitative decline. Sixty-three-year-old engineer Antonina was very clear about the moral and cultural decay the post-Soviet society has experienced. She directly linked both developments and perceived the television as a kind of societal mirror:

> Yes, I have many favorite movies. . . . Well, today you sometimes watch "Retro" on Channel 5. . . . Or we download something from the internet. But to watch something contemporary. . . . Our movies of the 1960s, 1970s, they are like a drink of fresh water among that, what flows on the screen today; this is a nightmare of blood, brutality, violence and debauchery. I can only say it this way. [. . .] Oh my God, how to remember all these movies now. Well, such a movie like *Letiat zhuravli* (The Cranes Are Flying), they are classics [. . .] *Semnadtsat' mgnovenii vesny* (Seventeen Moments of Spring)—that's our family movie, we watched it endlessly. Well, and then all the comedies. . . . Bright movies, rather primitive for our times, but there was a purity of human relations, that *Karnaval'naia noch'* (Carnival Night), it's the same. People were purer, people were simpler, in some respect, they were more primitive than todays. But they were nearer to the laws of God than todays. There is rather much cynicism, brutality. And people started to live each for themselves. But then—the life in our collective—we could not think of ourselves without the collective! Obviously, the collective had a very good aspect—you did not live unsociably, but you lived facing society, the people. This is what the West lacked, we had it plentifully. [. . .] That means our movies covered such dreams, strivings for life. And now? Who made love to whom? Who stole from whom . . . ? [. . .] I don't like these shaggy-dog stories, this mud they broadcast. I don't like it. How did all this horror begin? It's a nightmare! That means this negative, aggressive flow that you simply cannot watch; you simply become a psychopath. I think that the television causes much more harm than benefits.[9]

Antonina's narrative reveals much about the way traditional Soviet values were represented by Soviet TV. Obviously, they have survived the fall of the Soviet Union and provide ground for the current nostalgia for the lost Soviet society. The way Antonina compared Soviet and post-Soviet television highlighted her still-persistent emotional bonding to the Soviet TV. Although not all respondents were still as enthusiastic about Soviet television as Antonina, one can conclude that TV consumption had rather contributed to the stability of Soviet society than challenged it before perestroika changed the communication rules. Antonina made clear that in her view, Soviet movies and series symbolized the good human aspects of the lost Soviet life. "When it didn't make any difference to you, how your neighbor lives," Antonina deplored the current indifference and egoism of the modern society in contrast to the warmness of the Soviet life. This retrospective ascription points to the communicative space that television opened and that gave the viewers the opportunity to embed their private life into a broader societal context represented by television. Today, the viewers can link their personal narratives to their memories of those Soviet television programs that had evoked pleasure and entertainment.

It is worth noting that even retrospectively critical and reserved viewers reveal an affirmative emotional attitude to the Soviet lifestyle. At least, they did not simply identify media contents with false propaganda and deceptive ideology. As television was embedded in a complex communication structure, we could hardly conclude from people's watching habits that it turned out to have been a nail in the coffin of the Soviet regime. On the contrary, it provided a specific entertainment culture that successfully addressed the majority of the Soviet society. Television often reduced complex questions of living together, of the political, economic, and social order to concrete figures and stories presented on the screen. It shaped not only specific Soviet chronotopes but also particular Soviet lifestyles by presenting faces and relating authoritative discourse to the everyday life of many people.

Thus, the increasing TV consumption turned television into a symbol of the new Soviet way of life. Referring to game shows, serials, and movies, viewers debated the normalized daily life of the 1960s and 1970s. This communication on values, attitudes, and emotions should be assessed against the background of the increasing interest of the party in the medium and the rising financial investment in the technical infrastructure. It reflected the rise of TV consumption as a highly popular leisure habit and officially recognized it as an important mediator of the late Soviet way of life. Hard work for the benefit of the Soviet society on the one hand and watching TV and living a Soviet life one the other hand became ideologically deeply entangled.

Nevertheless, television opened a space for a wide range of TV uses and interpretations that the interviews reflect. Media use and watching TV started to differ with increasing media offerings. The interviews show that quantitative and material aspects of watching habits considerably converged, though striking cultural differences between a small group and the majority of the viewers remained. People with secondary education tended to consume media rather equally, whereas people with higher education at least retrospectively revealed more differences in their media usage, their assessments of the programs, and their motives to switch on the TV.

Those respondents who did not openly reject Soviet television suggested the way people got involved and established "shared" emotional bonds toward a broadly accepted popular culture and the integration of TV into the hierarchy of traditional performing arts. The television shifted the cultural boundaries, meanings, and uses that promoted its perception as "window to the world." Needless to say, TV as a "window" had opened onto two directions: inward and outward. This becomes clear from the way viewers addressed Soviet television. Viewers' letters touched upon a wide variety of topics, such as TV programs, ideological conformity, airtime, aesthetics, technological developments, and pedagogical aspects of how to use the new medium. Being in touch with the TV audience became increasingly important, for the TV staff as well as for the party that regarded television as the most powerful propaganda tool after the late 1960s. Letters that praised party politics and the Soviet system referred to the television as a reliable information source. Television sustained the letter communication by using letters in documentaries and entertainment shows. By steadily suggesting its close contact with the audience, the television positioned itself in the field of political communication not only as a medium of entertainment but also as agent of societal interests and improvements.

Letters and criticism obeyed specific sociopolitical language conventions. The belief in possible improvements gave moral grounds to disagree if nobody contested the regime in its legitimacy. The regime, in turn, rewarded the soft skill of criticizing within the official rhetoric boundaries. Thereby, viewers could play an active part in the communication by reproducing authoritative discourse. The screening of authoritative discourse made its form more flexible and allowed for changing or adding meanings that were not completely prescribed by official discourses. Viewers were fascinated by the impression of immediacy that television created. From the beginning, they requested to be directly addressed by the announcer and criticized them for looking down and reading out a manuscript instead of speaking without notes.[10]

All these issues became part of the negotiations about a normalized private life. The television played an important part by normalizing consumer issues, by interlacing home, consumption, and leisure in a public-private continuum. Reproducing the genre of consumer advice and information, it interlaced authoritative discourse with tangible questions of lifestyle and consumer taste, with personal experiences and local events in a more interactive, perhaps even intrusive way compared with print media and radio. Television did not simply replicate authoritative discourse. The mass media challenged and promoted the commitment of its users but could hardly control its popularity.

Viewers could easily switch off boring television programs. As the television required more attention, it needed to stage authoritative discourse much more according to the viewers' preferences than did the radio, which rather constituted a background medium. The television therefore tried to involve the audience by inviting viewer-consumers to communicate about their experiences, ideas, and suggestions.

We could suppose that TV at first accelerated the normalization of authoritative discourse, as it easily disseminated ideological language to a broad audience. However, the inherent dynamic of television was not very suitable for the circular model of language the late Soviet system entailed. Whereas, as Alexei Yurchak argues, *Pravda* was easily able to publish anonymous, hyperindividual editorials without identifiable authors or voices of eyewitnesses, TV could not. It essentially needed voices, faces, and perspectives. Yurchak demonstrates that the narrative structures of authoritative texts referred to the past and future but not to the present.[11] Television, however, bridged time and space by presenting voices, faces, and perspectives. It was a very current medium and subsisted on promptness and speed and on immediacy and authenticity. The creation of knowledge and meaning was much more insecure due to viewers' uses of the medium compared to the reading of newspapers. Formulaic speech on television quite quickly risked losing any literal meaning, which could have caused rejection and a feeling of absurdity.[12] All this challenged television to interlace authoritative discourse with the viewers' everyday life, to involve them and benefit from their input. Therefore, television involved their audience much more directly than the other media.

The fact that viewers "voluntarily" maintained their communication with television about consumer topics demonstrates that TV supported the normalization of the late Soviet lifestyle. The audience welcomed these new strategies. Internal reports on letters document that topics and expectations with which the viewers addressed Soviet television up to 1985 generally remained the same, although we can observe a certain change in the letters presented in the early 1980s. With the decline of audience response to the programs, the reproduction of authoritative discourse became more frozen and shallow. Up until this point, we observe the space the televisual reproduction of authoritative discourse opened for establishing emotional bonds to Soviet material and medial culture. We therefore should assume that television probably created lasting bonds for a certain number of viewers, and not only the less educated, as the interviews document.

The interpretation that television intensified the dialogue with the audience was at first propelled by the perception of a rather revolutionizing immediacy and the global reach of the medium. Both literally expanded the space of communication. Television became a "third space" in Bhabha's sense as it filtered representations, interlaced them, and initiated new ones. In several respects, television conveyed the impression that it responded to viewers' private concerns. Television staged viewers' requests more vividly than any other medium and bonded the solution of issues to concrete faces of local officials. The impression of a new dialogue faded away in the early 1980s with the habituation effect of watching television and the realization that television would not fundamentally resolve any systemic shortcomings. This becomes clear in a longer perspective that includes the communicative changes perestroika entailed. Perestroika removed many taboos of the former official language so that viewers started to criticize the regime and high officials. Viewers' response to television became rather ambiguous. On the one hand, television staged the political transformation, recording people's voices on the street. On the other, even though television tried to update its formats, people now explicitly identified it as a medium of the corrupt communist elite. In January 1989, twenty-six-year-old veteran of labor A. A. Matveev, who lived in a small village in Irkutsk region, indignantly addressed the producers of the local TV program *Dialog* (Dialog):

> I watched your program, it's a shame and disgusting to hear and watch all these ladies and gentlemen (*ledi i dzhentl'menov*). They have meetings every day although there's no sausage, no milk, no trousers in our shops. Not chatting, not appealing for, but all need to work, thus you don't shiver, don't pose in front of the camera, do you feel some shame?[13]

In the same year, Irkutsk's seventy-seven-year-old war veteran S. A. Stanitskii criticized the same TV program: "I watched *Dialog po pis'mam* (Dialog on letters). A dialogue did not emerge. Your invited dialog partners do not understand. Their brain is subjugated by instructions and administration, the soul is missing. [. . .] All hardships of the country that we are going through and endure hurt and are a shame."[14]

The changes perestroika brought about are obvious. We know very well that many people experienced them, if not as a catastrophe, then as a watershed in their life story.[15] Many interviewees considered the reforms to have quite negatively impacted on their personal life and thus did not take the opportunity to criticize the Soviet regime and its values as such. They certainly criticized the uneven distribution of privileges, the way the elites exploited their political power to secure its own, separate way of life. The interviews give evidence of the contradictory trajectories of development of the late Soviet Union. The autobiographical narratives challenge the contemporary attribution of stagnation that historiography had reproduced until recently, when it has finally started to debate the significance of the Brezhnev era. The retrospective perception of perestroika as a radical change brings up the question of longer continuities in attitudes, values, and norms up to the present day. At a minimum, the narratives of the interviewees challenge the idea of the inevitable collapse of the Soviet Union.[16]

Such retrospective compliant assessments of the Soviet social and political system and its longevity are influenced by nostalgic assumptions about the lost order and one's own social situation. This way of making sense of one's past life was particularly influenced by general social and political developments after perestroika. As we saw in chapters 4 and 5, contemporary viewers' letters to Soviet television let us trace points of reference for post-Soviet nostalgia.[17] The letters point to the mechanisms with which television promoted the emotional bonding to the sociopolitical system. Television offered interpretative frames for Soviet life that could easily be revived after the demise of the Soviet Union, as many people considered them to be familiar and reliable.

Television, however, had to cope with the radical changes that not only called for basic structural reforms but also for the elimination of ideological restrictions. Although Soviet television functioned rather well in the context of authoritative discourse, it failed to adapt to the new communication rules perestroika brought about. Ironically, it tried to be the spearhead of a new critical journalism that wanted to spontaneously engage people on the streets. The dimension of the political problems and people's loss of confidence toward the old political elites became overwhelming in the late 1980s. The ideological breakdown let the capacities of television as pleasure-giving medium almost vanish. Television also lost the appreciation of many viewers, as Zinaida Mikhailovna recalled. She denied that satirical or humorous programs conveyed any political messages and agreed that television created confidence in Soviet times: "In its early stage television generally brought up confidence. Everything was taken for absolute truth. Although, I do not say that we had such an informative 'Voice of America.'"

Perhaps American correspondent Hedrick Smith did not expect such a positive assessment from a Soviet viewer when he depicted Moscow's Central Television as "leaden and immobile" during the 1970s. In his book about the changing Soviet society during perestroika and glasnost, he highlighted the unfreezing of the Soviet mass media. Before perestroika, Central Television appeared in his view as having been "the most politically orthodox of all the media" and accordingly "a dull wasteland in the early 1970s." According to Smith, Soviet TV was so dull that allegedly even the kids of a Soviet diplomat in Washington found it only boring after having become used to American TV.[18]

Looking back, Zinaida Mikhailovna would have probably agreed, like many other respondents, with Hedrick Smith that Central Television ideologically strongly impacted on the perceptions of the Soviet audience. However, she would have very likely refused the image of a wasteland and would have painted a much more colorful and diversified picture of Soviet television. Therefore, the anecdote about the bored son of the Soviet diplomat, who was raised in Washington, may be right. It certainly appeals to the Western reader and matches her or his clichés, but it tells only half of the story before perestroika came up.

Also, the Soviet mass media, the readers, the radio listeners, and the TV viewers did not reach the ideal of the sophisticated and effective thought control George Orwell envisioned in I because Central Television's programming strategies took far too little advantage of the television's capacities. Somewhat ironically, this negligence especially applied to propaganda programs. The respondents usually considered the broadcasts of Khrushchev's and Brezhnev's speeches as boring. This was because the presentations not only presented only seemingly predictable information but basically lacked any specific televisual staging. Therefore, the viewers' ignorance of these broadcasts suggested that politics should and probably could have been visualized more captivatingly to match the ideas of the audience about attractive programs. Thus, the broadcasts of political events not only underperformed but also lagged behind the expectations of the audience whose viewing habits had gotten accustomed to all kinds of entertaining shows during the 1960s and 1970s.

The emotionalization and the good mood that many respondents recollected complicate the question of how confirming as well as rejecting attitudes impacted on the use and assessment of the different TV program genres. The variety of attitudes offered space for the deconstruction of "true and false" and for dissonances. The respondents' ambivalent assessments of Soviet television make clear that the medium highly impacted on the normalization of the Soviet life. It perhaps even opened a window not only to the outside world, as many respondents evoked, but, not least, also to its own "brave new world" until perestroika swept the familiar communication rules away. Then, a new story of everyday life and media use started to pave its way.

NOTES

1. Interview with Boris M. Firsov, October 2010. Narrating about Khrushchev the respondent shifted from the past tense to the present tense.

2. Bachtin, "Das Wort im Roman," 244–247.

3. Bhabha, *The Location*, 33, 56, 312; Sasse, *Michail Bachtin*, 136–140.

4. Daniel Dayan, "The Peculiar Public of Television," *Media, Culture & Society* 23, 6 (2001): 743–765; Leif Kramp, *Gedächtnismaschine Fernsehen. Vol. 1: Das Fernsehen als Faktor der gesellschaftlichen Erinnerung* (Berlin: Akademie Verlag, 2011), chapter 4.

5. Martha C. Nussbaum, *Political Emotions. Why Love Matters for Justice* (Cambridge, MA, and London: The Belknap Press of Harvard University Press, 2013), 200–203.

6. Kateryna Khinkulova, "Hello, Lenin? Nostalgia on Post-Soviet Television in Russia and Ukraine," *VIEW. Journal of European Television History and Culture* 1, 2 (2012): 94–104.

7. Ekaterina Kalinina, *Mediated Post-Soviet Nostalgia* (Stockholm: Södertörn University, 2014) 156–161.

8. Ekaterina Kalinina, "The Flow of Nostalgia: Experiencing Television from the Past," *International Journal of Communication* 10 (2016): 5324–5342.

9. Interview with Antonina (St. Petersburg, December 2011).

10. GARF, f. 6903, op. 31, d. 2, 1956, l. 35.

11. Yurchak, *Everything Was Forever*, 62.

12. Many former Soviet TV viewers recall this response on watching Brezhnev's speeches on TV. Cf. the quote in Yurchak, *Everything Was Forever*, 97.

13. TsNSIO, Arkhiv neopublikovannykh pisem, papka No. 3 "Irkutskoe televidenie," d. 162, l. 2.

14. TsNSIO, Arkhiv neopublikovannykh pisem, papka No. 3 "Irkutskoe televidenie," d. 153, l. 1; ibid., d. 163, l. 1.

15. Regarding post-Soviet biographical narratives, see Bönker, "*Perestroika.*"

16. Historians seek, for example, to investigate the mechanisms, communications, and practices that contributed to the stability of the late socialist order. Gorbachev's influential catchword of "stagnation" (*zastoi*) not only legitimized the need for reform, but also caricatured the relative security and confidence many Soviet people felt until Brezhnev died. On the question of periodization and the causes for the collapse, see Boris Belge and Martin Deuerlein (eds.), *Goldenes Zeitalter der Stagnation? Perspektiven auf die sowjetische Ordnung der Brežnev-Ära* (Tübingen: Mohr Siebeck, 2014); Jörg Baberowski, "Criticism as Crisis, Or Why the Soviet Union Still Collapsed," *Journal of Modern European History* 9 (2011): 148–166; Manfred Hildermeier, "'Well said is half a lie.' Observations on J. Baberowski's 'Criticism as Crisis, or Why the Soviet Union Still Collapsed,'" 289–297.

17. Cf. for example: *Obzor pisem telezritelei za 1981 god*, 167; GARF, f. 6903, op. 36, d. 1, 1971: Godovoi obzor pisem i obzory pisem telezritelei za ianvar'-dekabr' 1971, 461.

18. Hedrick Smith, *The New Russians* (New York: Random House, 1990), 162–163.

Bibliography

ARCHIVES

GARF (Gosudarstvennyi Arkhiv Rossiiskoi Federatsii)

f. 6903: Gosudarstvennyi komitet SSSR po televideniiu i radioveshchaniiu

op. 1: Sekretariat predsedatelia i otdel kadrov, 1933–1970 g.g.

d. 499, 1956: Stenogramma soveshchaniia u nachal'nika upravleniia po voprosam razvitiia televideniia ot 1 fevralia 1956 goda.

d. 542, 1958: Pis'ma v TsK KPSS i SM SSSR o rabote Komiteta.

d. 578, 1959: Pis'ma v TsK KPSS i SM SSSR o rabote Komiteta, t. 1.

d. 612, 1959: Stenogramma zasedaniia obshchestvennogo soveta telezritelei pri tsentral'noi studii televideniia.

d. 613, 1959: Protokoly zasedanii obshchestvennogo soveta telezritelei pri tsentral'noi studii televideniia.

d. 623, 1960: Pis'ma v TsK KPSS o rabote komiteta.

d. 666, 1960: Protokoly zasedanii obshchestvennogo soveta telezritelei pri tsentral'noi studii televideniia.

d. 667, 1960: Dokumenty o rabote obshchestvennogo soveta telezritelei pri tsentral'noi studii televideniia (spravki, predlozheniia).

d. 724, 1961: Protokoly zasedanii obshchestvennogo soveta telezritelei pri tsentral'noi studii televideniia.

d. 869, 1965: Stenogramma zasedaniia sektscii televideniia konferentsii „Televidenie i kritika" g. Moskva.

d. 883, 1965: Spravka otdela pisem Komiteta o rabote s pis'mami zritelei, poluchennykh Tsentral'nym televideniem v 1965 g.

d. 894, 1966: Protokoly NoNo 17, 18 zasedanii Komiteta i materialy k nim.

op. 3: Nauchno-technicheskii otdel, 1917–1970

d. 130, 1959: Protokol i stenogrammy konferentsii radioslushatelei i telezritelei g. Permi, Sverdlovska.

d. 147, 1960: Stenogramma vstrechi rabotnikov Tsentr. studii televideniia s telezriteliami g. Kaluga.
d. 178, 1962: Stenogramma vstrechi rabotnikov radio i televideniia s trudjashchimisia zavodov imeni Il'icha, gidroagregatov "Kauchuk," "Ėlektrosvet."
d. 184, 1962: Itogovaia spravka o rezul'tatakh anketnogo oprosa telezritelei o dnevnykh programmakh tsentral'nogo televidenia.
d. 185, 1962: Spravka NMO o neobkhodimosti izucheniia auditorii radioslushatelei i telezritelei Vsesoiuznogo radio i televideniia.
d. 208, 1963: Itogovaia spravka o rezul'tatakh anketnogo oprosa radioslushatelei i telezritelei "O roli radioveshchaniia i televideniia v kul'turnom i ėsteticheskom vospitanii naseleniia" i dokumenty k nei.
d. 277, 1965: Itogovaia spravka o rezul'tatakh sots. issledovaniia populiarnosti radio i telev. program, provedennogo na predpr. g. Moskvy.
d. 289, 1965, Dokumenty o provedenii ezhednevnogo operativnogo oprosa zhitelej Moskvy i Mosk. obl. o peredachakh Vsesoiuznogo radio i TsT, t. 2.
d. 306, 1965: Referat "Zritel' – o televidenii, televidenie – zriteliu," podgot. NMO dlia seminara v g. Aleksandrii (ARE).
d. 319, 1966: Polozhenie o NMO. Kopiia.
d. 320, 1966: Stenogramma sovmestnogo soveshchaniia NMO i sotsiologicheskoi gruppy Akademii obshchestvennykh nauk pri TsK KPSS na temu "Izuchenie radio i televizionnoi auditorii i effektivnosti veshchaniia."
d. 403, 1967: Programma sotsiologicheskogo issledovaniia NMO po teme "Struktura zaprosov i trebovanii radiosl. i telezr. po printsipu intellektual'noi vertikali."
d. 404, 1967: Otchet o sotsiologicheskom issledovanii na temu "Ob effektivnosti informatsii i propagandy na vneshnepoliticheskie temi po radio i televideniiu" dlia naseleniia Sovetskogo soiuza.
d. 417, 1967: Statisticheskie tablitsy o rezul'tatach issledovanii televizionnoi auditoria v 1967 godu.
d. 563, 1970: Otchet NMO ob issledovanii temu: "Effektivnost' tsentral'nykh i mestnykh programm radioveshchaniia i televideniia po chetyrem oblastiam. Zaprosy i trebovaniia radioslushatelei i telezritelei v zavisimosti ot urovnia intellektual'nogo razvitiia."
op. 10: Otdely pisem. Tsentral'nogo televideniia. Tsentral'nogo vnutrisoiuznogo radioveshchaniia. Tsentral'nogo Radioveshchaniia na zarubezhnye strany, 1952–1970 gg.
d. 18, 1959: Obzory pisem telezritelei i spravki o rabote s pis'mami za 1958–1959 gg.
d. 34, 1963: Godovoi obzor pisem i obzory pisem za ianvar'-mai radioslushatelei Tsentral'nogo Vnutrisoiuznogo Radioveshchaniia, Tsentral'nogo Radioveshchaniia na zarubezhnye strany i Tsentral'nogo televideniia za 1963.
d. 44, 1963: Obzor pisem radioslushatelei i telezritelei, postupivshikh v dni podgotovki k plenumu TsK KPSS 18 iiunia 1963 g.
d. 46, 1963: Obzor pisem telezritelei ob uluchshenii programm Tsentral'nogo televideniia, podgotovlennyi nauchno-metodicheskim otdelom.
d. 58, 1964: Obzory pisem radioslushatelei (sovetskikh i zarubezhnykh) i telezritelei za ianvar'-mart, iiul'-dekabr' 1964.

d. 64, 1965: Obzor pisem telezriteli za ianvar‘ 1965.
d. 66, 1965: Obzor pisem telezritelei za aprel’-avgust, oktiabr’ 1965.
d. 70, 1966: Analiz pisem telezritelei za period s 1964 po 1966.
d. 79, 1968: Godovoi obzor pisem telezritelei za 1968 g.
d. 89, 1969: Obzory pisem telezritelei za ianvar’-dekabr’ 1969 g.
d. 101, 1970: Godovoi obzor pisem telezritelei za 1970 g.
op. 28, 1963–1970 gody: Gosudarstvennyi komitet SM SSSR po televideniiu i radioveshchaniiu (1963 – dekabr’ 1965 g.), Komitet po radioveshchaniiu i televideniiu pri SM SSSR (dekabr’ 1965 – iiul’ 1970 g.)
d. 31, 1967, t. 3, Pis’ma v TsK KPSS, Verkhovnyi Sovet SSSR i RSFSR i TsK VLKSM po voprosam deiatel’nosti Gosudarstvennogo Komiteta Soveta Ministrov SSSR po televideniiu i radioveshchiiu.
op. 31, 1955–1971 gg.: Tsentral’noe televidenie. Glavnaia direktsiia programm. Tsentral’naia studiia televideniia (TsST), 1951–1960 gg. Tsentral’noe televidenie s 1960 goda
d. 2, 1956: Stenogramma konferentsii telezritelei zavoda “Optika” g. Krasnogorska.
d. 14, 1959: Stenogramma zasedaniia Obshchestvennogo soveta telezritelei pri Tsentral’noi studii televideniia.
d. 61, 1968: Zamechaniia otdela retsenzirovaniia po peredacham TsT.
op. 36: Tsentral’noe televidenie. Otdel pisem, 1971–1975 gg.
d. 1, 1971: Godovoi obzor pisem i obzory pisem telezritelei za ianvar’-dekabr’ 1971.
d. 18, 1973: Obzor pisem telezritelei za Aprel’ 1973 g.
d. 40, 1975: Obzor pisem telezritelei. Ianvar’ 1975 g.
op. 48 ch. 1: Tsentr nauchnogo programmirovaniia
d. 10, 1971: Osveshchenie podgotovki k XXIV s’’ezdu KPSS v informatsionno-publitsisticheskikh programmakh TsT za dekabr’ 1970.
d. 21, 1971: “Chto govorilos’ o televidenii i radioveshchanii na XX, XXI, XXII, XXIII s’’ezdach KPSS.” Spravka.
d. 22, 1971: Svedeniia anketnogo oprosa auditorii teleperedachi “Leninskii universitet millionov.”
d. 48, 1971: Kuznecov V. G. Obzor peredach TsT SSSR s uchastiem delegatov XXIV s’’ezda KPSS (31 marta – 7 aprelia 1971).
d. 55, 1971: Mironovich G. M. Rubriki obshchestvenno-politicheskikh peredach TsT v 1970.
d. 175, 1973: Mironovich G. A. “Programmy TsT, osveshchaiushchie khod i razvitie sotsialisticheskogo sorevnovaniia v strane.”

GARO (Gosudarstvennyi Arkhiv Rostovskoi Oblasti)

f. R-4237: Rostovskii oblastnoi komitet po radioveshchaniiu i televideniiu
op. 1 (prod.): Televidenie. Redaktsiia pisem
d. 876, 1963: Materialy redaktsii pisem, tom II.
d. 877, 1963: Materialy redaktsii pisem, tom III.

d. 1051, 1964: Materialy televizionnykh peredach redaktsii otdela pisem, tom III.
op. 2 (prod.): Televidenie. Redaktsiia pisem
d. 677, 1969: Dokumenty televizionnykh peredach promyshlennoi redaktsiia.
d. 754, 1969: Dokumenty teleperedach redaktsii pisem.
d. 759, 1969: Dokumenty teleperedach redaktsii pisem.

RGANI (Rossiiskii gosudarstvennyi arkhiv noveishei istorii)

f. 5: Apparat tsentral'nogo komiteta KPSS (1949–1991 gg.)
op. 33: Otdely TsK KPSS
d. 106.
op. 67: Otdely TsK KPSS (1974)
d. 133, 1974/75.
op. 90: Otdely TsK KPSS (1984)
d. 90, 1984/85.

RGAKFD (Rossiiskii gosudarstvennyi arkhiv kinofotodokumentov)

Fotodokumenty:
0-212658 ch/b: S. Tartakovskii. Mashinist depo im. Il'icha Zapadnoi zheleznoi dorogi Korolev s sem'ei v novoi kvartire v vysotnom zdanii u Krasnykh vorot za prosmotrom televizora, Moskva, dekabr' 1952.
0-229243 ch/b: Bez avtora. Sem'ia prepodavatelia Egor'evskogo stankostroitel'nogo tekhnikuma I. V. Firsova u televizora, Moskovskaia oblast', iiun' 1952.
1-4232 tsv: V. S. Tarasevich. Brigadiry stroitel'nogo kombinata Madalin Mamadaliev (sleva) i Abitbai Tokhtabaev s sem'iami smotriat televizor, p'iut chai, Uzbekistan, Syrdar'inskaia obl., 1957–1959.
1-104926 ch/b: D. Pirogov. Rabotnitsa miasokombinata A. V. Kostina u televizora v svoei novoi kvartire, Moskva, avril' 1963.
1-88353 ch/b: I. Krasutskii. Kolkhoznik kolkhoza im. Frunze V. L. Atroshkin s zhenoi v svoem dome v derevne Zabrod'e, Mogilevskaia oblast'/Belorussiia, oktiabr' 1966.

TsAOPIM (Tsentral'nyi arkhiv obshchestvenno-politicheskoi istorii Moskvy)

f. 2930: Partkom Gosudarstvennogo komiteta soveta ministrov SSSR po televideniiu i radioveshchaniiu, Moskvoretskij raion g. Moskvy
op. 1, 1957–1970 gg.
d. 158, 1964: Protokoly partsobranij, zasedanie partsobraniia redaktsii program, otdela pisem, redaktsii informatsii.
d. 383, 1966: Protokoly zasedanii partkoma pervichnoi partorg. TsT.

d. 527, 1967: Protokoly partsobranij, zasedanie partsobraniia glavnykh red. TsT: informatsii.
d. 1105, 1970: Protokoly partsobranii, zasedanie partsobraniia Glavnoi Redaktsii Informatsii.
op. 2, 1971–1975 gg.
d. 232, 1972: Protokoly zasedanii partkoma.
d. 239, 1972: Protokoly partsobranij, zasedanie partsobraniia Tsentra nauchn. Programmirovaniia.
op. 3, 1976–1978 gg.
d. 260, 1977: Protokoly partsobranij, zasedanie partsobraniia Tsentra nauchn. Programmirovaniia.
d. 480, 1978: Protokoly partsobranij, zasedanie partsobraniia Tsentra nauchn. Programmirovaniia.
op. 5, 1979–1982 gg.
d. 213, 1980: Stenogramma plenuma partkoma "O rabote s pis'mami telezritelei."
d. 564, 1982: Protokol plenuma partkoma o zadachach kommunistov Gosteleradio SSSR po vypolneniiu Postanovleniia TsK KPSS "O merakh po dal'neishemu ulushcheniiu raboty s pis'mami i predlozheniiami trudiashchikhsia v svete reshenii XXVI s''ezda KPSS."

TsGA SPb (Tsentral'nyi gosudarstvennyi arkhiv Sankt-Peterburga)

f. R-7384: Sankt Peterburgskii gorodskoi sovet narodnykh deputatov, 1917–1993.
op. 42: Arkhivnaia opis' del postoiannogo khraneniia, 1963–1967.
d. 332, 1963: Zhaloby i zaiavleniia trudiashchikhsia i otvety na nikh Ispolkoma Lengorsoveta o nepravil'nosti deistvii uchrezhdenii, organizatsii, predpriiatii i otdel'nykh rukovoditelei, t. 1.

TsGAIPD SPb (Tsentral'nyi gosudarstvennyi arkhiv istoriko-politicheskikh dokumentov Sankt-Peterburga)

f. R-24: Leningradskii oblastnoi komitet KPSS (1927–1991), Smol'ninskii raion, Leningrad
op. 145: Obshchee deloproizvodstvo, 1968–1973.
d. 20, 1972: O rabote s pis'mami i izuchenii ėffektivnosti peredach Leningradskogo televideniia v 1971 godu.

TsGALI SPb (Tsentral'nyi gosudarstvennyi arkhiv literatury i iskusstva Sankt-Peterburga)

f. R-293: Gosudarstvennaia teleradiokompaniia "Peterburg – 5-yi kanal." Sankt-Peterburg. 1933–1998
op. 3–1: Opis' del postoiannogo khraneniia, 1956–1968.

d. 132, 1963: Obzory o rabote s pis'mami radioslushatelei i telezritelei.
d. 151, 1964: Obzor o rabote s pis'mami telezritelei za 1964 god.

TsNSIO (Tsentr nezavisimykh sotsial'nykh issledovanii, Irkutsk)

Arkhiv neopublikovannykh pisem:
papka No. 3: "Irkutskoe televidenie"
d. 153, 162, 163.

PUBLISHED SOURCES

Arlazorov, Michail. "Ekran zhizni." *Znanie – sila: ezhemesiachnyi nauchnopopuliarnoi i nauchno-khudozhestvennyi zhurnal dlia molodezhi* (1956), 1: 41–42.

Baiar, Ol'ga G., and Raisa N. Blashkevich. *Kvartira i ee ubranstvo.* Moscow: Gosudarstvennoe izdatel'stvo literatury po stroitel'stvu, arkhitekture i stroitel'nym materialam, 1962.

Curtis, Charlotte. "The Way People Live." In *Anatomy of the Soviet Union*, edited by Harrison E. Salisbury, 33–58. London et al.: Nelson, 1967.

Efremov, Nikolai V. *Domovodstvo.* Kiev: Izdatel'stvo Urozhai, 1965.

Firsov, Boris. "Srednego zritelia net." *Zhurnalist* (1967), 12: 42–45.

Gal'perina, Elena V., ed. *KVN raskryvaet sekrety.* Moscow: Molodaia gvardiia, 1967.

Gordon, Leonid A., and Ėduard V. Klopow. *Der Mensch in seiner Freizeit.* Moskau: Verlag Progreß, 1976.

———, Ėduard V. Klopov, and Leon A. Onikov. *Cherty sotsialisticheskogo obraza zhizni: Byt gorodskikh rabochikh vchera, segodnia, zavtra.* Moskau: Izdatel'stvo Znanie, 1977.

Goriachev, Vladimir F. "K voprosu o tipologii detskoi auditorii." In *Televidenie i deti*, edited by Gosteleradio, Tsentr nauchnogo programmirovaniia and Nauchno-issledovatel'skii institut obshchikh problem vospitaniia APN SSSR, 51–58. Moscow: Iskusstvo, 1976.

Goriaeva, Tatiana M., ed. *"Velikaia kniga dnia . . .". Radio v SSSR. Dokumenty i materialy.* Moscow: Rosspen, 2007.

Grushin, Boris A. et al. *Die freie Zeit als Problem. Soziologische Untersuchungen in Bulgarien, Polen, Ungarn und der Sowjetunion.* Berlin: Berlin Verlag, 1970.

Higgins, Marguerite. *Red Plush and Black Bread.* Garden City, NY: Doubleday & Company, Inc., 1955.

Hollander, Gayle Durham. *Soviet Political Indoctrination. Developments in Mass Media and Propaganda since Stalin.* New York, Washington, London: Praeger, 1972.

Inkeles, Alex. *Public Opinion in the Soviet Union. A Study in Mass Persuasion.* Cambridge, MA: Harvard University Press, 1951.

Iurovskii, Aleksandr Ia. *Televidenie – poiski i resheniia. Ocherki istorii i teorii sovetskoi telezhurnalistiki*. Moscow: Iskusstvo, 1983.

Kaiser, Robert G. *Russia. The People & The Power*. New York: Atheneum, 1976.

Kogan, Lev N. "Sblizhenie sotsial'nykh grupp v sfere sotsialisticheskoi kul'tury." *Sotsiologicheskie issledovaniia* (1977), 2: 34–41.

———. "Buch und audio-visuelle Medien." In *Massenkommunikation in der UdSSR. Sowjetische Beiträge zur empirischen Soziologie der Journalistik*, edited by Hansjürgen Koschwitz, 148–166. Freiburg and München: Verlag Karl Alber, 1979.

Kolpakov, Boris T., and Vasilii D. Patrushev. *Biudzhety vremeni gorodskogo naseleniia*. Moskau: Statistika, 1971.

Lapin, Sergei. "Televizionnoe veshchanie." In *Bol'shaia Sovetskaia Ėntsiklopediia*, edited by A. M. Prokhorov, 378–380. 3rd ed. Moscow: Sovetskaia ėntsiklopedia, 1976.

———. "Tribuna liudei truda." *Zhurnalist* (1972), 5, 14–16.

Levine, Irving R. *Main Street, U.S.S.R.* Garden City, NY: Doubleday & Company, Inc., 1959.

Likhachev, Boris T. "Rol' televideniia v ėsteticheskom vospitanii detei." In *Televidenie i deti*, edited by Gosteleradio, Tsentr nauchnogo programmirovaniia, and Nauchno-issledovatel'skii institut obshchikh problem vospitaniia APN SSSR, 6–13. Moscow: Iskusstvo, 1976.

Narodnoe khoziaistvo SSSR, 1922–1972. Moscow: Statistika, 1972.

Narodnoe khoziaistvo SSSR v 1988 g. Statisticheskii ezhegodnik. Moscow: Statistika, 1989.

Narodnoe khoziaistvo SSSR za 60 let. Iubileinyi statisticheskii ezhegodnik. Moscow: Statistika, 1977.

Obzor pisem radioslushatelei za 1970 god. Moscow: Gosudarstvennyi komitet po televideniiu i radio, 1971.

Obzor pisem telezritelei za 1970 god. Moscow: Gosudarstvennyi komitet po televideniiu i radio, 1971.

Obzor pisem telezritelei za 1971 god. Moscow: Gosudarstvennyi Komitet SSSR po televideniiu i radioveshchaniiu, 1972.

Obzor pisem telezritelei, ianvar' 1972 god. Moscow: Gosudarstvennyi komitet po televideniiu i radio, 1972.

Obzor pisem telezritelei, fevral' 1972 god. Moscow: Gosudarstvennyi komitet po televideniiu i radio, 1972.

Obzor pisem telezritelei za 1972 god. Moscow: Gosudarstvennyi komitet po televideniiu i radio, 1973.

Obzor pisem telezritelei za 1973 god. Moscow: Gosudarstvennyi komitet po televideniiu i radio, 1974.

Obzor pisem telezritelei za iiul' 1974 goda (i 1 polugodie). Moscow: Gosudarstvennyi komitet po televideniiu i radio, 1974.

Obzor pisem telezritelei za oktiabr' 1975 g. Moscow: Gosudarstvennyi komitet po televideniiu i radio, 1975.

Obzor pisem telezritelei za 1975 god. Moscow: Gosudarstvennyi Komitet SSSR po televideniiu i radioveshchaniiu, 1976.
Obzor pisem telezritelei za 1976 g. Moscow: Gosudarstvennyi komitet po televideniiu i radio, 1977.
Obzor pisem telezritelei za fevral' 1977 g. Moscow: Gosudarstvennyi komitet po televideniiu i radio, 1977.
Obzor pisem telezritelei za 1977 g. Moscow: Gosudarstvennyi komitet po televideniiu i radio, 1978.
Obzor pisem telezritelei za 1979 god. Moscow: Gosudarstvennyi komitet po televideniiu i radio, 1980.
Obzor pisem telezritelei za 1980 god. Moscow: Gosudarstvennyi komitet po televideniiu i radio, 1981.
Obzor pisem telezritelei za 1981 god. Moscow: Gosudarstvennyi komitet po televideniiu i radio, 1982.
Obzor pisem telezritelei za oktiabr' 1981 god. Moscow: Gosudarstvennyi komitet po televideniiu i radio, 1982.
Obzor pisem telezritelei za 1982 god. Moscow: Gosudarstvennyi komitet po televideniiu i radio, 1983.
Obzor pisem telezritelei za ianvar' 1983 g. Moscow: Gosudarstvennyi komitet po televideniiu i radio, 1983.
Obzor pisem telezritelei za sentiabr' 1983 g. Moscow: Gosudarstvennyi komitet po televideniiu i radio, 1983.
Obzor pisem telezritelei za 1984 g. Moscow: Gosudarstvennyi komitet po televideniiu i radio, 1985.
Obzor pisem telezritelei za fevral' 1985 g. Moscow: Gosudarstvennyi komitet po televideniiu i radio, 1985.
Robinson, John P., Vladimir G. Andreyenkov, and Vasily D. Patrushev. *The Rhythm of Everyday Life. How Soviet and American Citizens Use Time.* Boulder, CO, San Francisco, CA, London: Westview Press, 1989.
Rogers, Rosemarie. "Normative Aspects of Leisure Time Behavior in the Soviet Union." *Sociology and Social Research* (1974), 58: 369–379.
Roth, Paul. *Die kommandierte öffentliche Meinung. Sowjetische Medienpolitik.* Stuttgart: Seewald, 1982.
Ryvkina, Rozalina V. *Obraz zhizni sel'skogo naseleniia. Metodologiia, metodika i rezul'taty izucheniia sotsial'no-ėkonomicheskich aspektov zhiznedejatel'nosti.* Novosibirsk: Izdatel'stvo "Nauka," Sibirskoe otdelenie, 1979.
Salisbury, Harrison E. "What Russians See on TV." *The New York Times Magazine*, July 11, 1954, 14.
———. "Fifty Years that Shook the World." In *Anatomy of the Soviet Union*, edited by idem., 3–32. London et al.: Nelson, 1967.
Sappak, Vladimir. "Televidenie, 1960. Iz pervykh nabliudenii." *Novyi mir,* no. 10, October 1960, 177–203.
———. *Televidenie i my: chetyre besedy.* Moscow: Iskusstvo, 1963.
———. *Televidenie i my: chetyre besedy.* In Vladimir Sappak and Vera Shitova, *Sem' let v teatre* and *Televidenie i my: chetyre besedy.* Moscow: Iskusstvo, 1968.

Sapunov, Boris. "Televidenie v sisteme socialisticheskoi kul'tury." In *Televidenie vchera, segodnja, zavtra-'86*, edited by Ėduard Efimov. Moscow: Iskusstvo, 1986.

Savastjuk Anton I. et al., eds. *Sovetskaia obraz zhizni: Soderzhanie, struktura, dinamika*. Minsk: Izdatel'stvo Narodnaia Asveta, 1980.

Smith, Hedrick. *The New Russians*. New York: Random House, 1990.

Televidenie i deti, edited by Gosteleradio, Centr nauchnogo programmirovaniia and Nauchno-issledovatel'skii institut obshchich problem vospitaniia APN SSSR. Moscow: Iskusstvo, 1976.

Tovarnyi slovar', vol. 7: Plavki-svekla, edited by Ivan A. Pugachev. Moskva: Gosudarstvennoe izdatel'stvo torgovoj literatury, 1959.

Vorob'eva, Liudmila I. et al., eds. *Domovodstvo: VII i VIII klassy*. Minsk: Izdatel'stvo Narodnaia Asveta, 1964.

CONTEMPORARY PERIODICALS

Iskusstvo kino
Krokodil
Literaturnaia gazeta
Novyi mir
Ogonek
Pravda
Sovetskaia kul'tura
Sovetskaia pechat'
Sovetskoe radio i televidenie
Zhurnalist

EXPLICITLY QUOTED INTERVIEWS

Experts

Boris M. Firsov, October 2010, St. Petersburg by Elena Bogdanova. He was born in 1929 in Saransk, was professor of sociology, and was director of Leningrad TV station 1962–1966.

Ėvelina E., October 2010, Rostov-on-Don by Galina Orlova and Sof'ia Kontorovich. Ėvelina E. was born in Rostov-on-Don in 1935, was graduated from Moscow Institute of Culture, and served as television director of Rostov television station from 1958 until the early 1990s.

Liudmila, October 2010, Irkutsk by Michail Rozhanskii. Liudmila was born in Irkutsk in 1955, television journalist, graduated from university, and started to work at the local TV station on the basis of an internship in 1974.

Roman R., October 2010, Rostov-on-Don by Galina Orlova and Sof'ia Kontorovich. Roman was born in 1934, and is a graduate from Moscow Library Institute, a director, and a documentary filmmaker.

V. V., November 2010, Samara by Oksana Zaporozhets. V. V., male, was born in 1948, graduated from Kuibyshev (Samara) Pedagogical Institute, and worked at Kuibyshev television from 1973 to 1999.

Valentina V., November 2010, Samara by Elena Zhidkova. Valentina V. was born in Orenburg region in 1937 and graduated from the Alma-Ata university department of journalism. She first worked for several newspapers before she was hired as editor at Kuibyshev (Samara) television in late 1965.

FORMER VIEWERS

Anonymous participants of a table talk: The talk was conducted by Ol'ga Galanova with some of her relatives in a small city in Nizhnii Novgorod region in August 2014.

Al'bina, January 2012, St. Petersburg by Elena Bogdanova. Al'bina was born in 1941, graduated from a technical university, and worked as engineer.

Aleksandr Nikolaevich, April 2010, Iaroslavl' by Anna Tikhomirova. Aleksandr was born in 1949 and graduated with a PhD from university.

Aleksandra, November 2010, St. Petersburg by Elena Bogdanova. Aleksandra was born in Leningrad in 1931. She has a doctorate in biology.

Aleksandra Apollonovna, November 2011, Tyumen by phone by Elena Mingaleva. Aleksandra Apollonovna was born in Rostov-on-Don in 1938 and lives in Tyumen since 1967. She graduated from a philological faculty and worked as translator and language teacher for English and German. Her father was a White Army emigrant to Paris who returned to the Soviet Union and was shot in 1941. Her mother was an accountant.

Anastasija, February 2012, St. Petersburg by Elena Bogdanova. Anastasija was born in Leningrad in 1954. She graduated from a technological university.

Anatolii, November 2011, Moscow by Kirsten Bönker. Anatolii was born in 1957 and graduated from university.

Anatolii Ivanovich, December 2011, Dzerzhinsk, Nizhnyi Novgorod region by Elena Mingaleva via phone. Anatolii was born in 1950 and finished secondary education.

Anna B., March 2012, Germany by Vadim Korolik. Anna B. is of German origin and was born in Kazakhstan in 1955. She finished a secondary school, and went to a pedagogical institute that she did not complete. She worked as kindergarten teacher and lived in Kazakhstan and in Altai region for a while before the family moved back to Kazakhstan in 1963.

Antonina, December 2011, St. Petersburg by Elena Bogdanova. Antonina was born in 1947, graduated from a technical university, and worked as engineer.

Elena, November 2010, St. Petersburg by Elena Bogdanova. Elena was born in Leningrad in 1961 and graduated from the faculty of economics of Leningrad university.

Elena Sergeevna, October 2010, in Samara by Oksana Zaporozhets. She grew up in a small village in Astrakhan region, moved to Kuibyshev (Samara), and graduated from the local history department. She worked as teacher and today works as saleswoman.

Georgii, November 2010, St. Petersburg by Elena Bogdanova. Georgii was born in Leningrad in 1953, graduated from university, and works as a public employee in St. Petersburg. He was a member of the Communist Party. His father was a scientist and doctor. He was also a member of the Communist Party and worked a while for the KGB.

Igor, March 2011, Moscow by Kirsten Bönker. Igor was born near Moscow in 1949 and finished secondary school.

Irina Alekseevna, March 2012, Omsk by Elena Mingaleva by phone. Irina Alekseevna was born in the village of Kochenevo/Novosibirsk region in 1958. She finished secondary school and worked as a cook.

Irina N., July 2010, Samara by Elena Zhidkova. Irina was born in Chernjakhovsk in 1953, Kaliningrad region; her family moved to Kuibyshev (Samara) region in 1957. She finished secondary school and worked as accountant.

Ivan, March 2012, St. Petersburg by Elena Bogdanova. Ivan was born in the city of Efremov, Tula region, in 1955. He studied at the workers' faculty in Leningrad, obtained a PhD in philosophy, and worked as an assistant professor before he switched over to a private enterprise offering business consulting in the 1990s. He was a member of the Communist Party.

Katia, July 2010, St. Petersburg by Elena Bogdanova. Katia was born in Leningrad in 1955. She graduated from the faculty of biology Leningrad State University. She worked for years as a translator in Soviet times and as an editor of St. Petersburg 5-ii Kanal from 2004 to 2009.

Konstantin, January 2012, St. Petersburg by Elena Bogdanova. Konstantin was born in Kokchetav (Kokshetau)/Kazakhstan in 1941, is higher educated, and graduated from Leningrad university, faculty of history.

Liubov' Aleksandrovna, August 2012, Lilienthal (Germany) by E. Peters. Liubov' was born in 1950 and lived in Leninsk-Kuznetskii, Kemerovo region. She finished secondary school.

Liudmila, September 2010, Samara by Oksana Zaporozhets. Liudmila was born in Chapaevsk in 1951, Samara region, and finished secondary school.

Liudmila, November 2010, St. Petersburg by Elena Bogdanova. Liudmila was born in Penza region in 1942 and moved to Leningrad in 1962. She finished secondary school.

Liudmila, January 2012, St. Petersburg by Elena Bogdanova. Liudmila was born in a small village in Arkhangel'sk region in 1951 and moved to Leningrad in 1964. She graduated from the polytechnic and worked as metallurgical engineer.

Liudmila Iur'evna, October 2010, Rostov-on-Don by Galina Orlova. The interview was conducted together with Liudmila Vasil'evna. Liudmila Iur'evna was born in North Caucasus in 1958. She was raised for some years in an orphanage and lived with her teacher's family in Rostov-on-Don from 1970 on. She is a radio cutter.

Liudmila Vasil'evna, October 2010, Rostov-on-Don by Galina Orlova. The interview was conducted together with Liudmila Iur'evna. Liudmila Vasil'evna born in the small town Matveev-Kurgan, Rostov-on-Don region in 1946, and is higher educated.

Masha, September 2010, St. Petersburg by Elena Bogdanova. Masha was born in Leningrad in 1946 and graduated from university.

Mikhail, December 2011, St. Petersburg by Elena Bogdanova. Mikhail was born in Leningrad in 1952; he holds a doctoral degree in meteorology.

Mikhail G., July 2010, Samara by Elena Zhidkova. Mikhail G. was born in Samara region in 1948, graduated from university, and worked as engineer of telecommunication.

Nadezhda Ivanovna, August 2010, Usol'e-Sibirskoe, Irkutsk region by Alexander Ermakov. Nadezhda was born in 1950, and is higher educated.

Natal'ia G., February 2011, St. Petersburg by Elena Bogdanova. Natal'ia was born in Leningrad in 1951 and graduated from university.

Natal'ia Petrovna, October 2010, Samara by Oksana Zaporozhets. Natal'ia was born in Samara region in 1946, finished secondary school, and managed a trade company.

Nikolai, January 2011, St. Petersburg by Elena Bogdanova. Nikolai was born in Briansk region in 1940 and moved to Leningrad in the 1960s. He graduated from university and was member of the Communist Party.

Oleg Igor'evich, November 2010, Samara by Oksana Zaporozhets. Oleg was born in a village in Kuibyshev (Samara) region in 1958, and is higher educated.

Ol'ga, March 2012, Germany by Julia Schmidt. Ol'ga is of German origin and was born in Karaganda in 1957. She completed a technical secondary school.

Ol'ga Nikolaevna, October 2010, Irkutsk by E. Ivanova. Ol'ga Nikolaevna was born in Irkutsk in 1964. She graduated both from a technical and in addition from an economic college. Afterward, she worked in a housing trust.

Rudol'f N., July 2010, Samara by Elena Zhidkova. Rudol'f N. was born in Tashkent in 1940. He graduated from a technical institute and worked as engineer in Iran at times.

Ruslan, October 2010, Samara by Oksana Zaporozhets. Ruslan was born in Kuibyshev (Samara) in 1968. He had secondary education and works as driver at the time of the interview.

Svetlana Vladimirovna, June 2010, Samara by Elena Zhidkova. Svetlana was born in Kuibyshev (Samara) in 1947, has specialized secondary education/technical college, and is member of CPSU.

Tamara, December 2011, St. Petersburg by Elena Bogdanova. Tamara was born in Kazakhstan in 1944, and lived in the Far North before she moved to Leningrad. She graduated from the faculty of physics and mathematics of the Pedagogical University of Astana.

Tat'iana, March 2012, St. Petersburg by Elena Bogdanova. Tat'iana was born in a village in Gor'kii (Nizhnii Novgorod) region in 1949, and moved to Leningrad in 1972. She has secondary education and worked in a woodworking factory.

Tat'iana Iakovlevna, October 2010, Irkutsk by E. Ivanova. Tat'iana was born in Irkutsk in 1964, and graduated from the faculty of law of Irkutsk university. She works in the state administration.

Valerii Nikolaevich, August 2010, Usol'e-Sibirskoe, Irkutsk region by Alexander Ermakov. Valerii was born in 1958, and has secondary education.

Vladimir Ivanovich, April 2012, St. Petersburg by Elena Bogdanova. Vladimir Ivanovich was born in Leningrad in 1958, and has specialized secondary education.

Vladimir Jakovlevich, October 2010, Irkutsk by E. Ivanova. Vladimir was born in 1947, and graduated from university.

Waldemar, February 2012, Germany by Vadim Korolik. Waldemar was born in Novosibirsk in 1946. He is of German origin; his parents were deported to a NKVD sovkhoz near Novosibirsk, where his father worked as a transport worker. Waldemar is married to Zoia; he completed a technical secondary school.

Yuri, October 2010, St. Petersburg by Elena Bogdanova. He was born in Leningrad in 1938 and graduated from university with a diploma in engineering.

Yuri Vasil'evich, July 2010, Samara region by Elena Zhidkova. Yuri was born in Kuibyshev (Samara) in 1951, and graduated from university. He was member of the CPSU and deputy of the district soviet.

Zina, March 2012, St. Petersburg by Elena Bogdanova. Zina was born in Leningrad in 1949 and completed secondary education.

Zinaida Michailovna, July 2010, Tol'iatti, Samara region by Elena Zhidkova. Zinaida was born in Riazan' region in 1937, graduated from Moscow pharmaceutical institute, and worked as teacher.

Zoia, February 2012, Germany by Vadim Korolik. She was born in Novosibirsk in 1947. She is married to Waldemar and of German origin. She completed a technical secondary school.

INTERNET SOURCES AND WEBSITES

http://www.ceae.ru/SSSR.htm.

http://cccp.tv.

http://www.euscreen.eu/.

http://gtrf.ru/.

http://www.mfk.ch/fernsehen.html.

http://www.mojgorod.ru/pskovsk_obl/pskov/index.html.

http://www.TeachingAmericanHistory.org/library/index.asp?documentprint=176.

http://tv-80.ru/.

http://tvmuseum.ru.

http://www.wdr.de/bilder/mediendb/Fotostrecken/wdr5/themenwochen_und_specials/2012/Sommer_1962/Bilder_zugeschnitten/0724_film_fernseher_akg-images_990_m.jpg.

http://www.welt.de/fernsehen/article13396067/Huberty-die-Hausfrauen-und-das-Eckenverhaeltnis.html.

https://www.youtube.com/watch?v=z6RLCw1OZFw.

»Utrenniaia pochta« (Morning post): http://www.youtube.com/watch?v=e5s_-MdQrhI.

LITERATURE

Anders, Günther. "Die Antiquiertheit des Menschen (1956)." In *Texte zur Theorie und Geschichte des Fernsehens*, edited by Michael Grisko, 101–121. Stuttgart: Reclam, 2009.

Anderson, Kathryn, and Dana C. Jack. "Learning to Listen: Interview Techniques and Analyses." In *The Oral History Reader*, edited by Robert Perks, and Alistair Thomson, 129–142. 3rd ed. Abingdon/Oxon and New York: Routledge 2016.

Arutunyan, Anna. *The Media in Russia*. Maidenhead/Berkshire: McGraw Hill Open University Press, 2009.

Attwood, Lynne. *Creating the New Soviet Woman: Women's Magazines as Engineers of Female Identity, 1922–53*. New York: St. Martin's Press, 1999.

Baberowski, Jörg. "Criticism as Crisis, or Why the Soviet Union Still Collapsed." *Journal of Modern European History* (2011), 9: 148–166.

Bachtin, Michail M. *Die Ästhetik des Wortes*, edited by Rainer Grübel. Frankfurt a. M.: Suhrkamp, 1979.

Badenoch, Alexander, Fickers, Andreas, and Henrich-Franke, Christian. "Airy Curtains in the European Ether: Introduction." In *Airy Curtains: European Broadcasting during the Cold War*, edited by Alexander Badenoch, Andreas Fickers, and Christian Henrich-Franke, 9–26. Baden-Baden: Nomos-Verlag, 2013.

Belge, Boris, and Martin Deuerlein, eds. *Goldenes Zeitalter der Stagnation? Perspektiven auf die sowjetische Ordnung der Brežnev-Ära*. Tübingen: Mohr Siebeck, 2014.

Beliaev, Edward, and Pavel Butorin. "The Institutionalization of Soviet Sociology. Its Social and Political Context." *Social Forces* (1982), 61: 418–435.

Beumers, Birgit. *Pop Culture Russia! Media, Arts, and Lifestyle*. Santa Barbara, CA: ABC CLIO, 2005.

———, Stephen Hutchings, and Natalia Rulyova, eds. *The Post-Soviet Russian Media: Conflicting Signals*. New York: Routledge, 2009.

Beutelschmidt, Thomas, and Richard Oehmig. "Connected Enemies? Programming Transfer between East and West during the Cold War and the Example of East German Television." *VIEW. Journal of European Television History and Culture* (2014), 3(5): 60–67.

Bhabha, Homi K. *The Location of Culture*. London: Routledge, 1994.

Bignell, Jonathan, and Andreas Fickers, eds. *A European Television History*. Malden, MA: Wiley-Blackwell, 2008.

Bönker, Kirsten. "Depoliticalisation of the Private Life? Reflections on Private Practices and the Political in the Late Soviet Union." In *Writing Political History Today*, edited by Willibald Steinmetz, Ingrid Gilcher-Holtey, and Heinz-Gerhard Haupt, 207–234. Frankfurt: Campus, 2013.

———. "Der sowjetische Fernsehzuschauer als ‚neuer Mensch': Die sozialwissenschaftliche Freizeitforschung und das Problem der Rückständigkeit." In *Die Zukunft der Rückständigkeit. Festschrift für Manfred Hildermeier*, edited by David Feest, and Lutz Häfner, 373–397. Köln, Weimar, Wien: Böhlau, 2016.

———. "*Perestroika* and the Loss of Certainties: The Post-Soviet Revaluation of Soviet Money Practices and Social Equality." *Journal of Modern European History* (2017), 15(3): 367–394.

———, Julia Obertreis, and Sven Grampp, eds. *Television Beyond and Across the Iron Curtain*. Cambridge: Cambridge Scholars Publishing, 2016.

Bösenberg, Jost-Arend. *Die Aktuelle Kamera. Medienpolitische Untersuchungen zu den Lenkungsmechanismen im Fernsehen der DDR, in der Zeit von 1952 bis 1990*. PhD manuscript. Freie Universität, Berlin, 2002.

Bogdanova, Elena A. "Gazetnye zhaloby kak strategii zashchity potrebitel'skikh interesov: Pozdnesovetskii period." *Teleskop: Nabliudeniia za povsednevnoi zhizn'iu peterburzhtsev* (2002), 6: 44–48.

———. "Konstruirovanie problem zashchity prav potrebitelei: Retrospektivnyi analiz." *Rubezh* (2003), 18: 167–177.

Bogdanowa, Olga S. *Freizeitgestaltung*. Ost-Berlin: Volk und Wissen Verlag, 1962.

Bolgov, Vladimir I. *Biudzhet vremeni pri sotsializme. Teoriia i metody issledovaniia*. Moskau: Nauka, 1973.

Boretskii, Rudol'f. *Televizionnaia programma: ocherk teorii propagandy*. Moscow: Komitet po radioveshchaniiu i televideniiu pri sovete ministrov SSSR. Redaktsionno-izdatel'skii otdel, 1967.

Boškovska, Nada, Strobel, Angelika, and Daniel Ursprung, eds. *'Entwickelter Sozialismus in Osteuropa'. Arbeit, Konsum und Öffentlichkeit*. Berlin: Duncker & Humblot, 2016.

Boym, Svetlana. *Common Places: Mythologies of Everyday Life in Russia*. Cambridge, MA: Harvard University Press, 1994.

———. *The Future of Nostalgia*. New York: Basic Books, 2001.

———. http://monumenttotransformation.org/atlas-of-transformation/html/n/nostalgia/nostalgia-svetlana-boym.html (accessed April 15, 2015).

Breckner, Roswitha. "Von den *Zeitzeugen* zu den *Biographen*. Methoden der Erhebung und Auswertung lebensgeschichtlicher Interviews [1994]." In *Oral History*, edited by Julia Obertreis, 131–151. Stuttgart: Franz Steiner, 2012.

Breitenborn, Uwe. *Wie lachte der Bär? Systematik, Funktionalität und thematische Segmentierung von unterhaltenden nonfiktionalen Programmformen im Deutschen Fernsehfunk bis 1969*. Berlin: Weißensee Verlag, 2003.

Bren, Paulina. *The Greengrocer and His TV: The Culture of Communism After the 1968 Prague Spring*. Ithaca, NY: Cornell University Press, 2010.

———, and Mary Neuberger, eds. *Communism Unwrapped: Consumption in Cold War Eastern Europe*. New York: Oxford University Press, 2012.

Bublitz, Hannelore. "Diskurs und Habitus. Zentrale Kategorien der Herstellung gesellschaftlicher Normalität." In *›Normalität‹ im Diskursnetz soziologischer Begriffe*, edited Jürgen Link, Thomas Loer, and Hartmut Neuendorff, 151–162. Heidelberg: Synchron, 2003.

Burds, Jeffrey, "A Culture of Denunciation: Peasant Labor Migration and Religious Anathematization in Rural Russia, 1860–1905." In *Accusatory Practices: Denunciation in Modern European History, 1789–1989*, edited by Sheila Fitzpatrick, and Robert Gellately, 40–72. Chicago, IL: University of Chicago Press, 1997.

Bushnell, John. "The 'New Soviet Man' Turns Pessimist." In *The Soviet Union Since Stalin,* edited by Stephen F. Cohen, Alexander Rabinowitch, and Robert Sharlet, 179–199. Bloomington, IN: Indiana University Press, 1980.

Bussemer, Thymian. *Propaganda und Populärkultur. Konstruierte Erlebniswelten im Nationalsozialismus*. Wiesbaden: Deutscher Universitäts-Verlag, 2000.

Cassiday, Julie A., and Emily D. Johnson. "Putin, Putiniana and the Question of a Post-Soviet Cult of Personality." *The Slavonic and East European Review* (2010), 88: 681–707.

Castillo, Greg. *The Cold War on the Home Front. The Soft Power of Midcentury Design*. Minneapolis, MN, London: University of Minnesota Press, 2010.

Chartier, Roger. *Forms and Meanings: Texts, Performances, and Audiences from Codex to Computer*. Philadelphia, PA: University of Pennsylvania Press, 1995.

———. *Lesewelten. Buch und Lektüre in der frühen Neuzeit*. Frankfurt a. M., New York, Paris: Campus, 1990.

———. "The Chimera of the Origin: Archaeology, Cultural History, and the French Revolution." In *Foucault and the Writing of History*, edited by Jan Goldstein, 167–186. Oxford, Cambridge, MA, 1994.

———. "Le Monde comme representation." *Annales ESC* (1989) 6: 1505–1520.

———. "Die Welt als Repräsentation." In *Alles Gewordene hat Geschichte. Die Schule der ANNALES in ihre Texten* 1929–1992, edited by Matthias Middell, and Steffen Sammler, 320–347. Leipzig: Reclam, 1994.

———. "Mémoire et oubli. Lire avec Ricoeur." In *Paul Ricoeur et les sciences humaines*, edited by Christian Delacroix, François Dosse, and Patrick Garcia, 231–248. Paris: La Découverte, 2007.

Chernyshova, Natalya. *Soviet Consumer Culture in the Brezhnev Era*. New York: Routledge, 2013.

Collaborative Research Centre (Sonderforschungsbereich) 584. *The Political as Communicative Space in History. Research Programme*. Manuscript: University of Bielefeld, 2007. http://www.uni-bielefeld.de/%28en%29/geschichte/forschung/sfb584/research_program/index.html (accessed June 12, 2012).

Crowley, David, and Susan E. Reid. "Introduction: Pleasures in Socialism?" In *Pleasures in Socialism. Leisure and Luxury in the Eastern Bloc*, edited by David Crowley, and Susan E. Reid, 3–53. Evanston, IL: Northwestern University Press, 2010.

Daniel Dayan. "The Peculiar Public of Television." *Media, Culture & Society* (2001), 23(6): 743–765.

———, and Elihu Katz. *Media Events: The Live Broadcasting of History*. Cambridge, MA: Harvard University Press, 1992.

Darby, Joseph J. "Product Liability in the Soviet Union." *International Lawyer* (1977), 11: 179–192.

Davis, Fred. *Yearning for Yesterday. A Sociology of Nostalgia*. New York, London: Free Press, 1979.

Dingley, James. "Soviet Television and Glasnost'." In *Culture and the Media in the USSR Today*, edited by Julian Graffy, and Geoffrey Hosking, 6–25. London, New York: St. Martin's Press, 1989.

Dittmar, Claudia. *Feindliches Fernsehen: Das DDR-Fernsehen und seine Strategien im Umgang mit dem westdeutschen Fernsehen*. Bielefeld: Transcript, 2010.

———, and Susanne Vollberg, eds. *Zwischen Experiment und Etablierung. Die Programmentwicklung des DDR-Fernsehens 1958 bis 1963*. Leipzig: Leipziger Universitätsverlag, 2007.

———. *Alternativen im DDR-Fernsehen? Die Programmentwicklung 1981 bis 1985*. Leipzig: Leipziger Universitätsverlag, 2005.

Döveling, Katrin. *Emotionen — Medien — Gemeinschaft. Eine kommunikationssoziologische Analyse*. Wiesbaden: VS Verlag, 2005.

———, Christian von Scheve, and Elly A. Konijn, eds. *The Routledge Handbook of Emotions and Mass Media*. London, New York: Routledge, 2010.

Downey, John, and Mihelj, Sabina, eds. *Central and Eastern European Media in Comparative Perspective. Politics, Economy and Culture*. Farnham/Surrey, Burlington, NJ: Ashgate, 2012.

Duhamel, Luc. *The KGB Campaign against Corruption in Moscow, 1982–1987*. Pittsburgh, PA: University of Pittsburgh Press, 2010.

Eco, Umberto. *Das offene Kunstwerk*. Frankfurt a. M.: Suhrkamp, 1996.

Egorov, Vilen V. *Televidenie: Stranitsy istorii*. Moscow: Aspekt Press, 2004.

Emeliantseva, Ekaterina. "The Privilege of Seclusion: Consumption Strategies in the Closed City of Severodvinsk." *Ab Imperio: Studies of New Imperial History and Nationalism in the Post-Soviet Space* (2011), 2: 238–259.

Esquenazi, Jean-Pierre. "Télévision et espace public sous De Gaulle." *Cahiers d'histoire. Revue d'histoire critique* (2002), 86: 49–61.

Etzemüller, Thomas. "*Social engineering* als Verhaltenslehre des kühlen Kopfes: Eine einleitende Skizze." In *Die Ordnung der Moderne: Social Engineering im 20. Jahrhundert*, edited by Thomas Etzemüller, 11–39. Bielefeld: Transcript Verlag, 2009.

Evans, Alfred Jr. "The Decline of Developed Socialism? Some Trends in Recent Soviet Ideology." *Soviet Studies* (1986), 38(1): 1–23.

Evans, Christine Elaine. *From Truth to Time: Soviet Central Television, 1957–1985*. PhD manuscript University of Berkeley, 2010. http://escholarship.org/uc/item/2702x6wr.

———. "Song of the Year and Soviet Culture in the 1970s." *Kritika: Explorations in Russian and Eurasian History* (2011), 12(3): 617–645.

———. "The Soviet Way of Life as a Way of Feeling: Emotion and Influence on Soviet Central Television in the Brezhnev Era." *Cahiers du monde russe* (2015), 56(2–3): 543–69.

———. *Between Truth and Time. A History of Soviet Central Television*. New Haven, CT and London: Yale University Press, 2016.

Fickers, Andreas, and Catherine Johnson. "Transnational Television History: A Comparative Approach. Introduction." *Media History* (2010), 16(1): 1–11.

Field, Deborah. *Private Life and Communist Morality in Khrushchev's Russia*. New York: Peter Lang, 2007.

Firsov, Boris. *Raznomyslie v SSSR 1940–1960–e gody. Istoriia, teoriia i praktika*. St. Petersburg: Izdatel'stvo Evropeiskogo Universiteta v Sankt-Peterburge, Evropeiskii dom, 2008.

Fiske, John. *Television Culture*. London, New York: Routledge, 1987.

Fitzpatrick, Sheila. "Supplicants and Citizens: Public Letter – Writing in Soviet Russia in the 1930s." *Slavic Review* (1996), 55(1): 78–105.

Fulbrook, Mary. *The People's State. East German Society from Hitler to Honecker*. New Haven, CT: Yale University Press, 2005.

Gatejel, Luminita. *Warten, hoffen und endlich fahren. Auto und Sozialismus in der Sowjetunion, in Rumänien und der DDR (1956–1989/91)*. Frankfurt/New York: Campus, 2014.

Genis, Aleksandr, and Petr Vail'. *60-e. Mir sovetskogo cheloveka.* 2nd ed. Moscow: Novoe literaturnoe obozrenie, 1998.
Golubev, Alexey, and Olga Smolyak. "Making Selves through Making Things: Soviet Do-It-Yourself Culture and Practices of Late Soviet Subjectivation." *Cahiers du monde russe* (2014), 53(3–4): 517–541.
Gorsuch, Anne E., and Diane P. Koenker, eds. *The Socialist Sixties. Crossing Borders in the Second World.* Bloomington, IN: Indiana University Press, 2013.
Goscilo, Helena. "Luxuriating in Lack: Plenitude and Consuming Happiness in Soviet Paintings and Posters, 1930–1953." In *Petrified Utopia: Happiness Soviet Style*, edited by Marina Balina, and Evgeny Dobrenko, 53–78. London: Anthem Press, 2009.
Grushin, Boris A. *Chetyre zhizni Rossii v zerkale oprosov obshchestvennogo mneniia. Zhizn' 2-aia: Epokha Brezhneva.* Moscow: Progress-Traditsiia, 2001.
Gumbert, Heather. *Envisioning Socialism: Television and the Cold War in the German Democratic Republic.* Ann Arbor, MI: University of Michigan Press, 2014.
———. "Exploring the Transnational Media Exchange in the 1960s." *VIEW. Journal of European Television History and Culture* (2014), 3(5): 50–59.
Habscheid, Stephan. "Ratgebersendungen: Gesundheitsmagazine." In *Der sprechende Zuschauer: Wie wir uns Fernsehen kommunikativ aneignen*, edited by Werner Holly, Ulrich Püschel, and Jörg Bergmann, 173–186. Wiesbaden: Springer, 2001.
Hall, Stuart. "Encoding/Decoding." In *Media and Cultural Studies: Key Works,* edited by Meenakshi Gigi Durham, and Douglas M. Kellner, 163–173. Malden, MA: Blackwell, 2006.
———. "Die strukturierte Vermittlung von Ereignissen." In *Grundlagentexte zur Fernsehwissenschaft. Theorie – Geschichte – Analyse*, edited by Ralf Adelmann et al., 344–375. Konstanz: UVK, 2001.
Hannah, Gayle Durham. "Legale und dissidente Formen politischer Kommunikation in der Sowjetunion nach Stalin." *Osteuropa* (1976), 7: 491–512.
Hartmann, Tilo. "Mass Communication and Para-Social Interaction. Observations on Intimacy at a Distance von Donald Horton und R. Richard Wohl." In *Schlüsselwerke der Medienwirkungsforschung*, edited by Matthias Potthoff, 75–84. Wiesbaden: Springer, 2016.
Havens, Timothy, Anikó Imre, and Katalin Lustyik, eds. *Popular Television in Eastern Europe During and Since Socialism.* New York: Routledge, 2013.
Hepp, Andreas. *Medienkultur. Die Kultur mediatisierter Welten.* 2nd ed. Wiesbaden: VS Verlag, 2013.
Hickethier, Knut. *Geschichte des deutschen Fernsehens.* Stuttgart: J. B. Metzler, 1998.
———. "Die kulturelle Bedeutung medialer Emotionserzeugung." In *Audiovisuelle Emotionen. Emotionsdarstellung und Emotionsvermittlung durch audiovisuelle Medienangebote*, edited by Anne Bartsch, Jens Eder, and Kathrin Fahlenbrach, 104–122. Köln: Herbert von Halem Verlag, 2007.
———. "Das Fernsehen der DDR." In *Wie im Westen, nur anders: Medien in der DDR*, edited by Stefan Zahlmann, 119–130. Berlin: Panama-Verlag, 2010.
Hildermeier, Manfred. "'Well Said is Half a Lie'. Observations on J. Baberowski's 'Criticism as Crisis, or Why the Soviet Union Still Collapsed'." *Journal of Modern European History* (2011), 9: 289–297.

Hoffmann, David L. *Stalinist Values: The Cultural Norms of Soviet Modernity, 1917–1941*. Ithaca, NY: Cornell University Press, 2003.

Holzweißig, Gunter. *Zensur ohne Zensor. Die SED-Informationsdiktatur*. Bonn: Bouvier Verlag, 1997.

———. *Die schärfste Waffe der Partei. Eine Mediengeschichte der DDR*. Köln, Weimar, Wien: Böhlau Verlag, 2002.

Hopkins, Mark W. *Mass Media in the Soviet Union*. New York: Pegasus, 1970.

Horton, Donald, and R. Richard Wohl. "Mass Communication and Para-Social Interaction. Observations on Intimacy at a Distance." *Psychiatry* (1956), 19(3): 215–229.

Hutchings, Stephen. *Russian Literary Culture in the Camera Age. The Word as Image*. London and New York: Routledge, 2004.

———, and Natalia Rulyova. *Television and Culture in Putin's Russia: Remote Control*. New York: Routledge, 2009.

Huxtable, Simon. "The Problem of Personality on the Soviet Screen, 1950s–1960s." *View: Journal of European Television History and Culture* (2014), 3(5): 119–130.

———. "In Search of the Soviet Reader: The Kosygin Reforms, Sociology, and Changing Concepts of Soviet Society, 1964–1970." *Cahiers du monde russe* (2013), 54(3): 623–642.

Idrizi, Idrit. "Magic Apparatus" and "Window to the Foreign World"? The Impact of Television and Foreign Broadcasts on Society and State-Society Relations in Socialist Albania." In *Television Beyond and Across the Iron Curtain*, edited by Kirsten Bönker, Julia Obertreis, and Sven Grampp, 227–256. Cambridge: Cambridge Scholars Publishing, 2016.

Imre, Anikó. "Television for Socialist Women." *Screen* (2013), 54(2): 249–255.

———, Timothy Havens, and Katalin Lustyk, eds. *Popular Television in Eastern Europe During and Since Socialism*. Abingdon, New York: Routledge, 2012.

———. "Audience History as a History of Ideas: Towards a Transnational History." *European Journal of Communication* (2015), 30(1): 22–35.

Janco, Andrew. "*KVN:* Live Television and Improvised Comedy in the Soviet Union, 1957–71." In *Popular Television in Authoritarian Europe,* edited by Peter Goddard, 124–140. Manchester: Manchester University Press, 2013.

Kalinina, Ekaterina. "Multiple Faces of the Nostalgia Channel in Russia." *VIEW. Journal of European Television History and Culture* (2014), 3(5): 108–118.

———. *Mediated Post-Soviet Nostalgia*. Stockholm: Södertörn University, 2014.

———. "The Flow of Nostalgia: Experiencing Television from the Past." *International Journal of Communication* (2016), 10: 5324–5342.

Kashin, Michail P. "Uchebnomu televideniiu – dorogu v shkolu." In *Televidenie v shkole. Trudy 1-go Vsesoiuznogo seminara po uchebnomu televideniiu*, edited by idem, and Lev P. Pressman, 2–8. Moscow: Akademiia pedagogicheskikh nauk, 1967.

Katz, Elihu. "Mass Communication Research and the Study of Popular Culture: An Editorial Note on a Possible Future for This Journal." *Studies in Public Communication* (1959), 2: 1–6.

Keghel, Isabelle de. "Konsum im Blick: Visualisierungsstrategien in sowjetischen und ostdeutschen Printmedien (1953–1964)." *Comparativ* (2011), 21(3): 79–96.

Keilina, Zinaida A. *Televidenie v semeinom vospitanii (Metodicheskoe posobie k lektsii)*. Voronezh: Izdatel'stvo Kommuna, 1969.

Kelly, Catriona, and Vadim Volkov. "Directed Desires: *Kul'turnost'* and Consumption." In *Constructing Russian Culture in the Age of Revolution: 1881–1940*, edited by Catriona Kelly, and David Shepherd, 291–313. Oxford: Oxford University Press, 1998.

———. *Refining Russia: Advice Literature, Polite Culture, and Gender from Catherine to Yeltsin*. Oxford: Oxford University Press, 2001.

———. "Making a Home on the Neva: Domestic Space, Memory, and Local Identity in Leningrad and St. Petersburg, 1957-present." *Laboratorium: Russian Review of Social Research* (2011), 3(3): 53–96.

Kenez, Peter. *The Birth of the Propaganda State. Soviet Methods of Mass Mobilization, 1917–1929*. Cambridge et al.: Cambridge University Press, 1985.

Khakhulina, Liudmila. "The Persistence of Mass Conceptions of Justice." *Sociological Research* (2015), 54(2): 71–90.

Khinkulova, Kateryna. "Hello, Lenin? Nostalgia on Post-Soviet Television in Russia and Ukraine." *VIEW. Journal of European Television History and Culture* (2012), 1(2): 94–104.

Kivelson, Valerie A. *Autocracy in the Provinces: The Muscovite Gentry and Political Culture in the Seventeenth Century*. Stanford, CA: Stanford University Press, 1996.

Klaus, Elisabeth. "Der Gegensatz von Information ist Desinformation, der Gegensatz von Unterhaltung ist Langeweile." In *Medien – Politik – Geschlecht. Feministische Befunde zur politischen Kommunikationsforschung*, edited by Johanna Dorer, Brigitte Geiger, and Regina Köpl, 51–64. Wiesbaden: Springer, 2008.

Klumbytė, Neringa, and Gulnaz Sharafutdinova, eds. *Soviet Society in the Era of Late Socialism, 1964–1985*. Lanham, MD: Lexington, 2013.

Koenig, Mark Charles. *Media and Reform: The Case of Youth Programming on Soviet Television (1955–1990)*. Ph.D. diss., Columbia University, 1995.

Korin, A. *Fenomen "Chto? Gde? Kogda?"* Moscow: Eksmo, 2002.

Kozlov, Denis. *The Readers of Novyi Mir: Coming to Terms with the Stalinist Past*. Cambridge, MA: Harvard University Press, 2013.

Kracauer, Siegfried. *Totalitäre Propaganda*, edited by Bernd Stiegler. Berlin: Suhrkamp Verlag, 2013.

Kramp, Leif. *Gedächtnismaschine Fernsehen. Vol. 1: Das Fernsehen als Faktor der gesellschaftlichen Erinnerung*. Berlin: Akademie Verlag, 2011.

Kutyrev, Boris P. *Budzhet vremeni. Voprosy izucheniia i ispol'zovaniia*. Novosibirsk: Nauka, 1977.

Lebina, Natal'ia B., and Aleksandr N. Chistikov. *Obyvatel' i reformy. Kartiny povsednevnoi zhizni gorozhan v gody NEPa i khrushchevskogo desiatiletiia*. St. Petersburg: Dmitrii Bulanin, 2003.

Lehmann, Maike. "The Local Reinvention of the Soviet Project. Nation and Socialism in the Republic of Armenia after 1945." *Jahrbücher für Geschichte Osteuropas* (2011), 59(4): 481–508.

———. "Apricot Socialism: The National Past, the Soviet Project, and the Imagining of Community in Late Soviet Armenia." *Slavic Review* (2015), 74(1): 9–31.

Lehr, Stefan. "'Pište nám!' Dopisy diváků a posluchačů Československé televizi a rozhlasu." *Marginalia Historica: Časopis pro dějiny vzdělanosti a kultury* (2012), 3(2): 71–82.

———. "Eingaben von Bürgern an das Tschechoslowakische Fernsehen und Radio: Ein Beitrag zur Kommunikation zwischen Staat, Bürgern und Medien im Staatssozialismus." *Bohemia* (2019), 59(2) (forthcoming).

Lemon, Alaina. "Sympathy for the Weary State?: Cold War Chronotopes and Moscow Others." *Comparative Studies in Society and History* (2009), 51(4): 832–864.

Lenoe, Matthew. *Closer to the Masses: Stalinist Culture, Social Revolution, and Soviet Newspapers.* Cambridge, MA: Harvard University Press, 2004.

Lepp, Annika, and Mervi Pantti. "Window to the West. Memories of Watching Finnish Television in Estonia During the Soviet Period." *VIEW. Journal of European Television History and Culture* (2013), 2(3): 76–86.

Lindenberger, Thomas. "Geteilte Welt, geteilter Himmel? Der Kalte Krieg und die Massenmedien in gesellschaftsgeschichtlicher Perspektive." In *Zwischen Pop und Propaganda. Radio in der DDR*, edited by Klaus Arnold, and Christoph Classen, 27–44. Berlin: Christoph Links, 2004.

Lovell, Stephen. "In Search of an Ending: *Seventeen Moments* and the Seventies." In *The Socialist Sixties: Crossing Borders in the Second World,* edited by Anne E. Gorsuch, and Diane P. Koenker, 303–21. Bloomington, IN: Indiana University Press, 2013.

———. *Russia in the Microphone Age: A History of Soviet Radio, 1919–1970.* Oxford: Oxford University Press, 2015.

Lüthi, Lorenz M. "The People's Republic of China and the Warsaw Pact Organization, 1955–63." *Cold War History* (2007), 7(4): 479–494.

Luhmann, Niklas. *Die Realität der Massenmedien.* 4th ed. Wiesbaden: VS Verlag für Sozialwissenschaften, 2009.

———. *Die Wissenschaft der Gesellschaft.* Frankfurt a. M.: Suhrkamp, 1990.

———. *Soziale Systeme. Grundriß einer allgemeinen Theorie.* Frankfurt a. M.: Suhrkamp, 1987.

Lustyik, Katalin. "From a Socialist Endeavour to a Commercial Enterprise: Children's Television in East-Central Europe." In *Popular Television in Eastern Europe During and Since Socialism,* edited by Anikó Imre, Timothy Havens, and Katalyn Lustyik, 105–122. New York, London: Routledge, 2013.

Luthar, Breda, and Pušnik, Maruša, eds. *Remembering Utopia: The Culture of Everyday Life in Socialist Yugoslavia.* Washington, DC: New Academia Publishing, 2010.

Martin, Terry. *The Affirmative Action Empire: Nations and Nationalism in the Soviet Union, 1923–1939.* Ithaca, NY: Cornell University Press, 2001.

Menzel, Birgit, and Ulrich Schmid. "Der Osten im Westen. Importe der Populärkultur." *Osteuropa* 2007, 5: 3–22.

Merkel, Ina, ed., *"Wir sind doch nicht die Meckerecke der Nation". Briefe an das Fernsehen der DDR.* Berlin: Schwarzkopf & Schwarzkopf, 2000.

——— and Felix Mühlberg. "Eingaben und Öffentlichkeit." In "*Wir sind doch nicht die Meckerecke der Nation". Briefe an das Fernsehen der DDR*, edited by Ina Merkel, 11–32. Berlin: Schwarzkopf & Schwarzkopf, 2000.

Merl, Stephan. *Politische Kommunikation in der Diktatur. Deutschland und die Sowjetunion im Vergleich*. Göttingen: Wallstein, 2012.

———. "Staat und Konsum in der Zentralverwaltungswirtschaft. Rußland und die ostmitteleuropäischen Länder." In *Europäische Konsumgeschichte, Zur Gesellschafts – und Kulturgeschichte des Konsums (18.-20. Jahrhundert)*, edited by Hannes Siegrist, Hartmut Kaelble, and Jürgen Kocka, 205–241. Frankfurt a.M., New York: Campus, 1997.

———. "Konsum in der Sowjetunion: Element der Systemdestabilisierung?" *Geschichte in Wissenschaft und Unterricht* (2007), 58: 519–536.

Meyen, Michael. *Einschalten, Umschalten, Ausschalten? Das Fernsehen im DDR-Alltag*. Leipzig: Leipziger Universitätsverlag, 2003.

———. "Haben die Westmedien die DDR stabilisiert? Zur Unterhaltungsfunktion bundesdeutscher Rundfunkangebote." *SPIEL* (2001), 20(1): 117–133.

———. *Denver Clan und Neues Deutschland. Mediennutzung in der DDR*. Berlin: Christoph Links Verlag, 2003.

Mickiewicz, Ellen Propper. *Changing Channels: Television and the Struggle for Power in Russia.* 2nd ed. Durham, NC: Duke University Press, 1999.

———. *Media and the Russian Public.* New York: Praeger, 1981.

———. "Excavating Concealed Tradeoffs: How Russians Watch the News." *Political Communication* (2005), 22(3): 355–380.

———. *Split Signals: Television and Politics in the Soviet Union.* New York: Oxford University Press, 1988.

Mihelj, Sabina. "The Politics of Privatization. Television Entertainment and the Yugoslav Sixties." In *The Socialist Sixties. Crossing Borders in the Second World*, edited by Anne E. Gorsuch, and Diane P. Koenker, 251–267. Bloomington, IN: Indiana University Press, 2013.

———. "Audience History as a History of Ideas: The 'Active Audience' of the Socialist Sixties." *European Journal of Communication* (2015), 30(1): 22–35.

———. "Negotiating Cold War Culture at the Crossroads of East and West: Uplifting the Working People, Entertaining the Masses, Cultivating the Nation." *Comparative Studies in Society and History* (2011), 53(3): 509–539.

Miklóssy, Katalin, and Melanie Ilic, eds. *Competition in Socialist Society*. New York: Routledge, 2014.

Milenkovitch, Michael. *The View from Red Square: A Critique of Cartoons from* Pravda *and* Izvestiia, *1947–1964*. New York, Buenos Aires: Hobbs, Dorman, 1966.

Millar, James. "The Little Deal: Brezhnev's Contribution to Acquisitive Socialism." *Slavic Review* (1985), 44(4): 694–706.

Mironova, Anna A. "Trust, Social Capital, and Subjective Individual Well-Being." *Sociological Research* (2015), 54(2): 121–133.

Mommsen, Margareta. *Hilf mir, mein Recht zu finden. Russische Bittschriften von Iwan dem Schrecklichen bis Gorbatschow*. Frankfurt a.M.: Propyläen Verlag, 1987.

Moskvichev, Lev N. et al., eds. *Sotsiologiia i vlast'. Sbornik 1. Dokumenty 1953–1968.* (Sociology and power. Collection 1. Documents 1953–1968). Moscow: Academia, 1997.

Murray, John. *The Russian Press from Brezhnev to Yeltsin: Behind the Paper Curtain.* Aldershot: Elgar, 1994.

Mustata, Dana. "'The Revolution Has Been Televised . . .'. Television as Historical Agent in the Romanian Revolution." *Journal of Modern European History* (2012), 10(1): 76–97.

———. "Within Excess Times and a Deficit Space: Cross-border Television as a Transnational Phenomenon in 1980s Romania." In *Transnational Television History. A Comparative Approach*, edited by Andreas Fickers, and Catherine Johnson, 89–102. Abingdon, New York: Routledge 2012.

———. "Geographies of Power: The Case of Foreign Broadcasting in Dictatorial Romania." In *Airy Curtains: European Broadcasting during the Cold War*, edited by Alexander Badenoch, Andreas Fickers, and Christian Henrich-Franke, 151–179. Baden-Baden: Nomos-Verlag, 2013.

Neller, Katja. *DDR-Nostalgie. Dimensionen der Orientierungen der Ostdeutschen gegenüber der ehemaligen DDR, ihre Ursachen und politischen Konnotationen.* Wiesbaden: Springer, 2006.

Nepomnyashchy, Catharine Thiemer. "The Blockbuster Miniseries on Soviet TV: Isaev-Stirlitz, the Ambiguous Hero of *Seventeen Moments of Spring.*" *Soviet and Post-Soviet Review* (2002), 29(3): 257–276.

Nordenstreng, Kaarle, and Tapio Varis. *Television Traffic—A One-Way Street? A Survey and Analysis of the International Flow of Television Material.* Paris: UNESCO, 1974.

Nordenstreng, Kaarle, Vartanova, Elena, and Yassen Zassoursky, eds. *Russian Media Challenge.* Helsinki: Kikimora Publications, 2002.

Nussbaum, Martha C. *Political Emotions. Why Love Matters for Justice.* Cambridge, MA, and London: The Belknap Press of Harvard University Press, 2013.

———. *Politische Emotionen.* Berlin: Suhrkamp, 2014.

Oates, Sarah. "The Neo-Soviet Model of the Media." *Europe-Asia Studies* (2007), 59(8): 1279–1297.

Öhner, Vrääth. "Die Dystopie Fernsehen." In *Klassenproduktion. Fernsehen als Agentur des Sozialen*, edited by Andrea Seier, and Thomas Waitz, 101–110. Hamburg, Münster: Lit 2014.

Oushakine, Serguei Alex. "'Against the Cult of Things': On Soviet Productivism, Storage Economy, and Commodities with No Destination." *Russian Review* (2014), 73(2): 198–236.

———. "'We're Nostalgic but We're Not Crazy': Retrofitting the Past in Russia." *Russian Review* (2007), 66: 451–482.

Paasilinna, Reino. *Glasnost' and Soviet Television: A Study of the Soviet Mass Media and Its Role in Society from 1985–1991.* Helsinki: YLE Finnish Broadcasting, 1995.

Pashkov, Aleksei S. *Obraz zhizni naseleniia krupnogo goroda. Opyt kompleksnogo sotsial'nogo issledovaniia* (Chelovek i obshchestvo, vol. 23). Leningrad: Izdatel'stvo Leningradskogo Universiteta, 1988.

Patrushev, Vasilii D. "Osnovnye izmeneniia v sfere svobodnogo vremeni gorodskogo naseleniia v posledniuiu tret' XX veka." In *Biudzhet vremeni i peremeny v zhiznedeiatel'nosti gorodskikh zhitelei v 1965–1998 godakh*, edited by Tat'iana M. Karakhanova, 34–44. Moscow: Izdatel'stvo Instituta sotsiologii RAN, 2001.

Patterson, Patrick Hyder. *Bought and Sold. Living and Losing the Good Life in Socialist Yugoslavia*. Ithaca, NY: Cornell University Press, 2011.

Pehe, Veronika. "Responses to *The Thirty Cases of Major Zeman* in the Czech Republic." *VIEW. Journal of European Television History and Culture* (2014), 3(5): 100–107.

Penati, Cecilia. "'Remembering Our First TV Set'. Personal Memories as a Source for Television Audience History." *VIEW. Journal of European Television History and Culture* (2013), 2(3): 4–12.

Platt, Kevin M. F. "Russian Empire of Pop: Post-Socialist Nostalgia and Soviet Retro at the 'New Wave' Competition." *Russian Review* (2013), 72: 447–469.

Pressman, Lev P. "Uchebnaia teleperedacha i trebovaniia k nei." In *Televidenie v shkole. Trudy 1-go Vsesoiuznogo seminara po uchebnomu televideniiu*, edited by Michail P. Kashin, and Lev P. Pressman, 45–72. Moscow: Akademiia pedagogicheskikh nauk, 1967.

Prokhorov, Alexander. "Size Matters: The Ideological Functions of the Length of Soviet Feature Films and Television Mini-Series in the 1950s and 1960s." *KinoKultura* (2006), 12. http://www.kinokultura.com/2006/12 – prokhorov.shtml.

Prokhorova, Elena. *Fragmented Mythologies: Soviet TV Miniseries of the 1970s*. Ph.D. diss., University of Pittsburgh, 2003.

Rajagopalan, Sudha. *Leave* Disco Dancer *Alone! Indian Cinema and Soviet Movie-Going After Stalin*. Bloomington, IN: Indiana University Press, 2008.

Raleigh, Donald J. *Soviet Baby Boomers. An Oral History of Russia's Cold War Generation*. Oxford: Oxford University Press, 2012.

Razzakov, Fedor. *Gibel' sovetskogo TV. Tainy televideniia: Ot Stalina do Gorbacheva, 1930–1991*. Moscow: Eksmo, 2009.

Reid, Susan E. "The Khrushchev Kitchen: Domesticating the Scientific-Technological Revolution." *Journal of Contemporary History* (2005), 40: 289–316.

———. "Khrushchev Modern. Agency and Modernization in the Soviet Home." *Cahiers du monde russe* (2006), 47(1–2): 227–268.

———. "Cold War in the Kitchen: Gender and the De-Stalinization of Consumer Taste in the Soviet Union under Khrushchev." *Slavic Review* (2002), 61: 211–252.

———. "Communist Comfort: Socialist Modernism and the Making of Cosy Homes in the Khrushchev Era." *Gender & History* (2009), 21(3): 465–498.

———. "The Meaning of Home: 'The Only Bit of the World You Can Have to Yourself'." In *Borders of Socialism. Private Spheres of Soviet Russia*, edited by Lewis Siegelbaum, 145–170. New York, Houndmills, Basingstoke: Palgrave Macmillan, 2006.

Reifová, Irena. "A Study in the History of Meaning-Making: Watching Socialist Television Serials in the Former Czechoslovakia." *European Journal of Communication* (2015), 30(1): 79–94.

Requate, Jörg. "Kommunikation und Propaganda in den Medien der DDR. Möglichkeiten und Grenzen des kommunikativen Austausches zwischen den audiovisuellen Medien der DDR und ihrem Publikum." *Siegener Periodicum zur Internationalen Empirischen Literaturwissenschaft (SPIEL)* (1995), 14(2): 230–243.

———. "Öffentlichkeit und Medien als Gegenstände historischer Analyse." *Geschichte und Gesellschaft* (1999), 25: 5–32.

Rosenthal, Gabriele. *Erlebte und erzählte Lebensgeschichte. Gestalt und Struktur biographischer Selbstbeschreibungen*, Frankfurt a. M., New York: Campus, 1995.

Roth-Ey, Kristin. *Moscow Prime Time: How the Soviet Union Built the Media Empire that Lost the Cultural Cold War.* Ithaca, NY: Cornell University Press, 2011.

———. "Playing for Cultural Authority: Soviet TV Professionals and the Game Show in the 1950s and 1960s." In *Pleasures in Socialism. Leisure and Luxury in the Eastern Bloc*, edited by David Crowley, and Susan E. Reid, 147–176. Evanston, IL: Northwestern University Press, 2010.

Rozov, A., ed. *Shabolovka 53: stranitsy istorii televideniia.* Moscow: Iskusstvo, 1988.

Rutland, Peter, and Victoria Smolkin-Rothrock. "Introduction: Looking Back at Brezhnev." *Russian History* (2014), 41: 299–306.

Safire, William. "The Cold War's Hot Kitchen." *New York Times*, July 23, 2009. http://www.nytimes.com/2009/07/24/opinion/24safire.html?_r=0 (accessed October 15, 2015).

Sahadeo, Jeff. "Soviet 'Blacks' and Place Making in Leningrad and Moscow." *Slavic Review* (2012), 71(2): 331–358.

Sasse, Silvia. *Michail Bachtin zur Einführung*. Hamburg: Junius, 2010.

Saxer, Ulrich, and Martina Märki-Koepp. *Medien-Gefühlskultur. Zielgruppenspezifische Gefühlsdramaturgie als journalistische Produktionsroutine*. München: Oelschläger, 1992.

Schanze, Helmut, and Bernhard Zimmermann eds. *Das Fernsehen und die Künste.* München: Wilhelm Fink Verlag, 1994.

Schmid, Ulrich M. *Technologien der Seele. Vom Verfertigen der Wahrheit in der russischen Gegenwartskultur*. Berlin: Suhrkamp Verlag, 2015.

Schmidt, Siegfried J. "Medien, Kultur: Medienkultur. Ein konstruktivistisches Gesprächsangebot." In *Kognition und Gesellschaft. Der Diskurs des Radikalen Konstruktivismus*, edited by idem, 425–450. Frankfurt a. M.: Suhrkamp, 1992.

Schnell, Rüdiger. *Haben Gefühle eine Geschichte? Aporien einer History of emotions*, part 1. Göttingen: V&R, 2015.

Schubert, Markus, and Hans-Jörg Stiehler. "Programmentwicklung im DDR-Fernsehen zwischen 1953 und 1963. Programmstrukturanalytische Betrachtungen zu den Anfängen des DDR-Fernsehens." In *Zwischen Experiment und Etablierung. Die Programmentwicklung des DDR-Fernsehens 1958 bis 1963*, edited by Claudia Dittmer, and Susanne Vollberg, 25–63. Leipzig: Leipziger Universitätsverlag, 2007.

Schwoch, James. *Global TV: New Media and the Cold War, 1946–1969.* Urbana-Champaign, IL: University of Illinois Press, 2009.

Scuteri, Lucia Gaja. "TV as a Linguistic Issue in Yugoslavian Slovenia: A Brief Chronology from the 1960s to the 1980s." In *Television Beyond and Across the*

Iron Curtain, edited by Kirsten Bönker, Julia Obertreis, and Sven Grampp, 196–226. Cambridge: Cambridge Scholars Publishing, 2016.

Shaw, Tony, and Denise Youngblood. *Cinematic Cold War: The American and Soviet Struggle for Hearts and Minds*. Lawrence: University Press of Kansas, 2010.

Shlapentokh, Vladimir. *The Politics of Sociology in the Soviet Union*. Boulder, CO: Westview, 1987.

Shonin, Nikolai. "Radio and Television as Art Forms in the Soviet Union." *Cultures* (1977), 4: 83–92.

Smoliak, Ol'ga. "'Abazhur,' 'abort,' 'abrikos': Reprezentatsiia sovetskoi povsednevnosti v knigakh po domovodstvu." In *Razryvy i konventsii v otechestvennoi kul'ture*, edited by Oleg Leibovich, 138–184. Perm: PGIIK, 2011.

Sorescu-Marinković, Annemarie. "Serbian Language Acquisition in Communist Romania." *Balcanica* (2010), XLI: 7–31.

———. "The World through the TV Screen. Everyday Life under Communism on the Western Romanian Border." *Martor – The Museum of the Romanian Peasant Anthropology Review* (2012), 17: 173–187.

Spigel, Lynn. "Media Homes: Then and Now." *International Journal of Cultural Studies* (2001), 4: 385–411.

———. *Welcome to the Dreamhouse: Popular Media and Postwar Suburbs*. Durham, NC: Duke University Press, 2001.

———. *Make Room for TV: Television and the Family Ideal in Post-War America*. Chicago, IL: University of Chicago Press, 1992.

Stepanov, Anatolii A. "Psikhologicheskie problemy uchebnogo televideniia." In *Televidenie v shkole. Trudy 1-go Vsesoiuznogo seminara po uchebnomu televideniiu*, edited by Michail P. Kashin, and Lev P. Pressman, 104–120. Moscow: Akademiia pedagogicheskikh nauk, 1967.

Stephan, Anke. "Erinnertes Leben: Autobiographien, Memoiren und Oral-History-Interviews als historische Quellen." In *Digitales Handbuch zur Geschichte und Kultur Russlands und Osteuropas*. https://epub.ub.uni-muenchen.de/627/1/Stephan-Selbstzeugnisse.pdf (accessed February 14, 2015).

Stites, Richard. *Russian Popular Culture: Entertainment and Society Since 1900*. Cambridge: Cambridge University Press, 1992.

Stroeva, Adel' S. *Deti, kino i televidenie*. Moskva: Znanie, 1962.

Suny, Ronald Grigor. "The Contradictions of Identity: Being Soviet and National in the USSR and After." In *Soviet and Post-Soviet Identities*, edited by Mark Bassin, and Catriona Kelly, 17–36. Cambridge: Cambridge University Press, 2012.

Thomaß, Barbara, and Tzankoff, Michaela, eds. *Medien und Transformation in Osteuropa*. Wiesbaden: Westdeutscher Verlag, 2001.

Tikhomirova, Anna. "Soviet Women and Fur Consumption in the Brezhnev Era." In *Pleasures in Socialism: Leisure and Luxury in the Eastern Bloc*, edited by David Crowley, and Susan E. Reid, 283–308. Evanston, IL: Northwestern University Press, 2010.

Tomlinson, John. *The Culture of Speed: The Coming of Immediacy*. Los Angeles, CA: Sage Publications, 2007.

Tsipursky, Gleb. "'As a Citizen, I Cannot Ignore These Facts'. Whistleblowing in the Khrushchev Era." *Jahrbücher für Geschichte Osteuropas* (2010), 58: 52–69.

Vassallo, Aude. *La télévision sous de Gaulle. Le contrôle gouvernemental de l'information (1958/1969)*. Bruxelles: De Boeck Supérieur, 2005.

Veenis, Milena. *Material Fantasies. Expectations of the Western Consumer World among East Germans*. Amsterdam: Amsterdam University Press, 2012.

Vogel, Meike. *Unruhe im Fernsehen. Protestbewegung und öffentlich-rechtliche Berichterstattung in den 1960er Jahren*. Göttingen: Wallstein Verlag, 2010.

Vollberg, Susanne. "'Wiederholungssender,' Russenprogramm' oder alternatives Massenprogramm? Zur Konzeption und Realisation des zweiten Programms des DDR-Fernsehens." In *Das Ende der Langeweile? Programmgeschichte des DDR-Fernsehens*, edited by Claudia Dittmar, and Susanne Vollberg, 147–182. Leipzig: Leipziger Universitätsverlag, 2003.

Vonnard, Philippe, Nicola Sbetti, and Grégory Quin eds., *Beyond Boycotts. Sport during the Cold War in Europe*. Berlin, Boston, MA: De Gruyter/Oldenbourg, 2018.

Vorderer, Peter. "Unterhaltung durch Fernsehen: Welche Rolle spielen parasoziale Beziehungen zwischen Zuschauern und Fernsehakteuren?" In *Fernsehforschung in Deutschland. Themen – Akteure – Methoden*, vol. 2, edited by Walter Klingler, Gunnar Roters, and Oliver Zöllner, 689–707. Baden-Baden: Nomos, 1998.

Wasiak, Patryk. "The Great Époque of the Consumption of Imported Broadcasts. West European Television Channels and Polish Audiences during the System Transition." *VIEW Journal of European Television History and Culture* (2014), 3(5): 68–78.

Weiss, Daniel. "Prolegomena zur Geschichte der verbalen Propaganda in der Sowjetunion." In *Slavistische Linguistik 1994. Referate des XX. Konstanzer Slavistischen Arbeitstreffens, Zürich 20.-22.9.1994*, edited by Daniel Weiss, 343–391. München: Sagner 1995.

Wespe, Aglaia. *Alltagsbeobachtung als Subversion. Leningrader Dokumentarfilm im Spätsozialismus*. Göttingen: V&R unipress, 2014.

Wiehl, Anna. "ARTE: French-German Experiments in Crossing the Borders. 'One Media – Three Screens' Convergence and Interactivity at Its Full Potential?" *VIEW Journal of European Television History and Culture* (2014), 3(6): 78–94.

Wolfe, Thomas C. *Governing Socialist Journalism. The Press and the Socialist Person after Stalin*. Bloomington, IN: Indiana University Press, 2005.

Weinberg, Elizabeth A. *The Development of Sociology in the Soviet Union*. London and Boston, MA: Routledge & Kegan Paul, 1974.

Weingart, Brigitte. "Alles (McLuhans Fernsehen im *global village*)." In *Diskursgeschichte der Medien nach 1945: Medienkultur der 60er Jahre*, vol. 2, edited by Irmela Schneider, Torsten Hahn, and Christina Bartz, 215–240. Opladen, Wiesbaden: Westdeutscher Verlag, 2003.

Willems, Herbert. "Normalität, Normalisierung, Normalismus." In *›Normalität‹ im Diskursnetz soziologischer Begriffe*, edited by Jürgen Link, Thomas Loer, and Hartmut Neuendorff, 51–83. Heidelberg: Synchron, 2003.

Williams, Raymond. *Culture*. London: Fontana, 1981.

———. *Television. Technology and Cultural Form.* 2nd ed. London: Routledge, 1990.

Wojnowski, Zbigniew. "The Soviet People: National and Supranational Identities in the USSR After 1945." *Nationalities Papers. The Journal of Nationalism and Ethnicity* (2015), 43(1): 1–7.

Wolff, Franka. *Glasnost erst kurz vor Sendeschluss. Die letzten Jahre des DDR-Fernsehens (1985–1989/90).* Köln, Weimar, Wien: Böhlau, 2002.

Yurchak, Alexei. *Everything Was Forever, Until It Was No More: The Last Soviet Generation.* Princeton, NJ: Princeton University Press, 2005.

———. "Soviet Hegemony of Form: Everything Was Forever, Until it Was No More." *Comparative Studies in Society and History* (2003), 45(3): 480–510.

Yurtaeva, Yulia. "Intervision. Searching for Traces." *VIEW Journal of European Television History and Culture* (2014), 3(5): 23–34.

Zakharova, Larissa. *S'habiller à la soviétique: La mode et le Dégel en URSS.* Paris: CNRS, 2011.

Zasurskii, Ia. N., ed. *Teleradioėfir: istoriia i sovremennost'.* Moscow: Aspekt, 2005.

Zeller, Manfred. *Das sowjetische Fieber. Fußballfans im poststalinistischen Vielvölkerreich.* Stuttgart: *ibidem*-Verlag, 2015.

Zerubavel, Eviatar. *Time Maps: Collective Memory and the Social Shape of the Past.* Chicago, IL: University of Chicago Press, 2003.

Zhuk, Sergei I. *Rock and Roll in the Rocket City. The West, Identity, and Ideology in Soviet Dnipropetrovsk*, 1960–1985. Baltimore, MD: Johns Hopkins University Press, 2010.

Zhukova, Maria. "Rashid Nugmanov's film *Igla* (*The Needle*): 'Televisionised' Cinema?" In *Television Beyond and Across the Iron Curtain*, edited by Kirsten Bönker, Julia Obertreis, and Sven Grampp, 162–192. Cambridge: Cambridge Scholars Publishing, 2016.

Zok, Michael. "Das polnische Fernsehen in den 1980 Jahren. Polska Telewizja als Gegenstand und Austragungsort politischer Konflikte." *Rundfunk und Geschichte* (2013), 39(3–4): 25–34.

———. *Die Darstellung der Judenvernichtung in Film, Fernsehen und politischer Publizistik der Volksrepublik Polen 1968–1989.* Marburg: Verlag Herder-Institut, 2015.

Zoonen, Liesbet van, and Jan Wieten. "It Wasn't Exactly a Miracle. The Arrival of Television in Dutch Family Life." *Media Culture & Society* (1994), 16(4): 641–659.

Index

About the Author

Kirsten Bönker is research fellow of the Gerda Henkel-Foundation. She was visiting professor of East European History at the Georg-August-University of Göttingen in 2019/20. From 2014 to 2018, she was visiting professor of East European History, the History of Modern Societies, and Contemporary History at Bielefeld University and at the Carl von Ossietzky University of Oldenburg (Germany). She holds a PhD in history from Bielefeld University in 2007. She is the author of *Jenseits der Metropolen. Öffentlichkeit und Lokalpolitik im Gouvernement Saratov (1890–1914)* (Böhlau 2010) and the editor of the book series *Rethinking the Cold War*.

www.ingramcontent.com/pod-product-compliance
Ingram Content Group UK Ltd.
Pitfield, Milton Keynes, MK11 3LW, UK
UKHW041841031026
15587UKWH00002B/932

9 781498 526883